Student Solutions Manual

Physical Chemistry

SECOND EDITION

David W. Ball

Prepared by

Jörg C. Woehl

University of Wisconsin, Milwaukee

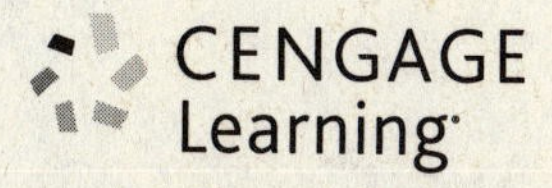

Australia • Brazil • Mexico • Singapore • United Kingdom • United States

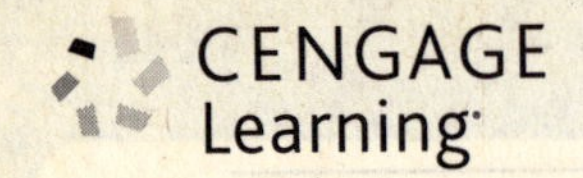

For product information and technology assistance, contact us at
Cengage Learning Customer & Sales Support, 1-800-354-9706.

For permission to use material from this text or product, submit all requests online at **www.cengage.com/permissions**
Further permissions questions can be emailed to **permissionrequest@cengage.com**.

ISBN-13: 978-1-285-07478-8
ISBN-10: 1-285-07478-5

Cengage Learning
200 First Stamford Place, 4th Floor
Stamford, CT 06902
USA

Cengage Learning is a leading provider of customized learning solutions with office locations around the globe, including Singapore, the United Kingdom, Australia, Mexico, Brazil, and Japan. Locate your local office at: **www.cengage.com/global**.

Cengage Learning products are represented in Canada by Nelson Education, Ltd.

To learn more about Cengage Learning Solutions, visit **www.cengage.com**.

Purchase any of our products at your local college store or at our preferred online store **www.cengagebrain.com**.

Printed in the United States of America
1 2 3 4 5 6 7 18 17 16 15 14

Table of Contents

Preface to the Student

When working with this manual, it is helpful to understand how results were rounded. Following best practices in problem solving, all calculated results (whether they are intermediate results or final results) are given with the correct number of significant figures:

- In operations other than addition and subtraction (multiplication, division, exponentials, logarithms, etc.), the result is rounded so as to have the same number of significant figures as the value with the fewest number of significant figures. For example, in exercise 2.1 we calculate work by multiplying 30 N by 30 m. Both values have two significant figures, so the result has two significant figures as well. However, "900 J" would imply three significant figures, so we use scientific notation and write the result as "9.0×10^2 J". (This makes sense because the real work could be as low as (29 N) (29 m) = 841 J or as high as (31 N) (31 m) = 961 J, so "900 J" is clearly too accurate). On some rare occasions, I have deviated from this rule to avoid unusual notation (for example, writing "20,560 cm^{-1}" is more common than "2.056×10^4 cm^{-1}"), but in those cases the number of significant figures is explicitly stated.
- In additions and subtractions, the accuracy of the result is dictated by the least accurate term (having the fewest number of decimal places). For example, in exercise 1.21(a) we are adding 0.25 atm to 1.0 atm. Of the two, "1.0 atm" has the lower accuracy, because the pressure is accurately known to the tenth place only (meaning that the real pressure could be as low as 0.9 atm or as high as 1.1 atm). Therefore, the result of the addition can only be given to the tenth place as well, so we have to round it to "1.3 atm".

However, the fact that these rules are applied to intermediate results does not imply that the rounded value is used in subsequent calculations - this would lead to what is known as truncation errors. Instead, we store the exact value of the intermediate result in the calculator's memory and substitute it for the rounded value in all subsequent calculations. Rounding then occurs only once the final result is calculated. For example, even though the pressure in 1.21(a) is given as "1.3 atm", the exact value of 1.25 atm is actually used wherever the placeholder "1.3 atm" appears. This means that in part (c) of the exercise, the result of (0.25 atm) / (1.3 atm) is exactly 0.2 (or 0.20 when rounded to two significant figures), not 0.1923... (which, when rounded, would yield 0.19). Therefore, should you get a different result when repeating the calculations in this manual on your calculator, it's probably because you are using rounded values for intermediate results instead of the exact values obtained in previous steps.

To make transitioning between the main text and this manual as seamless as possible, the values that are used for physical constants are those given in the main text. For example, we use R = 8.314 J/(mol K) rather than the more accurate value of 8.314472 J/(mol K) from the front cover. Atomic masses are used as given in the front cover. Isotope masses are taken from the NIST website on "Atomic Weights and Isotopic Compositions for All Elements" and rounded to two decimal places (except for H and D, which are rounded to three decimal places).

In constructing this Student Solutions Manual, I have followed the philosophy of the main text and tried to make the solutions as clear, accurate, and explicit as possible. I hope you will find it to be a helpful resource for studying physical chemistry.

Jörg C. Woehl
Milwaukee, Wisconsin

CHAPTER 1

Gases and the Zeroth Law of Thermodynamics

1.1. The drawing is left to the student. The calorimeter, water bath, and associated equipment (thermometers, ignition system, and so forth) are the system, while the surroundings are everything outside the apparatus.

1.3. A closed system has boundaries that prevent matter from moving in and out, although energy can transfer across them. An example is a soda can that hasn't been opened yet; neither gas nor liquid can escape, but energy can be transferred into and out of it (*e.g.*, by cooling it down in a refrigerator).

1.5. **(a)** 0 °C **(b)** 300 K **(c)** −20 °C (which is 253 K)

1.7. The pressure difference between the top and bottom of a column of liquid of height h is ρgh. In SI units, the density of water is

$$\left(1.0\ \frac{\text{g}}{\text{cm}^3}\right)\left(\frac{1\ \text{kg}}{1000\ \text{g}}\right)\left(\frac{10^6\ \text{cm}^3}{1\ \text{m}^3}\right)=1.0\times10^3\ \text{kg/m}^3\text{, and}$$

the maximum height is then $\dfrac{101325\ \text{Pa}}{\left(1.0\times10^3\ \text{kg/m}^3\right)\left(9.80\ \text{m/s}^2\right)}=10\ \text{m}\cong 34\ \text{ft}$.

[1 Pa = 1 kg/(m·s^2)]

1.9. A temperature difference will usually result in heat flow from the hot to the cold system if they are in thermal contact with each other. Under certain circumstances, heat can also flow in the inverse direction, from cold to hot, if external work is put into that process (*e.g.*, in a refrigerator, where heat is extracted from the inside and dumped into the surrounding air)

1.11. For this sample of gas under these conditions,

$F(p)=\dfrac{V}{T}=\dfrac{0.0250\ \text{L}}{(-33.0+273.15)\ \text{K}}=1.04\times10^{-4}\ \text{L/K}$. (Note the conversion to kelvin for the temperature.) If the volume is going to be $\left(66.9\ \text{cm}^3\right)\left(\dfrac{1\ \text{L}}{1000\ \text{cm}^3}\right)=0.0669\ \text{L}$:

$$T=\frac{V}{F(p)}=\frac{0.0669\ \text{L}}{1.04\times10^{-4}\ \text{L/K}}=643\ \text{K}.$$

1.13. Because $p_1V_1=p_2V_2$ at fixed T, the new volume is $\left(\dfrac{1.04\ \text{atm}}{0.047\ \text{atm}}\right)(67\ \text{L})=1.5\times10^3\ \text{L}$.

1.15. Because $p_1/T_1=p_2/T_2$ at fixed V, the pressure increases to $(4.5\ \text{atm})\left(\dfrac{317\ \text{K}}{298\ \text{K}}\right)=4.8\ \text{atm}$.

1.17. There are many possible conversions. For example, using the fact that 1 cal = 4.184 J:

$$\left(8.314\frac{\text{J}}{\text{mol}\cdot\text{K}}\right)\left(\frac{1\text{ cal}}{4.184\text{ J}}\right)=1.987\frac{\text{cal}}{\text{mol}\cdot\text{K}}.$$

1.19. Knowing that $R=0.08205\ \frac{\text{L}\cdot\text{atm}}{\text{mol}\cdot\text{K}}=8.314\frac{\text{J}}{\text{mol}\cdot\text{K}}$, we multiply by (mol·K)/0.08205 to find: 1 L·atm = 101.3 J (the exact conversion is 1 L·atm = 101.325 J).

1.21. **(a)** If 1.00 L of He gas at 0.75 atm is expanded to 3.00 L at fixed temperature, the new pressure would be $(0.75\text{ atm})\left(\frac{1.00\text{ L}}{3.00\text{ L}}\right)=0.25\text{ atm}$ ($p_1V_1=p_2V_2$ at fixed temperature). Similarly, the new pressure for Ne would be $(1.5\text{ atm})\left(\frac{2.00\text{ L}}{3.00\text{ L}}\right)=1.0\text{ atm}$. The total pressure from combining the two ideal gases into the same volume is then p_{tot} = 0.25 atm + 1.0 atm = 1.3 atm.

(b) p_{He} = 0.25 atm; p_{Ne} = 1.0 atm

(c) $x_{He}=\frac{n_{He}}{n_{tot}}=\frac{p_{He}V/(RT)}{p_{tot}V/(RT)}=\frac{p_{He}}{p_{tot}}=\frac{0.25\text{ atm}}{1.3\text{ atm}}=0.20$; $x_{Ne}=1-x_{He}=0.80$.

1.23. p_{CO_2} = (0.96) (90 bar) = 86 bar. The density of CO_2 is $\rho=\frac{m}{V}=\frac{n\cdot M}{V}$, where M is its molar mass. Assuming ideal gas behavior, we can substitute $\frac{n}{V}=\frac{p}{RT}$:

$$\rho=\frac{(86\text{ bar})(44.010\text{ g/mol})}{(0.08314\ \frac{\text{L}\cdot\text{bar}}{\text{mol}\cdot\text{K}})(730\text{ K})}=\left(63\frac{\text{g}}{\text{L}}\right)\left(\frac{1\text{ L}}{1000\text{ cm}^3}\right)=0.063\text{ g/cm}^3.$$

1.25. $n=m/M$ = 100.0 g / (36.4606 g/mol) = 2.743 mol HCl produce half the amount of hydrogen gas (1.371 mol). At (273.15 + 47.5) K = 320.65 K, this results in a volume of

$$V=\frac{nRT}{p}=\frac{(1.371\text{ mol})\left(0.08205\frac{\text{L}\cdot\text{atm}}{\text{mol}\cdot\text{K}}\right)(320.65\text{ K})}{1.02\text{ atm}}=35.4\text{ L}.$$

1.27. **(a)** This is an equation for a straight line with slope 5, so at x = 5 and x = 10, the slope is simply 5.

(b) The slope of this function is given by its first derivative: $\frac{d}{dx}(3x^2-5x+2)=6x-5$.

At x = 5, the slope is (6)(5) – 5 = 25. At x = 10, the slope is (6)(10) – 5 = 55.

(c) $\frac{d}{dx}\left(\frac{7}{x}\right)=-\left(\frac{7}{x^2}\right)$. At x = 5, the slope is $-\frac{7}{25}=-0.28$; at x = 10, the slope is $-\frac{7}{100}=-0.07$.

1.29. **(a)** $\left(\dfrac{\partial V}{\partial p}\right)_{T,n} = \dfrac{\partial}{\partial p}\left(\dfrac{nRT}{p}\right) = -\dfrac{nRT}{p^2}$ **(b)** $\left(\dfrac{\partial V}{\partial n}\right)_{T,p} = \dfrac{\partial}{\partial n}\left(\dfrac{nRT}{p}\right) = \dfrac{RT}{p}$

(c) $\left(\dfrac{\partial T}{\partial V}\right)_{p,n} = \dfrac{\partial}{\partial V}\left(\dfrac{pV}{nR}\right) = \dfrac{p}{nR}$ **(d)** $\left(\dfrac{\partial p}{\partial T}\right)_{V,n} = \dfrac{\partial}{\partial T}\left(\dfrac{nRT}{V}\right) = \dfrac{nR}{V}$

1.31. R is a constant, not a variable, and one cannot define a function of a constant (per definition, a function is a relation between a set of inputs and corresponding outputs, but becomes meaningless if the set of inputs is just a single number), even less so its derivative.

The derivative of the "function" R (a function of constant value equal to R) with respect to any variable, however, is well defined and equal to 0.

1.33. $\left[\dfrac{\mathrm{d}}{\mathrm{dV}}\left(\dfrac{\mathrm{dT}}{\mathrm{dp}}\right)_{\mathrm{n,V}}\right]_{\mathrm{n.p}}$ or $\left[\dfrac{\mathrm{d}}{\mathrm{dp}}\left(\dfrac{\mathrm{dT}}{\mathrm{dV}}\right)_{\mathrm{n,p}}\right]_{\mathrm{n,V}}$

1.35. The van der Waals constant a represents the pressure correction and is related to the magnitude of the interactions between gas particles. The van der Waals constant b is the volume correction and is related to the size of the gas particles.

1.37. One cylinder holds $n = \dfrac{m}{M} = \dfrac{\rho V}{M} = \dfrac{\left(0.840\ \text{g/cm}^3\right)\left(1000\ \text{cm}^3/\text{L}\right)\left(120\ \text{L}\right)}{28.0134\ \text{g/mol}} = 3.60 \times 10^3\ \text{mol}$

of liquid nitrogen. Neglecting the an^2/V^2 term, we have $V = \dfrac{nRT}{p} + nb = n\left(\dfrac{RT}{p} + b\right)$,

which evaluates to

$$V = \left(3.60 \times 10^3\ \text{mol}\right)\left[\frac{\left(0.08205\dfrac{\text{L}\cdot\text{atm}}{\text{mol}\cdot\text{K}}\right)(77\ \text{K})}{1.0000\ \text{atm}} + 0.03913\frac{\text{L}}{\text{mol}}\right] = 22874\ \text{L}.$$

(Note that although the result can normally be given with only two significant digits, limited by the temperature reading of "77 K", we will deviate from this rule here and list 5 significant digits because we are looking for a mathematical solution through an iterative process.

The pressure correction at this volume is

$\dfrac{an^2}{V^2} = \dfrac{\left(1.390\dfrac{\text{atm}\cdot\text{L}^2}{\text{mol}^2}\right)(3.60 \times 10^3\ \text{mol})^2}{\left(22874\ \text{L}\right)^2} = 0.0344\ \text{atm}$. Including this correction

properly in the van der Waals equation by replacing p with $p + an^2/V^2$ results in

$$V = \left(3.60\times10^3\ \text{mol}\right)\left[\frac{\left(0.08205\,\frac{\text{L}\cdot\text{atm}}{\text{mol}\cdot\text{K}}\right)(77\ \text{K})}{1.0000\ \text{atm}+0.0344\ \text{atm}}+0.03913\frac{\text{L}}{\text{mol}}\right]=22118\ \text{L}.$$

If we continue in the same way, we obtain the series 22874 L, 22118 L, 22068 L, 22064 L, 22064 L; the volume remains unchanged thereafter. Therefore, the value 22064 L solves the van der Waals equation; the properly rounded value would be 2.2×10^3 L.

1.39. We first rewrite the van der Waals equation as $p=\frac{nRT}{V-nb}-\frac{an^2}{V^2}$. The partial derivative with respect to volume is then $\left[\frac{\partial}{\partial V}\left(\frac{nRT}{V-nb}-\frac{an^2}{V^2}\right)\right]_{T,n}=-\frac{nRT}{(V-nb)^2}+\frac{2an^2}{V^3}$.

The virial equation with respect to volume can be written as $p=\frac{RT}{\bar{V}}\left(1+\frac{B}{\bar{V}}+\frac{C}{\bar{V}^2}+\cdots\right)$, or with $\bar{V}=\frac{V}{n}$: $p=\frac{nRT}{V}\left(1+\frac{nB}{V}+\frac{n^2C}{V^2}+\cdots\right)$. The partial deriative with respect to volume is

$$\left[\frac{\partial}{\partial V}\left(\frac{nRT}{V}\left[1+\frac{nB}{V}+\frac{n^2C}{V^2}+\cdots\right]\right)\right]_{T,n}=-\frac{nRT}{V^2}\left(1+\frac{nB}{V}+\frac{n^2C}{V^2}+\cdots\right)+\frac{nRT}{V}\left(\frac{-nB}{V^2}-\frac{2n^2C}{V^3}-\cdots\right)$$

$=-\frac{nRT}{V^2}\left(1+\frac{2nB}{V}+\frac{3n^2C}{V^2}+\cdots\right)$. Note that we have applied the product rule for differentiation here.

1.41. Helium's value for B is decreasing but may never be negative again. As you go to higher and higher temperatures, helium (or any gas) will act more and more like an ideal gas. Thus, we expect that at higher and higher temperatures, B will approach zero.

1.43. For Ne, $a = 0.2107$ atm L^2/mol^2. Because 1 atm = 1.01325 bar and 1 L = 1000 cm^3, we can substitute for atm and L:

$$a=(0.2107)\left[\frac{(1.01325\ \text{bar})\left(1000\ \text{cm}^3\right)^2}{\text{mol}^2}\right]=2.135\times10^5\ \frac{\text{bar}\cdot\text{cm}^6}{\text{mol}^2}.$$

1.45. The van der Waals constant b is directly proportional to the proper volume of the gas particles and is therefore always positive.

1.47. The compressibility is $Z=1+\frac{B}{\bar{V}}=1+\frac{-0.021\ \text{L/mol}}{22.4\ \text{L/mol}}=1.00$ if we include only the second virial term at standard temperature and pressure. The third virial coefficient changes this by $\frac{C}{\bar{V}^2}=\frac{0.0012\ \text{L}^2/\text{mol}^2}{(22.4\ \text{L/mol})^2}=2.4\times10^{-6}$, or $\frac{2.4\times10^{-6}}{1.00}=0.00024\%$.

1.49. Nitrogen's Boyle temperature is 327 K and therefore very close to room temperature, which means that it should behave close to ideally at room temperature.

1.51. For a van der Waals gas, $\left(\frac{\partial p}{\partial V}\right)_{T,n} = -\frac{nRT}{(V-nb)^2} + \frac{2an^2}{V^3}$ (see exercise 1.39).

(a)

$$\left(\frac{\partial \text{p}}{\partial \text{V}}\right)_{\text{T,n}} = -\frac{(1\text{ mol})\left(0.08205\frac{\text{L}\cdot\text{atm}}{\text{mol}\cdot\text{K}}\right)(298\text{ K})}{\left[25.0\text{ L}-(1\text{ mol})(0.0428\text{ L/mol})\right]^2} + \frac{2\left(2.253\frac{\text{atm}\cdot\text{L}^2}{\text{mol}^2}\right)(1\text{ mol})^2}{(25.0\text{ L})^3} = -0.0395\frac{\text{atm}}{\text{L}}$$

(b) $\left(\frac{\partial p}{\partial V}\right)_{T,n} = -0.001314$ atm/L. Higher temperatures and lower densities (higher volumes for the same amount of gas) favor ideal behavior, so (b) should be closer to ideal behavior.

1.53. At high temperatures, a can be ignored because the pressure correction term an^2/V^2 is inversely proportional to the square of volume, which increases with increasing temperature. The van der Waals equation of state then becomes $p(V-nb) = nRT$.

1.55. $$\text{p} = \frac{\left(0.08205\frac{\text{L}\cdot\text{atm}}{\text{mol}\cdot\text{K}}\right)(273.15\text{ K})}{22.41\text{ L/mol}-0.0139\text{ L/mol}} - \frac{741.6\frac{\text{atm}\cdot\text{L}^2\cdot\text{K}}{\text{mol}^2}}{(273.15\text{ K})(22.41\text{ L/mol})^2} = 0.9953\text{ atm}.$$

The value for an ideal gas is $p = \frac{RT}{\overline{V}} = \frac{\left(0.08205\frac{\text{L}\cdot\text{atm}}{\text{mol}\cdot\text{K}}\right)(273.15\text{ K})}{22.41\text{ L/mol}} = 1.000\text{ atm}$.

1.57. No. Pressure is the force exerted by the gas particles per unit area of surface, and it does not matter whether this force is generated by an ideal or nonideal gas.

1.59. Applying the cyclic rule would result in $\left(\frac{\partial \text{p}}{\partial \text{p}}\right)_{\text{T}} = -\frac{\left(\frac{\partial \text{T}}{\partial \text{p}}\right)_{\text{p}}}{\left(\frac{\partial \text{T}}{\partial \text{p}}\right)_{\text{p}}}$. However, this equation is mathematically meaningless – the partial derivatives on the right-hand side do not exist, because one cannot take a derivative with respect to a variable that is held constant. (This results in the contradiction that the left-hand side, although not very informative or useful, yields 1, while the right-hand side, the ratio of two identical derivatives multiplied by -1, would seem to yield negative 1.) There is not much danger in encountering the original partial derivative, however, because we are usually interested in the variation of p with respect to variables other than p itself.

1.61. STP: $\alpha = \frac{1}{V}\left(\frac{\partial V}{\partial T}\right)_p = \frac{1}{V}\frac{nR}{p} = \frac{1}{T} = \frac{1}{273.15\text{ K}} = 3.6610\times10^{-3}\text{ K}^{-1}$

SATP: $\alpha = \frac{1}{298.15\text{ K}} = 3.3540\times10^{-3}\text{ K}^{-1}$

1.63. The van der Waals equation contains volume in several places (if multiplied out, one can see that it is a cubic equation in volume), and cannot be rewritten in a form so that volume is all by itself on one side of the equation. As a consequence, taking the derivative of volume with respect to temperature (as for the expansion coefficient) or with respect to pressure (as for the isothermal compressibility) is not straightforward.

1.65. Using the definition of α, we have $\left(\frac{\partial V}{\partial T}\right)_{p,n} = \alpha V$; isolating V from the definition of κ and substituting it in there leads to $\left(\frac{\partial V}{\partial T}\right)_{p,n} = \alpha\left[-\frac{1}{\kappa}\left(\frac{\partial V}{\partial p}\right)_{T,n}\right] = -\frac{\alpha}{\kappa}\left(\frac{\partial V}{\partial p}\right)_{T,n}$.

The left-hand side is positive, as volume increases with increasing temperature at fixed pressure. However, volume decreases with increasing pressure at fixed temperature, so the partial derivative on the right-hand side is negative, which (with the negative sign and the fact that both α and κ are positive) leads again to a positive value.

1.67. $$\frac{\alpha}{\kappa} = \frac{\frac{1}{V}\left(\frac{\partial V}{\partial T}\right)_p}{-\frac{1}{V}\left(\frac{\partial V}{\partial p}\right)_T} = -\frac{\left(\frac{\partial V}{\partial T}\right)_p}{\left(\frac{\partial V}{\partial p}\right)_T} = \left(\frac{\partial p}{\partial T}\right)_V$$

1.69. Assuming that (dry) air contains about 79% of nitrogen (M_{N2} = 28.0134 g/mol) and 21% of oxygen (M_{O2} = 31.9988 g/mol), we use the (weighted) average of 29 g/mol for the molar mass:

$$p = \exp\left[-\frac{(0.029 \text{ kg/mol})(9.81 \text{ m/s}^2)(1840 \text{ m})}{\left(8.314 \frac{\text{J}}{\text{mol}\cdot\text{K}}\right)(273.15+26.0)\text{ K}}\right] \text{atm} = 0.81 \text{ atm}.$$ Note that it is important to use SI units in the exponential so that all units cancel out:

$$\frac{\text{kg/mol}\cdot\text{m/s}^2\cdot\text{m}}{\text{J/(mol}\cdot\text{K)}\cdot\text{K}} = \frac{\text{kg}\cdot\text{m}^2/(\text{mol}\cdot\text{s}^2)}{\text{kg}\cdot\text{m}^2/(\text{mol}\cdot\text{s}^2)}$$

(we have made use of the fact that 1 J = 1 kg·m^2/s^2).

1.71. We rearrange the scores in order of increasing grade: 3, 3, 4, 5, 7, 7, 9, 10. Two scores (3 and 7) appear twice, so their probability is 2 out of 8 or ¼; for all other scores, it is 1/8.

Thus, we have: $$\text{average} = \frac{\left(\frac{1}{4}\right)\cdot 3+\left(\frac{1}{8}\right)\cdot 4+\left(\frac{1}{8}\right)\cdot 5+\left(\frac{1}{4}\right)\cdot 7+\left(\frac{1}{8}\right)\cdot 9+\left(\frac{1}{8}\right)\cdot 10}{\left(\frac{1}{4}\right)+\left(\frac{1}{8}\right)+\left(\frac{1}{8}\right)+\left(\frac{1}{4}\right)+\left(\frac{1}{8}\right)+\left(\frac{1}{8}\right)} = \frac{6}{1} = 6.$$

1.73. The probability of a particle being in a state is proportional to $e^{-\Delta E/(RT)}$, where ΔE represents the difference in energy from the energy minimum. The population ratio between two states with energies ΔE_1 and ΔE_2 (relative to the energy minimum) is

therefore $$\frac{\exp\left(\frac{-\Delta E_1}{RT}\right)}{\exp\left(\frac{-\Delta E_2}{RT}\right)} = \exp\left[\frac{-(\Delta E_1 - \Delta E_2)}{RT}\right].$$

(a) 200 K: $\exp\left[\frac{-500\text{ J}}{\left(8.314\frac{\text{J}}{\text{mol}\cdot\text{K}}\right)(200\text{ K})}\right] = 0.740$ **(b)** 0.887 **(c)** 0.942

With increasing temperature the higher energy state becomes more populated; the population ratio approaches 1:1.

1.75. **(a)** atom: $\langle E_{\text{trans}}\rangle = \frac{3}{2}RT$; $\langle E_{\text{rot}}\rangle = 0$ **(b)** linear: $\langle E_{\text{trans}}\rangle = \frac{3}{2}RT$; $\langle E_{\text{rot}}\rangle = RT$

(c) linear: $\langle E_{\text{trans}}\rangle = \frac{3}{2}RT$; $\langle E_{\text{rot}}\rangle = RT$ **(d)** non-linear: $\langle E_{\text{trans}}\rangle = \frac{3}{2}RT$; $\langle E_{\text{rot}}\rangle = \frac{3}{2}RT$

1.77. Vibrational degrees of freedom have relatively large energy gaps, so that one can no longer regard them as a continuous distribution of energies. Mathematically, this is means that the summation can no longer be transformed in a simple integration and leads to a more complicated expression for $\langle E_{\text{vib}}\rangle$. For high vibrational energy values, this simplifies to $\langle E_{\text{vib}}\rangle \approx 0$ (see section 1.9), whereas the low vibrational energy limit is $\langle E_{\text{vib}}\rangle \approx RT$.

CHAPTER 2

The First Law of Thermodynamics

2.1. $w = \vec{F} \cdot \vec{s} = |F| \cdot |s| \cdot \cos\theta$

(a) w = (30 N) (30 m) (cos 0°) = 900 N·m = 9.0×10^2 J (or about 900 J)

(b) w = (30 N) (30 m) (cos 45°) = (9.0×10^2 N·m) (0.7071) = 6.4×10^2 J.

2.3. The volume change from 50.00 mL to 450.00 mL equals +400.00 mL, so that

$$w = -p_{\text{ext}}\Delta V = -(2.33\text{ atm})(400.00\text{ mL})\left(\frac{1\text{ L}}{1000\text{ mL}}\right)\left(\frac{101.32\text{ J}}{1\text{ L}\cdot\text{atm}}\right) = -94.4\text{ J}$$

2.5. The volume change from 5 mL = 0.005 L to 3.350 L equals +3.345 L, so that

$$w = -p_{\text{ext}}\Delta V = -(1\text{ atm})(3.345\text{ L})\left(\frac{101.32\text{ J}}{1\text{ L}\cdot\text{atm}}\right) = -338.9\text{ J}$$

The work needed to expand the balloon is 338.9 J.

2.7. **(a)** $w = -p_{\text{ext}}\Delta V = -(0.455\text{ atm})(2.00\text{ L} - 0.77\text{ L})\left(\frac{101.32\text{ J}}{1\text{ L}\cdot\text{atm}}\right) = -56.7\text{ J}$

(b) Assuming an ideal gas, the work performed is

$$w_{\text{rev}} = -nRT\ln\frac{V_{\text{f}}}{V_{\text{i}}} = -(0.033\text{ mol})\left(8.314\frac{\text{J}}{\text{mol}\cdot\text{K}}\right)\left[(273.15+35.0)\text{ K}\right]\left(\ln\frac{2.00\text{ L}}{0.77\text{ L}}\right) = -80.7\text{ J}$$

2.9. According to Example 2.4, the reversible work performed by a van der Waals gas is

$w_{\text{rev}} = -nRT\ln\frac{V_{\text{f}}-nb}{V_i-nb} - an^2\left(\frac{1}{V_{\text{f}}}-\frac{1}{V_{\text{i}}}\right)$. For ethane, a = 5.489 atm·L²/mol² and b = 0.0638 L/mol: $w_{\text{rev}} = -8.68\times10^3\text{ J} - (-5.12\text{ L}\cdot\text{atm})\left(\frac{101.32\text{ J}}{\text{L}\cdot\text{atm}}\right) = -8.16\times10^3\text{ J}$.

2.11. Solving $q = m\cdot c\cdot\Delta T$ for ΔT using c = 14.304 J/(g·K) from Table 2.1 yields a temperature

change of $\Delta T = \frac{q}{m\cdot c} = \frac{3930\text{ J}}{(79.8\text{ g})\left(14.304\frac{\text{J}}{\text{g}\cdot\text{K}}\right)} = 3.44\text{ K (or °C)}$.

2.13. These three compounds experience hydrogen bonding between their molecules. Because it requires more energy to overcome the effects of hydrogen bonding, these compounds have higher specific heat capacities than other, similar-mass molecules.

2.15. The heat lost by the iron is $q = m_{Fe} \cdot c_{Fe} \cdot (T_f - T_{Fe})$, where q is negative because the final temperature T_f is lower than the initial temperature T_{Fe}, which is (273.15 + 100.0) K = 373.2 K.

The heat gained by water is $q' = m_{H_2O} \cdot c_{H_2O} \cdot (T_f - T_{H_2O})$, where q' is positive because the final temperature T_f is higher than the initial temperature T_{H2O}, which is (273.15 + 22.0) K = 295.2 K. Setting $q = -q'$ and solving the resulting equation $m_{Fe} \cdot c_{Fe} \cdot (T_f - T_{Fe}) = -m_{H_2O} \cdot c_{H_2O} \cdot (T_f - T_{H_2O})$ for T_f leads to $T_f = \dfrac{m_{Fe}c_{Fe}T_{Fe} + m_{H_2O}c_{H_2O}T_{H_2O}}{m_{Fe}c_{Fe} + m_{H_2O}c_{H_2O}}$. Substituting the numerical values (with temperatures in kelvin) yields a final temperature of 298 K (or about 25 °C).

2.17. The verification of equation 2.8 follows from Boyle's law, which says $p_i V_i = p_f V_f$. This can be rearranged to give $\dfrac{V_f}{V_i} = \dfrac{p_i}{p_f}$. Substituting this into equation 2.7: $w_{rev} = -nRT \ln\dfrac{V_f}{V_i}$ becomes $w_{rev} = -nRT \ln\dfrac{p_i}{p_f}$ (which is what we are supposed to verify).

2.19. An open system allows for matter and energy exchange with the surroundings (open beaker). A closed system does not allow matter to pass, but energy can be transferred into and out of the system (closed bottle). An isolated system does not allow for passage of matter or energy into or out of the system; perfectly isolated systems do not exist, but a closed, thermally insulated container is a good approximation.

2.21. For an isolated system, there is no passage of energy into or out of the system, which means that energy cannot be exchanged in the form of heat ($q = 0$) or work ($w = 0$). Thus, equation 2.10 is simply a special case of equation 2.11; the latter also applies to closed systems.

2.23. Using $w_{rev} = -nRT \ln(V_f/V_i)$:

$$w_{rev} = -(0.245 \text{ mol})\left(8.314\frac{\text{J}}{\text{mol}\cdot\text{K}}\right)[(273.15+95.0)\text{ K}]\left(\ln\frac{1.00 \text{ mL}}{1000 \text{ mL}}\right) = -5.18\times10^3 \text{ J}$$

2.25. To calculate ΔU from the first law, we need both w and q. In each of the two cases, the system *absorbs* heat, so $q = +155$ J:

(a) $w - p_{ext}\Delta V = -(0.455 \text{ atm})[(2.00-0.77)\text{ L}]\left(\dfrac{101.32 \text{ J}}{\text{L}\cdot\text{atm}}\right) = -56.7 \text{ J}$

$\Delta U = q + w = +155 \text{ J} - 56.7 \text{ J} = +98 \text{ J}$

(b)

$$w_{rev} = -nRT\ln\frac{V_f}{V_i} = -(0.033 \text{ mol})\left(8.314\frac{\text{J}}{\text{mol}\cdot\text{K}}\right)[(273.15+35.0)\text{ K}]\left(\ln\frac{2.00 \text{ L}}{0.77 \text{ L}}\right) = -81 \text{ J}$$

$\Delta U = +155 \text{ J} - 81 \text{ J} = +74 \text{ J}$

2.27. **(a)** From the description, we can infer that the volume does not change if the sample is burned in a bomb calorimeter. Therefore, the heat generated by the ignition of a sample is equal to the change in internal energy ΔU.

(b) This process occurs open to the atmosphere, *i.e.* at constant (air) pressure. Therefore, the heat taken up equals the change in enthalpy ΔH.

(c) The refrigerator volume remains constant (assuming that the door is well sealed), so the loss of heat during the cooldown equals ΔU.

(d) As in (b), a fire inside a fireplace is open to the atmosphere, so the heat generated is equal to ΔH.

2.29. Temperature is a state function because an overall change in temperature is determined solely by the initial temperature and the final temperature, not the path a series of temperature changes took.

2.31. To calculate ΔU (which is a state function and therefore not path-dependent), we can break the process down into two simpler steps:

(a) Isothermal expansion from 10.0 L to 15.0 L at a constant temperature of 295 K, followed by (b) heating to 350 K while keeping the volume constant.

For step (a), the change in internal energy ΔU_a is zero if we assume that H_2 behaves like an ideal gas. For step (b), we have $\Delta T = 350\text{ K} - 295\text{ K} = 55\text{ K}$ and therefore

$q = m \cdot c \cdot \Delta T = (2.02\text{ g})\left(14.304 \frac{\text{J}}{\text{g} \cdot K}\right)(55\text{ K}) = +1.6 \times 10^3\text{ J}$. Furthermore, $w = 0$ because the volume does not change, so $\Delta U_b = q + w = +1.6 \times 10^3$ J. The total change in internal energy for (a) and (b) is therefore $\Delta U = \Delta U_a + \Delta U_b = +1.6$ kJ.

2.33. Because $\Delta U = q + w$, the change in internal energy ΔU will be zero if $q = -w$, even for a process whose initial conditions are not the same as its final conditions.

2.35. The increase of the steam's temperature by 35 °C (or 35 K) requires a heat of

$q = m \cdot c \cdot \Delta T = (7.23\text{ g})\left(2.04 \frac{\text{J}}{\text{g} \cdot \text{K}}\right)(35\text{ K}) = +5.2 \times 10^2\text{ J}$. (We assume here that the specific heat of steam, given for 100 °C in Table 2.1, does not change between 100 °C and 145 °C.)

The expansion at constant pressure requires an energy of

$$w = -p_{\text{ext}}\Delta V = -(0.985\text{ atm})\left[(8.00 - 2.00)\text{ L}\right]\left(\frac{101.32\text{ J}}{\text{L} \cdot \text{atm}}\right) = -599\text{ J}.$$

According to the first law $\Delta U = q + w$, which evaluates to $\Delta U = +5.2 \times 10^2\text{ J} - 599\text{ J} = -0.8 \times 10^2\text{ J}$ (about -80 J).

$\Delta H = \Delta U + \Delta(pV)$: We know the initial volume ($V_i = 2.00$ L) and final volume ($V_f = 8.00$ L), but need to calculate the initial and final pressures to evaluate ΔH (note that the temperature does not stay constant here, so $\Delta(pV) \neq 0$). Assuming an ideal gas, the initial pressure is $p_{\mathrm{i}} = \dfrac{nRT_{\mathrm{i}}}{V_{\mathrm{i}}} = \dfrac{\left(\dfrac{7.23\ \mathrm{g}}{18.0152\ \mathrm{g/mol}}\right)\left(0.08205\,\dfrac{\mathrm{L\cdot atm}}{\mathrm{mol\cdot K}}\right)(383\ \mathrm{K})}{2.00\ \mathrm{L}} = 6.31\ \mathrm{atm}$, while the same calculation for 8.00 L and (273.15 + 110 + 35) K = 418 K yields a final pressure of $p_{\mathrm{f}} = \dfrac{nRT_{\mathrm{f}}}{V_{\mathrm{f}}} = \dfrac{\left(\dfrac{7.23\ \mathrm{g}}{18.0152\ \mathrm{g/mol}}\right)\left(0.08205\,\dfrac{\mathrm{L\cdot atm}}{\mathrm{mol\cdot K}}\right)(418\ \mathrm{K})}{8.00\ \mathrm{L}} = 1.72\ \mathrm{atm}$. This results in a pV change of $\Delta(pV) = \left[(1.72\ \mathrm{atm})(8.00\ \mathrm{L}) - (6.31\ \mathrm{atm})(2.00\ \mathrm{L})\right]\left(\dfrac{101.32\ \mathrm{J}}{\mathrm{L\cdot atm}}\right) = 117\ \mathrm{J}$.

Therefore, $\Delta H = \Delta U + \Delta(pV) = -0.8\times10^2\ \mathrm{J} + 117\ \mathrm{J} = +0.3\times10^2\ \mathrm{J}$ (or about 30 J).

2.37. An exothermic process is a process that releases heat to the surroundings. Typically, these processes occur at constant pressure, where the heat given off by the system ($q < 0$) is equal to its change in enthalpy ΔH. The statement is therefore true *if* the pressure remains constant; however, a process that does not occur at constant pressure can release heat (is exothermic) even if ΔH is positive, so the statement is in general false.

2.39. 480 L of an ideal gas at 22 °C and 1.00 atm contains an amount of substance of $n = \dfrac{pV}{RT} = \dfrac{(1.00\ \mathrm{atm})(480\ \mathrm{L})}{\left(0.08205\,\dfrac{\mathrm{L\cdot atm}}{\mathrm{mol\cdot K}}\right)(295\ \mathrm{K})} = 19.8\ \mathrm{mol}$. Using $\Delta U = n\bar{C}_V\Delta T$ and a temperature drop of 18 °C (or 18 K), we obtain

$$\Delta U = (19.8\ \mathrm{mol})\left(12.47\,\frac{\mathrm{J}}{\mathrm{mol\cdot K}}\right)(-18\ \mathrm{K}) = -4.4\times10^3\ \mathrm{J}.$$

2.41. $w = -p_{\mathrm{ext}}\Delta V = -(10.0\ \mathrm{atm})(5.00\ \mathrm{L} - 20.00\ \mathrm{L})\left(\dfrac{101.32\ \mathrm{J}}{\mathrm{L\cdot atm}}\right) = +15.2\times10^3\ \mathrm{J}$.

The compression is isothermal (the temperature does not change), which – assuming an ideal gas – means that $\Delta U = 0$. Because $\Delta U = q + w$, this in turn requires $q = -w = -15.2$ kJ (*i.e.*, all the energy provided for the compression is released to the surroundings to keep the temperature constant). In addition, $\Delta H = \Delta U + \Delta(pV) = 0$ because the process occurs at constant temperature and Boyle's law applies: pV is constant and thus $\Delta(pV) = 0$.

2.43. $C_V = 21.6\,\dfrac{\mathrm{J}}{\mathrm{K}} + \left(4.18\times10^{-3}\,\dfrac{\mathrm{J}}{\mathrm{K}^2}\right)T - \dfrac{1.67\times10^5\ \mathrm{J\cdot K}}{T^2}$

2.45. $\left(\frac{\partial H}{\partial p}\right)_T = \left[\frac{\partial(U+pV)}{\partial p}\right]_T = \left(\frac{\partial U}{\partial p}\right)_T + \left[\frac{\partial(pV)}{\partial p}\right]_T = 0 + \left[\frac{\partial(nRT)}{\partial p}\right]_T$

(we have used the ideal gas law $pV = nRT$ here). The second partial derivative is zero because nRT is a constant at constant temperature. Therefore $\left(\frac{\partial H}{\partial p}\right)_T = 0$ for an ideal gas.

2.47. This derivation is given explicitly in the text in section 2.7.

2.49. For He: $T = \frac{2a}{Rb} = \frac{(2)(0.03508\ \text{atm}\cdot\text{L}^2/\text{mol}^2)}{\left(0.08205\frac{\text{L}\cdot\text{atm}}{\text{mol}\cdot\text{K}}\right)(0.0237\ \text{L/mol})} = 36.1\ \text{K}$

For H_2: $T = \frac{2a}{Rb} = \frac{(2)(0.244\ \text{atm}\cdot\text{L}^2/\text{mol}^2)}{\left(0.08205\frac{\text{L}\cdot\text{atm}}{\text{mol}\cdot\text{K}}\right)(0.0266\ \text{L/mol})} = 224\ \text{K}$

These values are quite close to the corresponding values given in the text (40 K and 202 K, respectively), only about 10% off the given values. The implications with respect to liquefaction of these two gases is that they need to be substantially precooled (below their inversion points) before liquefaction in a Joule-Thomson apparatus can be attempted.

2.51. Because the pressure change isn't too drastic, our answer to exercise 2.50 is probably within a few degrees of being correct – if a truly isenthalpic process can be arranged.

2.53. According to Table 2.2, the Joule-Thomson coefficient for N_2 at 50°C and 20 atm is 0.1709 K/atm. Because $\mu_{JT} = -\frac{\left(\frac{\partial H}{\partial p}\right)_T}{C_p}$, we can solve this for $\left(\frac{\partial H}{\partial p}\right)_T$ if we know C_p under these conditions. For 1 mol of nitrogen gas: Assuming $C_p = 29.1$ J/K (from translational and rotational contributions for a linear, ideal gas), we have $\left(\frac{\partial H}{\partial p}\right)_T = -(0.1709\ \text{K/atm})(29.1\ \text{J/K}) = -4.97\frac{\text{J}}{\text{atm}}$ and therefore $\left(\frac{\partial p}{\partial H}\right)_T = -\frac{1}{4.97\ \text{J/atm}} = -0.201\frac{\text{atm}}{\text{J}}$.

2.55. Because strictly speaking, heat capacities are extensive properties; they depend on the amount of matter in the system. Thus, the form in equation 2.37 is the most general expression that relates the two quantities.

2.57. First, we calculate the initial pressure and assume that this is the external pressure that remains constant throughout the compression (a good approximation, since the gas contracts slowly):

$$p = \frac{nRT}{V} = \frac{(0.145 \text{ mol})\left(0.08205 \ \frac{\text{L} \cdot \text{atm}}{\text{mol} \cdot \text{K}}\right)(273.15 \text{ K})}{5.00 \text{ L}} = 0.650 \text{ atm}.$$

Now we can calculate the work from $w = -p_{ext}\Delta V$, which results in

$$w = -(0.650 \text{ atm})(3.92 \text{ L} - 5.00 \text{ L})\left(\frac{101.32 \text{ J}}{\text{L} \cdot \text{atm}}\right) = +71.1 \text{ J}.$$

To determine ΔU, we need to calculate q first. For this, we need the final temperature, which can be determined by Charles' law (since pressure is constant), $\frac{V_i}{T_i} = \frac{V_f}{T_f}$. Solving for T_f results in $T_f = \frac{V_f}{V_i} T_i = \left(\frac{3.92 \text{ L}}{5.00 \text{ L}}\right)(273.15 \text{ K}) = 214 \text{ K}$, which means that $\Delta T = (214 \text{ K} - 273.15 \text{ K}) = -59 \text{ K}$. Using $q = n \cdot \bar{C}_p \cdot \Delta T$ and assuming a monoatomic gas with

$$\bar{C}_p = \tfrac{5}{2}R = 20.79 \ \frac{\text{J}}{\text{mol} \cdot \text{K}},$$

we have $q = n \cdot \bar{C}_p \cdot \Delta T = (0.145 \text{ mol})\left(20.79 \ \frac{\text{J}}{\text{mol} \cdot \text{K}}\right)(-59 \text{ K}) = -1.8 \times 10^2 \text{ J}$

and therefore $\Delta U = q + w = -1.8 \times 10^2 \text{ J} + 71.1 \text{ J} = -1.1 \times 10^2 \text{ J}$ (about -110 J).

2.59. Starting with $-R \ln V\Big|_{V_i}^{V_f} = \bar{C}_V \ln T\Big|_{T_i}^{T_f}$, we evaluate the terms at the given boundaries: $-R(\ln V_f - \ln V_i) = \bar{C}_V(\ln T_f - \ln T_i)$. According to the rules of logarithms, $\ln a - \ln b = \ln \frac{a}{b}$, so this can be rewritten as $-R \ln \frac{V_f}{V_i} = \bar{C}_V \ln \frac{T_f}{T_i}$, which is equation 2.44.

2.61. Diatomic molecules are linear, so that the heat capacity at constant volume from translational and rotational contributions is $\bar{C}_V = \tfrac{5}{2}R$. (Vibrational contributions only come into play at very high temperatures, so we do not need to consider them here.) For ideal gases, $\bar{C}_p = \bar{C}_V + R$, so that $\bar{C}_p = \tfrac{7}{2}R$. Therefore, $\gamma = \frac{\bar{C}_p}{\bar{C}_V} = \frac{\tfrac{7}{2}R}{\tfrac{5}{2}R} = \frac{7}{5}$ for an ideal diatomic gas.

2.63. **(a)** The expected molar heat capacity at constant volume for the diatomic molecule H_2 (assuming ideal behavior) is 5/2 R = 20.8 J/(mol·K) at low temperatures.

(b) The necessary heat is given by

$$q = n\cdot\bar{C}_V\cdot\Delta T = (1.00\text{ mol})\left(20.8\frac{\text{J}}{\text{mol}\cdot\text{K}}\right)(40\text{ K} - 22\text{ K}) = +374\text{ J}.$$

2.65. **(a)** As described in example 2.14, we can use $\left(\frac{p_f}{p_i}\right)^{\frac{2}{7}} = \frac{T_f}{T_i}$. If the initial pressure doubles, $p_f/p_i = 2$, and we obtain $\frac{T_f}{T_i} = 2^{2/7} = 1.219$, which means that the absolute temperature increases 21.9% above its initial value.

(b) The only change here is the value of γ, which now also includes vibrational contributions. For a diatomic molecule, $N = 2$, and because it is necessarily linear, there is only $3N - 5 = 1$ possible vibration that contributes R to the molar heat capacity at constant volume. Therefore, $\bar{C}_V = \frac{5}{2}R + R = \frac{7}{2}R$ and $\bar{C}_p = \frac{7}{2}R + R = \frac{9}{2}R$, which yields $\gamma = \frac{\bar{C}_p}{\bar{C}_V} = \frac{\frac{9}{2}R}{\frac{7}{2}R} = \frac{9}{7}$. Substituting this into $\left(\frac{p_f}{p_i}\right)^{\frac{\gamma-1}{\gamma}} = \frac{T_f}{T_i}$, we have $\left(\frac{p_f}{p_i}\right)^{\frac{2/7}{9/7}} = \left(\frac{p_f}{p_i}\right)^{\frac{2}{9}} = \frac{T_f}{T_i}$. For twice the initial pressure, $\frac{T_f}{T_i} = 2^{2/9} = 1.167$ (which is a 16.7% increase in absolute temperature).

2.67. Starting from equation 2.46, $\left(\frac{V_i}{V_f}\right)^{\gamma-1} = \frac{T_f}{T_i}$, we substitute $V_i = \frac{nRT_i}{p_i}$ (and the corresponding expression for V_f) to obtain $\left(\frac{T_i}{T_f}\right)^{\gamma-1}\left(\frac{p_f}{p_i}\right)^{\gamma-1} = \frac{T_f}{T_i}$. We then multiply the T ratio from the left to the right side of the equation, $\left(\frac{p_f}{p_i}\right)^{\gamma-1} = \left(\frac{T_f}{T_i}\right)\cdot\left(\frac{T_f}{T_i}\right)^{\gamma-1} = \left(\frac{T_f}{T_i}\right)^{\gamma}$, and take the γth root on both sides to obtain $\left(\frac{p_f}{p_i}\right)^{\frac{\gamma-1}{\gamma}} = \frac{T_f}{T_i}$. This is the most general form of equation 2.48.

2.69. Because $\left(\frac{p_f}{p_i}\right)^{\frac{\gamma-1}{\gamma}} = \frac{T_f}{T_i}$ uses only ratios of pressure, the units do not matter (as long as they are the same) and we can use the given pressure values directly. However, we need to use absolute temperatures, so $T_i = (22 + 273.15)\text{ K} = 295\text{ K}$:

$$T_f = \left(\frac{p_f}{p_i}\right)^{\frac{\gamma-1}{\gamma}} T_i = \left(\frac{46.7\text{ psi}}{14.7\text{ psi}}\right)^{\frac{2}{7}}(295\text{ K}) = 411\text{ K}.$$

2.71. $\Delta H = +333.5$ J (from Table 2.3). Using the equation $\Delta H = \Delta U + \Delta(pV) = \Delta U + p\Delta V$ (for constant pressure), we can calculate ΔU if we know the change in volume. The volumes of exactly 1 g of water and ice at 0 °C are:

$$\text{water: } V_l = \frac{1\text{ g}}{0.99984\text{ g/mL}} = 1.0002\text{ mL}\text{ ; ice: } V_s = \frac{1\text{ g}}{0.9168\text{ g/mL}} = 1.091\text{ mL}.$$

Therefore, $\Delta V = 1.0002\text{ mL} - 1.091\text{ mL} = -0.091\text{ mL}$, so that

$$\Delta U = \Delta H - p\Delta V = +333.5\text{ J} - (1.000\text{ bar})\left(\frac{1\text{ atm}}{1.01325\text{ bar}}\right)\left(-0.091\times10^{-3}\text{ L}\right)\left(\frac{101.32\text{ J}}{\text{atm}\cdot\text{L}}\right).$$

This evaluates to $\Delta U = +333.5\text{ J} + 0.0091\text{ J} = +333.5$ J. This shows that ΔU and ΔH can be very close, if not virtually the same, for many condensed-phase processes.

2.73. Steam burns hurt more than hot water burns because steam contains more energy than hot water of the same temperature (the difference is given by the heat of vaporization).

2.75. The drawing is left to the student; it should resemble Figure 2.13a.

2.77. The $\Delta_f H°$ values for NO_2 (g) and NO (g) at 298 K (25°C) are +33.10 kJ/mol and +90.29 kJ/mol, respectively (from Appendix 2). So, for 1 mol of NO_2 (g) at this temperature:

$\Delta_{rxn}H = (+33.10\text{ kJ}) - (+90.29\text{ kJ}) = -57.19$ kJ.

2.79. Appendix 2 lists the $\Delta_f H°$ value for CO_2 (g) as −393.51 kJ/mol at 298 K. The enthalpy of combustion for diamond can be expressed in terms of $\Delta_f H°$ values as follows:

−395.4 kJ/mol = (−393.51 kJ/mol) – $\Delta_f H°$[C (s, dia)], so $\Delta_f H°$[C (s, dia)] = +1.9 kJ/mol.

(This is close to the actual value of 1.897 kJ/mol, which is also given in Appendix 2.)

2.81. $\Delta_{sub}H = H_{gas} - H_{sol} = (H_{gas} - H_{liq}) + (H_{liq} - H_{sol}) = \Delta_{vap}H + \Delta_{fus}H$

2.83. Since the process occurs at constant volume, $q_V = \Delta U = -31723$ J and $w = 0$. To determine ΔH, we first need to know the balanced chemical equation for the combustion of benzoic acid:

$$C_6H_5COOH\text{ (s)} + 15/2\ O_2\text{ (g)} \rightarrow 7\ CO_2\text{ (g)} + 3\ H_2O\text{ (l)}$$

For every mole of benzoic acid, there is a change of (7 – 15/2) mol = −0.5 mol of gas.

1.20 g of benzoic acid contain 1.20 g / (122.123 g/mol) = 0.00983 mol, so that the change in amount of gas is 0.00491 mol. We can now determine ΔH at 24.6°C (297.8 K) using

$$\Delta H = \Delta U + (\Delta n)RT = -31723\text{ J} + (-0.00491\text{ mol})\left(8.314\ \frac{\text{J}}{\text{mol}\cdot\text{K}}\right)(297.8\text{ K}) = -31735\text{ J}.$$

2.85. Since the pressure remains constant, we have $q_p = \Delta H = -890.9$ kJ. For each mol of CH_4, 3 mol of gas disappear (1 mol CH_4 and 2 mol O_2) and 3 mol of gas are formed (1 mol CO_2 and 2 mol of water vapor), so that there is no net change in volume and thus $w = 0$.

Therefore, $\Delta U = q + w = -890.9$ kJ.

2.87. The thermite reaction combines aluminum powder and iron oxide and ignites the mixture to make aluminum oxide and iron:

$$2\ \mathrm{Al\ (s)} + \mathrm{Fe_2O_3\ (s)} \rightarrow 2\ \mathrm{Fe\ (s)} + \mathrm{Al_2O_3\ (s)}$$

The $\Delta_f H°$ values for iron oxide and aluminum oxide at 298 K are −825.5 kJ/mol and −1675.7 kJ/mol, respectively (from Appendix 2), so that $\Delta H = -1675.7$ kJ/mol – (−825.5 kJ/mol) = −850.2 kJ/mol at this temperature.

Because pressure remains constant, 850.2 kJ per mol of Fe_2O_3 are generated by the chemical reaction, which increases the temperature of the reaction products Fe and Al_2O_3. The specific heat capacities for Fe and Al_2O_3 are 0.452 J/(g·K) and 1.275 J/(g·K), respectively (Table 2.1), so

$$850.2\times10^3\ \mathrm{J} = (2\ \mathrm{mol})\left(55.845\ \frac{\mathrm{g}}{\mathrm{mol}}\right)\left(0.452\frac{\mathrm{J}}{\mathrm{g\cdot K}}\right)\Delta T + (1\ \mathrm{mol})\left(101.9612\ \frac{\mathrm{g}}{\mathrm{mol}}\right)\left(1.275\frac{\mathrm{J}}{\mathrm{g\cdot K}}\right)\Delta T$$

This yields $\Delta T = 4711$ K, resulting in a final temperature of 298 K + 4711 K = 5009 K.

CHAPTER 3

The Second and Third Laws of Thermodynamics

3.1. **(a)** Not spontaneous, because the melting point of ice is 0 °C.

(b) Spontaneous, because the melting point of ice is 0 °C.

(c) Spontaneous, because potassium compounds are generally soluble in water.

(d) Not spontaneous, because an unplugged refrigerator will warm up.

(e) Spontaneous, because of the effect of gravity on the leaf.

(f) Spontaneous, because both lithium and fluorine are rather reactive elements.

(g) Not spontaneous, because water does not break apart into hydrogen and oxygen without some input of energy.

3.3. $$e = 1 + \frac{q_3}{q_1} = 1 + \frac{-623\text{ J}}{+850\text{ J}} = 0.267 = 26.7\%$$

3.5. From $e = 1 - \frac{T_{\text{low}}}{T_{\text{high}}}$ and $T_{\text{high}} = (273.15 + 150)\text{ K} = 423\text{ K}$, we obtain

$T_{\text{low}} = (1 - e) \cdot T_{\text{high}} = (1 - 0.440)(423\text{ K}) = 237\text{ K}$, which is -36 °C.

3.7. 0 °C is 273 K and 100 °C is 373 K, so that $e = 1 - \frac{T_{\text{low}}}{T_{\text{high}}} = 1 - \frac{273\text{ K}}{373\text{ K}} = 0.268 = 26.8\%$.

3.9. Superheated steam has the advantage of a higher temperature, and according to $e = 1 - \frac{T_{\text{low}}}{T_{\text{high}}}$, the efficiency of a heat engine will increase with increasing T_{high}.

3.11. A Carnot cycle can start at an adiabatic expansion process (or at any other point in the cycle), as long as it performs the other three processes in the proper order.

3.13. If the refrigerator operates at $T_{\text{low}} = ½\, T_{\text{high}}$, its coefficient of performance is $e = \frac{T_{\text{low}}}{T_{\text{high}} - T_{\text{low}}} = \frac{½ T_{\text{high}}}{T_{\text{high}} - ½ T_{\text{high}}} = \frac{½}{½} = 1$. This means that for each amount of heat removed from the reservoir, the same amount of work has to be put in, which is very costly and inefficient.

3.15. Entropy is a state function, so the entropy change for an ideal Carnot cycle (as any other closed cycle) is zero. This can be seen directly from equation 3.11, which states that the entropy changes during the reversible isothermal expansion and compression add up to zero (note that the entropy changes during the reversible adiabatic expansion and compression do not contribute because $q_2 = q_4 = 0$).

3.17. The enthalpy of fusion indicates that 12.55 kJ of heat is necessary to melt 1 mol (or 197.0 g) of solid gold, which means that in order to melt 28.3 g Au,

$12.55\text{ kJ}\cdot\dfrac{28.3\text{ g}}{197.0\text{ g}}=1.80\text{ kJ}$ are needed. So the amount of heat released when 28.3 g of liquid gold solidifies is q = -1.80 kJ, and the entropy change at T = (1064 + 273.15) K = 1337 K is $\Delta S=\dfrac{-1.80\times 10^{3}\text{ J}}{1337\text{ K}}=-1.35\text{ J/K}$.

3.19. 1.00 mol of H_2O is 18.02 g H_2O. 0 °C is 273 K and 100 °C is 373 K, so that

$$\Delta S = mc\ln\frac{T_{\text{f}}}{T_{\text{i}}}=(18.02\text{ g})\left(4.18\ \frac{\text{J}}{\text{g}\cdot\text{K}}\right)\left(\ln\frac{373\text{ K}}{273\text{ K}}\right)=23.5\text{ J/K}.$$

3.21. Recall that the heat capacity at constant volume of 1 mol of a monoatomic ideal gas is 12.471 J/K and temperature-invariant. 45 °C is 318 K and 55 °C is 328 K, so that the entropy change for the isochoric warm-up is

$$\Delta S = C_V\ln\frac{T_{\text{f}}}{T_{\text{i}}}=\left(12.471\frac{\text{J}}{\text{K}}\right)\left(\ln\frac{328\text{ K}}{318\text{ K}}\right)=0.386\text{ J/K}.$$

Entropy is a state function, so the entropy change of the system is always 0.386 J/K, whether the process is carried out reversibly or not.

3.23. We first calculate the final temperature from the fact that the pressure of the ideal gas is 2.45 atm when the volume has reached 5.00 L:

$$T=\frac{pV}{nR}=\frac{(2.45\text{ atm})(5.00\text{ L})}{(0.500\text{ mol})\left(0.08205\ \dfrac{\text{L}\cdot\text{atm}}{\text{mol}\cdot\text{K}}\right)}=299\text{ K}.$$

This means that the process is essentially isothermal, and that the entropy change of the system is simply

$$\Delta S_{\text{sys}}=nR\ln\frac{V_{\text{f}}}{V_{\text{i}}}=(0.500\text{ mol})\left(8.314\ \frac{\text{J}}{\text{mol}\cdot\text{K}}\right)\left(\ln\frac{5.00\text{ L}}{13.00\text{ L}}\right)=-3.97\text{ J/K}.$$

For an ideal gas undergoing an isothermal process, $\Delta U = q + w = 0$, so $q=-w=-(-p_{\text{ext}}\Delta V)$.

$q=+p_{ext}\Delta V=(2.45\text{ atm})(13.00\text{ L}-5.00\text{ L})\left(\dfrac{101.32\text{ J}}{1\text{ L atm}}\right)=1.99\text{ kJ}$. Because this heat came from the surroundings, we obtain $\Delta S_{surr}=\dfrac{1.99\text{ kJ}}{298\text{ K}}=6.66\text{ J/K}$. So

$\Delta S_{\text{univ}}=\Delta S_{\text{sys}}+\Delta S_{\text{surr}}=+2.69\text{ J/K}.$

3.25. At 22.0 °C (295.2 K), 1 liter of air at exactly 1 atm pressure contains

$$n=\frac{pV}{RT}=\frac{(1\text{ atm})(1\text{ L})}{\left(0.08205\ \frac{\text{L}\cdot\text{atm}}{\text{mol}\cdot\text{K}}\right)(295.2\text{ K})}$$ = 0.04 mol air. The entropy change of 0.04 mol of gas undergoing a pressure change from 760 torr to 758 torr is

$$\Delta S=-nR\ln\frac{p_\text{f}}{p_\text{i}}=-(0.04\text{ mol})\left(8.314\ \frac{\text{J}}{\text{mol}\cdot\text{K}}\right)\left(\ln\frac{758\text{ torr}}{760\text{ torr}}\right)=9\times10^{-4}\text{ J/K}.$$

3.27. The amount of air contained in 15.6 L at 46.0 psi (3.17 bar) and 22.0 °C (295.2 K) is

$$n=\frac{pV}{RT}=\frac{(3.17\text{ bar})(15.6\text{ L})}{\left(0.08314\ \frac{\text{L}\cdot\text{bar}}{\text{mol}\cdot\text{K}}\right)(295.2\text{ K})}$$ = 2.02 mol air. Changing the temperature from 22.0 °C (295.2 K) to 85.0 °C (358.2 K) at constant (tire) volume changes the entropy by

$$\Delta S=n\bar{C}\ln\frac{T_\text{f}}{T_\text{i}}=(2.02\text{ mol})\left(20.79\ \frac{\text{J}}{\text{mol}\cdot\text{K}}\right)\left(\ln\frac{358.2\text{ K}}{295.2\text{ K}}\right)=8.11\text{ J/K}.$$

Note that although the air pressure in the tire increases as well, this is already taken into account in the calculation because we have used the constant-volume heat capacity.

3.29. Start with equation 3.21, which describes the entropy change of an ideal gas during an isothermal process in which the volume changes: $\Delta S=nR\ln\frac{V_\text{f}}{V_\text{i}}$.

The change in volume is accompanied by a change in pressure (since T is constant), and we can use Boyle's law to substitute volumes by pressures: $p_\text{i}V_\text{i}=p_\text{f}V_\text{f}$, which can be rearranged into $\frac{V_\text{f}}{V_\text{i}}=\frac{p_\text{i}}{p_\text{f}}$. Substituting leads to the equation we wanted to derive:

$$\Delta S=nR\ln\frac{p_\text{i}}{p_\text{f}}=-nR\ln\frac{p_\text{f}}{p_\text{i}}.$$

(Recall that when you take the reciprocal of a fraction inside a logarithm, the value of the logarithm changes sign: $\ln\frac{a}{b}=-\ln\frac{b}{a}$.)

3.31. For one mole of air, the percentages given are numerically equal to both the number of moles of each gas n_i in the mixture *and* the mole fraction x_i. Therefore:

$$\Delta_\text{mix}S=-R\sum n_i\ln x_i=-\left(8.314\frac{\text{J}}{\text{mol}\cdot\text{K}}\right)(0.79\text{ mol}\cdot\ln0.79+0.20\text{ mol}\cdot\ln0.20+0.01\text{ mol}\cdot\ln0.01)$$

$$\Delta_\text{mix}S=+4.6\text{ J/K}$$

3.33. As with exercise 3.31, the percentages give the number of moles *and* the mole fractions when 1 mole of total gas is involved, as it is here. Therefore,

$$\Delta_{\text{mix}}S = -R\sum n_i \ln x_i = -\left(8.314\frac{\text{J}}{\text{mol}\cdot\text{K}}\right)(0.40\text{ mol}\cdot\ln 0.40 + 0.60\text{ mol}\cdot\ln 0.60) = +5.6\text{ J/K}$$

3.35. **(a)** The final temperature of both silver samples is simply the average temperature of 75 °C (348 K), because both samples have the exact same mass.

(b) The hot Ag sample has an initial temperature of 150 °C (423 K), so the entropy change is $\Delta S_{\text{hot}} = n\bar{C}\ln\frac{T_\text{f}}{T_\text{i}} = (1.00\text{ mol})\left(25.75\frac{\text{J}}{\text{mol}\cdot\text{K}}\right)\left(\ln\frac{348\text{ K}}{423\text{ K}}\right) = -5.02\text{ J/K}$.

(c) The cold Ag sample has an initial temperature of 0 °C (273 K), so the entropy change is $\Delta S_{\text{cold}} = n\bar{C}\ln\frac{T_\text{f}}{T_\text{i}} = (1.00\text{ mol})\left(25.75\frac{\text{J}}{\text{mol}\cdot\text{K}}\right)\left(\ln\frac{348\text{ K}}{273\text{ K}}\right) = +6.25\text{ J/K}$.

(d) The total ΔS of the system is the sum of both values:

$\Delta S_{\text{total}} = -5.02\text{ J/K} + 6.25\text{ J/K} = +1.22\text{ J/K}$.

(e) If we assume that the two silver samples only exchange heat with each other but not with the surroundings, the system is isolated and the process is therefore spontaneous ($\Delta S_{\text{total}} > 0$).

3.37. The molar mass of water is 18.0152 g/mol, so 2.22 mol of water have a mass of 40.0 g. With an initial temperature of 25.0 °C (298.2 K) and a final temperature of 100 °C (373 K), we obtain $\Delta S = mc\ln\frac{T_\text{f}}{T_\text{i}} = (40.0\text{ g})\left(4.18\frac{\text{J}}{\text{g}\cdot\text{K}}\right)\left(\ln\frac{373\text{ K}}{298.2\text{ K}}\right) = +37.5\text{ J/K}$.

3.39. We want to calculate the *molar* entropy change for an isothermal pressure change, which means we need to use $n = 1$ mol:

$$\Delta S = -nR\ln\frac{p_\text{f}}{p_\text{i}} = -(1\text{ mol})\left(8.314\frac{\text{J}}{\text{mol}\cdot\text{K}}\right)\left(\ln\frac{0.97\text{ atm}}{2.55\text{ atm}}\right) = +8.04\text{ J/K}.$$

The molar entropy change is therefore 8.04 J/(mol·K).

3.41. Because (for an isolated system) the first law of thermodynamics prohibits the creation of new energy, the concept of "you can't win" may be used to convey – if inaccurately – that fact. And because the second law of thermodynamics requires an efficiency of less than 100%, you will always get less energy out of a process than the energy going into that process. Thus, "you can't even break even" may be a way to convey that idea.

3.43. The drawing for such an (impossible) heat engine would be the same as Figure 3.1, but without the low-temperature reservoir ($q_{\text{out}} = 0$). Because all heat extracted from the (hot) reservoir is converted into work, its efficiency would be 100%. This, however, is impossible because no engine can ever be 100% efficient, as was shown in the text.

3.45. **(a)** Using the various values of R, it can be shown that 1 L·atm = 101.32 J, so

$$k=\left(1.381\times10^{-23}\ \frac{\text{J}}{\text{K}}\right)\left(\frac{1\ \text{L}\cdot\text{atm}}{101.32\ \text{J}}\right)=1.363\times10^{-25}\ \text{L}\cdot\text{atm/K}.$$

(b)

$$k=\left(1.381\times10^{-23}\ \frac{\text{J}}{\text{K}}\right)\left(\frac{1\ \text{L}\cdot\text{atm}}{101.32\ \text{J}}\right)\left(\frac{1000\ \text{cm}^3}{1\ \text{L}}\right)\left(\frac{760\ \text{mmHg}}{1\ \text{atm}}\right)=1.036\times10^{-19}\ \text{cm}^3\cdot\text{mmHg/K}$$

3.47. **(a)** 1 g of liquid Au, because more energy states are available in a liquid than in a solid at the same temperature.

(b) 1 mole of CO_2 at STP, because a three-atomic gas will have more energy states available than a diatomic gas under the same conditions.

(c) 1 mole of Ar at 0.01 atm. At identical temperature (which we assume to be the case), the gas at lower pressure occupies a greater volume, and more microstates are available to the gas molecules.

3.49. For a monoatomic ideal gas such as Kr, the constant-volume heat capacity is temperature-invariant and equal to 12.471 J/(mol·K). The change in entropy due to the temperature change at constant volume is then

$$\Delta S=n\bar{C}_V\ln\frac{T_\text{f}}{T_\text{i}}=(1\ \text{mol})\left(12.471\frac{\text{J}}{\text{mol}\cdot\text{K}}\right)\left(\ln\frac{200.00\ \text{K}}{298.15\ \text{K}}\right)=-4.9794\ \text{J/K}.$$

Thus, the absolute entropy of 1 mol of Kr is lower by 4.9794 J/K, which means that its absolute molar entropy at 200.00 K is (163.97 – 4.9794) J/(mol·K) = 158.99 J/(mol·K).

3.51. In order of increasing entropy:

C(diamond) < C(graphite) < Si (crystal) < Fe (solid) < NaCl (solid) < $BaSO_4$ (solid).

All of these substances are solids, so the ordering is determined by the degree of order in the crystal (diamond versus graphite) and mass (elements with higher masses have more energy states available, therefore C < Si < Fe). Likewise, a greater variety of atoms will increase the entropy, which explains why the salts NaCl and $BaSO_4$ have the greatest entropies.

3.53. Initially, the entropy per molecule is $S_1=k\ln\Omega_1$ and changes to $S_2=k\ln\Omega_2=k\ln(2\Omega_1)$ after the process. Recall that ln(a·b) = ln(a) + ln(b), so the latter expression can be rewritten as $S_2=k\ln2+k\ln\Omega_1=k\ln2+S_1$. The difference $\Delta S=S_2-S_1$ is then simply $\Delta S=k\ln2=\left(1.381\times10^{-23}\ \frac{\text{J}}{\text{K}}\right)(\ln2)=+9.572\times10^{-24}$ J/K per molecule. Or per mole of molecules: $\Delta\bar{S}=\left(9.572\times10^{-24}\ \frac{\text{J}}{\text{K}}\right)\left(\frac{6.022\times10^{23}}{1\ \text{mol}}\right)=+5.764\ \frac{\text{J}}{\text{mol}\cdot\text{K}}$.

3.55. The listings of $\Delta_f H$ and $\Delta_f G$ refer to the formation reaction of the tabulated product from its constituent elements in their standard states, and are therefore zero for elements (in their standard states). The entropies of the latter are non-zero because zero entropy is only obtained for a perfect crystal at 0 K and the tabulated thermodynamic refer to room temperature.

3.57. The balanced chemical reaction is: 2 Al (s) + Fe_2O_3 (s) → Al_2O_3 (s) + 2 Fe (s)

$\Delta_{rxn}S = [(50.92 + 2 \cdot 27.3) - (2 \cdot 28.30 + 87.4)]$ J/K = −38.5 J/K

3.59. For the formation of H_2O (l): $\Delta_{rxn}S = [69.91 - (130.68 + ½ \cdot 205.14)]$ J/K = −163.34 J/K

For the formation of H_2O (g): $\Delta_{rxn}S = [188.83 - (130.68 + ½ \cdot 205.14)]$ J/K = −44.42 J/K

The formation of water vapor has a $\Delta_{rxn}S$ that is higher by 118.92 J/K compared to liquid water, because the entropy of the gas phase is higher than that of the liquid phase.

3.61. The entropy change for the combustion of 2 moles of C_8H_{18} is

$\Delta_{rxn}S = [(16 \cdot 213.785 + 18 \cdot 69.91) - (2 \cdot 361.2 + 25 \cdot 205.14)]$ J/K = −1171.96 J/K,

which corresponds to −585.98 J/K per mole of gasoline.

The molar mass of C_8H_{18} is 114.230 g/mol, so 2653 g correspond to 23.23 mol. The entropy change for the combustion of this amount of gasoline is then -13.61×10^3 J/K.

3.63. The entropy change ΔS for this reaction should be positive, because two solid substances are transformed into a solid, a liquid, and a gas, and liquids and especially gases have a much higher entropy content. This increase in entropy is the main reason of why the reaction is spontaneous, even if it absorbs energy and decreases the entropy of the surroundings (the entropy gain of the system is so high that it outweighs the entropy loss of the surroundings).

CHAPTER 4

Gibbs Energy and Chemical Potential

4.1. A spontaneous process occurs if one of the following statements about the system is true:

(a) $dS \geq 0$ while U and V remain constant;

(b) $dU \leq 0$ while V and S remain constant;

(c) $dH \leq 0$ while p and S remain constant.

Note that while this list is not exhaustive, *i.e.* does not cover all possible scenarios for spontaneous processes, a process that *does* meet one of the listed criteria *is* indeed spontaneous.

4.3. The expression on the left is less than zero if $\frac{dU + p\,dV}{T} \leq dS$. There are many situations that fulfill this requirement (for example if both left-hand side and right-hand side are positive, but the right-hand side is greater than the left-hand side etc.), but it is *certainly* fulfilled if the left-hand side is negative and the right-hand side is positive. That means that a decrease in energy ($dU < 0$, which makes $dU + p\,dV$ more negative) and an increase in entropy ($dS > 0$) are generally favorable for a spontaneous change.

4.5. The expansion is adiabatic, so $q = 0$, and occurs agains zero external pressure (free expansion into vacuum), so $w = -p_{\text{ext}}\,dV = 0$; thus, the system is isolated ($\Delta U = q + w = 0$) and the spontaneity condition $\Delta S > 0$ applies.

We know from the last chapter and the equation $\Delta S = nR\ln\frac{V_{\text{f}}}{V_{\text{i}}}$ that increasing the volume of an ideal gas at constant T increases its entropy. This means that the adiabatic free expansion of an ideal gas (which is isothermal) satisfies the spontaneity condition $\Delta S > 0$, and is therefore spontaneous.

4.7. Use the definition of the Gibbs energy $G = H - TS$ and take the derivative:

$dG = dH - T\,dS - S\,dT$. Now solve for dH, $dH = dG + T\,dS + S\,dT$, and substitute this in the original spontaneity condition $dH - V\,dp - T\,dS \leq 0$: $dG + T\,dS + S\,dT - V\,dp - T\,dS \leq 0$.

The $T\,dS$ terms cancel out: $dG + S\,dT - V\,dp \leq 0$. Under conditions of constant T and p, the second and third terms are zero, so this spontaneity condition simplifies to $(dG)_{T,p} \leq 0$.

4.9. Since ΔA is always less than or equal to the maximum amount of work the system can do during an isothermal process (the temperature is constant here), it must be less than w in our case because the process is not reversible. The work done by the system is

$$w = -p_{\text{ext}}\,dV = -\left(880\text{ mmHg} \times \frac{1\text{ atm}}{760\text{ mmHg}}\right)\left[(3.5-1.0)\text{ L} \times \frac{101.32\text{ J}}{1\text{ L}\cdot\text{atm}}\right] = -2.9\times10^{2}\text{ J}.$$

(Note that we cannot use $w = -nRT\ln\frac{V_f}{V_i}$ here, because this formula is *only* valid for reversible processes.)

Therefore, ΔA must be less than this amount of energy: $\Delta A < -2.9\times10^{2}$ J.

4.11. Because this is a reversible, isothermal process, ΔA is equal to the reversible work:

$$\Delta A = w_{\text{rev}} = -nRT\ln\frac{V_f}{V_i} = -(0.0200\text{ mol})\left(8.314\frac{\text{J}}{\text{mol}\cdot\text{K}}\right)(1400\text{ K})\left(\ln\frac{10V_i}{V_i}\right) = -536\text{ J}.$$

4.13. Using $\Delta_f G°$ data from Appendix 2, which are tabulated for 298 K = 25 °C, $\Delta_{rxn}G°$ at that temperature and standard pressure can be calculated from Hess's law (products-minus-reactants):

$$\Delta_{\text{rxn}}G° = \left[(1\cdot26.7)-(1\cdot124.4+3\cdot0)\right]\text{ kJ/mol} = -97.7\text{ kJ/mol}.$$

The spontaneity condition at constant T and p is $(dG)_{T,p} \le 0$, and because the Gibbs energy decreases, the reaction is spontaneous under these conditions.

4.15. We first calculate $\Delta_{rxn}G$ at 298 K and standard pressure using $\Delta_f G°$ data from Appendix 2 using Hess's law (products-minus-reactants):

$$\Delta_{\text{rxn}}G = \left[(1\cdot97.79)-(2\cdot51.30)\right]\text{ kJ} = -4.81\text{ kJ}.$$

We can also determine ΔH and ΔS for the process first, then use the equation $\Delta G = \Delta H - T\,\Delta S$:

$$\Delta_{\text{rxn}}H = \left[(1\cdot9.08)-(2\cdot33.10)\right]\text{ kJ} = -57.12\text{ kJ}$$

$$\Delta_{\text{rxn}}S = \left[(1\cdot304.38)-(2\cdot240.04)\right]\text{ J/K} = -175.70\text{ J/K} = -0.17570\text{ kJ/K}$$

Therefore, $\Delta_{\text{rxn}}G = -57.12\text{ kJ} - (298\text{ K})\left(-0.17570\frac{\text{kJ}}{\text{K}}\right) = -4.76\text{ kJ}.$

The two values are close enough that they can be considered equal.

4.17. We use the equation $\Delta G = \Delta H - T\,\Delta S$ and solve for ΔS:

$$\Delta S = \frac{\Delta H - \Delta G}{T} = \frac{+1.897\text{ kJ} - 2.90\text{ kJ}}{298\text{ K}} = -0.00336\text{ kJ/K} = -3.36\text{ J/K}.$$

This means that the graphite-to-diamond conversion at 1 bar and 25 °C is not spontaneous (which also means that the conversion of diamond to graphite is spontaneous; the fact that diamond is stable under these conditions is only due to the fact that the reaction is extremely slow). The value of ΔS makes sense because diamond is a much more ordered crystalline system than graphite.

4.19. The maximum amount of non-pV work that a process can potentially yield at constant pressure and temperature is given by ΔG. We therefore calculate $\Delta_{rxn}G$ at 25 °C (298 K) and standard pressure using $\Delta_f G°$ data for 298 K (Appendix 2):

$$\Delta_{rxn}G = \left(\left[2\cdot(-237.14)+1\cdot(-394.35)\right]-\left[1\cdot(-50.8)+2\cdot 0\right]\right)\text{ kJ} = -817.8\text{ kJ}$$

The maximum amount of electrical work produced by the reaction of 1.00 mol of methane in a fuel cell at 25 °C and standard pressure is therefore (1.00 mol) (+817.8 kJ/mol) = 818 kJ.

4.21. No (at least not at constant p and T), because the maximum amount of non-pV work that can be obtained from an isothermal and isobaric process is limited by ΔG and is therefore 0.

4.23. We will assume that the battery operates at constant pressure (standard pressure) and temperature (25 °C = 298 K). Under these conditions, the maximum amount of non-pV work is equal to $\Delta_{rxn}G°$, which we can calculate according to

$$\Delta_{rxn}G = (1\text{ mol})\cdot\Delta_f G^\circ_{MX} - \left(1\cdot 0+\tfrac{1}{2}\cdot 0\right)\text{ kJ} = (1\text{ mol})\cdot\Delta_f G^\circ_{MX},$$

because the Gibbs energies of formation for the elements are zero. From the appendix:

$\Delta_f G°$ (kJ/mol)	MF	MCl	MBr	MI
M = Li	−587.7	−372.2	−342.0	−270.3
M = Na	−546.3	−365.7	−349.0	−286.1
M = K	−537.8	−408.5	−380.7	−324.9

On the basis of this chart, a battery utilizing the reaction between lithium and fluorine would provide the most energy. Lithium-ion batteries are commercially available but do not rely on the lithium-halogen reaction for producing energy. Alkaline batteries get their name from the use of an alkaline electrolyte (potassium hydroxide), but do not rely on the alkali-halogen reaction either.

4.25. 2808 °C is (273.15 + 2808) K = 3081 K. Phase changes are isothermal because they occur at a well-defined temperature, and we can therefore use $\Delta G = \Delta H - T\,\Delta S$ to solve for ΔS.

With $\Delta G = 0$, we obtain $\Delta S = \dfrac{\Delta H - \Delta G}{T} = \dfrac{\Delta H}{T} = \dfrac{343\text{ kJ}}{3081\text{ K}} = +0.111\dfrac{\text{kJ}}{\text{K}} = +111\text{ J/K}$.

The entropy change during a liquid-to-gas phase change is much larger than that for a solid-to-liquid conversion because the gas phase has a much larger entropy than either solid or liquid phase.

4.27. ΔA is zero for the complete Carnot cycle, since A is a state function and the system returns to its original state after a complete cycle.

4.29. We can express U as follows: $A = U - TS \rightarrow U = A + TS$. Substituting the entropy S in the second term by the temperature-dependence of A, $\left(\frac{\partial A}{\partial T}\right)_V = -S$, leads to

$$U = A - T\left(\frac{\partial A}{\partial T}\right)_V.$$

(The volume-dependence of A, namely $\left(\frac{\partial A}{\partial V}\right)_T = -p$, is not needed here.)

4.31. U, H, A, and G are energies and have units of J; therefore, changes in these energies are also described using J. The entropy S has units of J/K, and changes in entropies are also expressed in J/K. Therefore, the partial derivatives $\left(\frac{\partial U}{\partial S}\right)_V$ and $\left(\frac{\partial H}{\partial S}\right)_p$ have units of $\frac{\text{J}}{\text{J/K}} = \frac{1}{1/\text{K}} = \text{K}$. This is consistent with the fact that these derivatives equal the temperature T.

Similarly, $\left(\frac{\partial A}{\partial T}\right)_V$ and $\left(\frac{\partial G}{\partial T}\right)_p$ have units of J/K, which is consistent with the fact that they equal the negative entropy, $-S$, which has units of J/K.

The volume V has SI units of m^3 ($1\ \text{m}^3 = 1{,}000$ L), and volume changes are also expressed in m^3. This means that $\left(\frac{\partial U}{\partial V}\right)_S$ and $\left(\frac{\partial A}{\partial V}\right)_T$ have units of $\frac{\text{J}}{\text{m}^3} = \frac{\text{N}\cdot\text{m}}{\text{m}^3} = \frac{\text{N}}{\text{m}^2}$, where we have used the fact that 1 J = 1 N·m (recall that work energy is force times distance). N/m^2 are units of pressure (force per area; $1\ \text{N/m}^2 = 1$ Pa), which is consistent with the fact that these derivatives equal the negative pressure, $-p$.

Similarly, because changes in pressure have units of pressure, the partial derivatives $\left(\frac{\partial H}{\partial p}\right)_S$ and $\left(\frac{\partial G}{\partial p}\right)_T$ have units of $\frac{\text{J}}{\text{N/m}^2} = \frac{\text{N}\cdot\text{m}}{\text{N/m}^2} = \frac{\text{m}}{1/\text{m}^2} = \text{m}^3$, again consistent with the fact that these derivatives are equal to the volume V.

4.33. A and G are defined as $A = U - TS$ and $G = H - TS$, respectively, so at constant T the changes in these energies are $\Delta A = \Delta U - T\,\Delta S$ and $\Delta G = \Delta H - T\,\Delta S$, respectively. This means that ΔA and ΔG values have entropy componenets to them, which are not necessarily zero for all isothermal processes.

4.35. To derive equation 4.35, we start with equation 4.31: $\left[\frac{\partial}{\partial V}\left(\frac{\partial U}{\partial S}\right)_V\right]_S = \left[\frac{\partial}{\partial S}\left(\frac{\partial U}{\partial V}\right)_S\right]_V$.

Since $\left(\frac{\partial U}{\partial S}\right)_V = T$ and $\left(\frac{\partial U}{\partial V}\right)_S = -p$, we substitute to get $\left(\frac{\partial T}{\partial V}\right)_S = \left[\frac{\partial(-p)}{\partial S}\right]_V = -\left(\frac{\partial p}{\partial S}\right)_V$.

To derive equation 4.36, we start with equation 4.32: $\left[\frac{\partial}{\partial V}\left(\frac{\partial A}{\partial T}\right)_V\right]_T = \left[\frac{\partial}{\partial T}\left(\frac{\partial A}{\partial V}\right)_T\right]_V$.

Since $\left(\frac{\partial A}{\partial T}\right)_V = -S$ and $\left(\frac{\partial A}{\partial V}\right)_T = -p$, we substitute to get $\left(\frac{\partial S}{\partial V}\right)_T = \left(\frac{\partial p}{\partial T}\right)_V$. Note that the two negative signs cancel each other out in this case.

To derive equation 4.37, we start with equation 4.33: $\left[\frac{\partial}{\partial T}\left(\frac{\partial G}{\partial p}\right)_T\right]_p = \left[\frac{\partial}{\partial p}\left(\frac{\partial G}{\partial T}\right)_p\right]_T$.

Since $\left(\frac{\partial G}{\partial p}\right)_T = V$ and $\left(\frac{\partial G}{\partial T}\right)_p = -S$, we substitute to get $\left(\frac{\partial V}{\partial T}\right)_p = -\left(\frac{\partial S}{\partial p}\right)_T$, or $\left(\frac{\partial S}{\partial p}\right)_T = -\left(\frac{\partial V}{\partial T}\right)_p$.

4.37. Let's start with the right-hand side of the equation. Using the fact that the expansion coefficient is defined as $\alpha = \frac{1}{V}\left(\frac{\partial V}{\partial T}\right)_p$ (see Chapter 1), we have $-\alpha V = \frac{-V}{V}\left(\frac{\partial V}{\partial T}\right)_p = -\left(\frac{\partial V}{\partial T}\right)_p$. According to the Maxwell relation in equation 4.37, $\left(\frac{\partial S}{\partial p}\right)_T = -\left(\frac{\partial V}{\partial T}\right)_p$, so that $-\alpha V = \left(\frac{\partial S}{\partial p}\right)_T$.

4.39. Finite changes in a state variable such as volume V are simply the sum over an (infinite) number of infinitesimal changes. This means that infinitesimal changes dV can be converted into finite changes ΔV through the process of integration: $\int dV = \Delta V$.

Equations 4.14-4.17 can therefore be rewritten in terms of finite changes by taking the integral of both sides. For example, integrating $dU = T\,dS - p\,dV$ leads to $\int dU = \int (T\,dS - p\,dV) = \int T\,dS - \int p\,dV$.

If temperature remains constant, T can be factored out of the first integral, and if pressure remains constant as well, it can be factored out of the second integral, and we obtain $\int dU = T\int dS - p\int dV$, or $\Delta U = T\,\Delta S - p\,\Delta V$.

Similarly, we obtain for the other equations:

$\Delta H = T\,\Delta S + V\,\Delta p$ if T and V remain constant; $\Delta A = -S\,\Delta T - p\,\Delta V$ if S and p remain constant;

$\Delta G = -S\,\Delta T + V\,\Delta p$ if S and V remain constant.

4.41. From the previous problem, we can see that $\left[\frac{\partial(\Delta U)}{\partial V}\right]_S = -\Delta p$. Multiplying both sides by dV, we get $d(\Delta U) = -(\Delta p)\cdot(dV)$ for an isentropic process. While this equation is strictly valid for infinitesimal volume changes dV, it can be used as an approximation only for finite changes in volume ΔV. With $\Delta p = (1.00 - 7.33)$ atm = -6.33 atm and $\Delta V = (10.0 - 3.04)$ L = 7.0 L, the change in ΔU is then approximately

$$-(-6.33\text{ atm})(7.0\text{ L})\left(\frac{101.32\text{ J}}{1\text{ L}\cdot\text{atm}}\right) = +4.5\times10^3\text{ J} = 4.5\text{ kJ}.$$

4.43. Using $\left(\frac{\partial H}{\partial p}\right)_S = V$, the equation simplifies to

$$\left[C_p - \left(\frac{\partial U}{\partial T}\right)_p - \left(\frac{\partial H}{\partial p}\right)_S\left(\frac{\partial p}{\partial T}\right)_V\right] = C_p - \left(\frac{\partial U}{\partial T}\right)_p - V\left(\frac{\partial p}{\partial T}\right)_V.$$

For an ideal gas, $p = \frac{nRT}{V}$ and $\left(\frac{\partial p}{\partial T}\right)_V = \frac{nR}{V}$, which leads to

$C_p - \left(\frac{\partial U}{\partial T}\right)_p - V\frac{nR}{V} = C_p - \left(\frac{\partial U}{\partial T}\right)_p - nR$. We also recognize that $C_p = \left(\frac{\partial H}{\partial T}\right)_p$, so our equation becomes $\left(\frac{\partial H}{\partial T}\right)_p - \left(\frac{\partial U}{\partial T}\right)_p - nR$. The two partial derivatives can be related to each other because $H = U + pV$: $\left(\frac{\partial H}{\partial T}\right)_p = \left(\frac{\partial U}{\partial T}\right)_p + \left[\frac{\partial(pV)}{\partial T}\right]_p$. But since pressure is kept constant, $\left[\frac{\partial(pV)}{\partial T}\right]_p = p\left(\frac{\partial V}{\partial T}\right)_p$, and $\left(\frac{\partial V}{\partial T}\right)_p = \frac{nR}{p}$ for an ideal gas ($V = nRT/p$).

Substituting this into our equation leads to

$$\left(\frac{\partial H}{\partial T}\right)_p - \left(\frac{\partial U}{\partial T}\right)_p - nR = \left(\frac{\partial U}{\partial T}\right)_p + p\frac{nR}{p} - \left(\frac{\partial U}{\partial T}\right)_p - nR = 0.$$

4.45. Using the equation from example 4.11: $\left(\frac{\partial U}{\partial V}\right)_T = T\left(\frac{\partial p}{\partial T}\right)_V - p$. For an ideal gas, $p = nRT/V$, and the partial derivative with respect to temperature is therefore $\left(\frac{\partial p}{\partial T}\right)_V = \frac{nR}{V}$.

Substituting this into the equation gives $\left(\frac{\partial U}{\partial V}\right)_T = T\frac{nR}{V} - p = \frac{nRT}{V} - p = p - p = 0.$

This is expected, as the internal energy of an ideal gas does not depend on volume, only on temperature (see euqation 2.28). We assume that in an ideal gas the individual particles do not interact with each other, so an increase in the average distance between them will have no effect on their energy if the temperature remains constant.

4.47. The Berthelot equation of state is $p=\frac{RT}{\overline{V}-b}-\frac{a}{T\overline{V}^2}$. In order to evaluate $(\partial U/\partial V)_T$ using the expression from Example 4.11, $\left(\frac{\partial U}{\partial V}\right)_T=T\left(\frac{\partial p}{\partial T}\right)_V-p$, we need to calculate $(\partial p/\partial T)_V$:

$\left(\frac{\partial p}{\partial T}\right)_V=\frac{R}{\overline{V}-b}-\frac{a}{\overline{V}^2}\left[\frac{d(1/T)}{dT}\right]=\frac{R}{\overline{V}-b}-\frac{a}{\overline{V}^2}\left(-\frac{1}{T^2}\right)=\frac{R}{\overline{V}-b}+\frac{a}{\overline{V}^2T^2}$. Substituting this into our expression leads to

$\left(\frac{\partial U}{\partial V}\right)_T=T\left(\frac{R}{\overline{V}-b}+\frac{a}{\overline{V}^2T^2}\right)-p=\frac{RT}{\overline{V}-b}+\frac{aT}{\overline{V}^2T^2}-p=\frac{RT}{\overline{V}-b}+\frac{a}{T\overline{V}^2}-p$. Substituting p by the Berthelot equation, we get $\left(\frac{\partial U}{\partial V}\right)_T=\frac{RT}{\overline{V}-b}+\frac{a}{T\overline{V}^2}-\left(\frac{RT}{\overline{V}-b}-\frac{a}{T\overline{V}^2}\right)=\frac{2a}{T\overline{V}^2}$.

4.49. According to the equation 4.24, $\left(\frac{\partial G}{\partial T}\right)_p=-S$; therefore $\left(\frac{\partial \Delta G}{\partial T}\right)_p=-\Delta S$.

(This becomes immediately clear when you write ΔG as G_2-G_1, because

$$\left(\frac{\partial \Delta G}{\partial T}\right)_p=\left[\frac{\partial(G_2-G_1)}{\partial T}\right]_p=\left(\frac{\partial G_2}{\partial T}\right)_p-\left(\frac{\partial G_1}{\partial T}\right)_p=-S_2+S_1=-(S_2-S_1)=-\Delta S.)$$

In exercise 3.57, we showed that the entropy change for the reaction was –38.5 J/K assuming standard conditions. Under these conditions $\left(\frac{\partial \Delta G}{\partial T}\right)_p=-\Delta S=+38.5$ J/K.

4.51. Take the reciprocal of equation 4.44: $\left(\frac{\partial \frac{1}{T}}{\partial \frac{\Delta G}{T}}\right)_p=\frac{1}{\Delta H}$ (remember that partial derivatives obey some of the same algebraic rules as fractions, because they are just the ratio of two infinitesimal numbers). A plot of $1/T$ on the y axis versus $\Delta G/T$ on the x axis therefore has a slope of $1/\Delta H$.

4.53. Assuming that He behaves as an ideal gas during the isothermal compression at -188 °C (or 85 K), we can use $\Delta G=nRT\ln\frac{p_f}{p_i}$. The pressures are not given, but we have the volumes instead, and because $p_iV_i = p_fV_f$ we can replace p_f/p_i by V_i/V_f (note that this reverses the initial and final values):

$$\Delta G = nRT\ln\frac{V_i}{V_f} = (3.66\text{ mol})\left(8.314\frac{\text{J}}{\text{mol}\cdot\text{K}}\right)(85\text{ K})\left(\ln\frac{15.5\text{ L}}{2.07\text{ L}}\right) = 5.2\times10^3\text{ J} = 5.2\text{ kJ}.$$

4.55. Let $u = 1/T$. The derivative du/dT is then $\frac{du}{dT} = -\frac{1}{T^2}$, or after multiplying both sides by dT and T^2: $dT = -T^2\,du$. We can now replace the dT in the denominator of equation 4.43:

$\frac{\partial}{\partial T}\left(\frac{\Delta G}{T}\right)_p = -\frac{\Delta H}{T^2} \rightarrow \frac{\partial}{-T^2\partial u}\left(\frac{\Delta G}{T}\right)_p = -\frac{\Delta H}{T^2}$. If we mutiply both sides of this equation with $-T^2$, we obtain equation 4.44 by substituting back $u = 1/T$:

$$\frac{\partial}{\partial u}\left(\frac{\Delta G}{T}\right)_p = +\Delta H \text{ or } \frac{\partial}{\partial(1/T)}\left(\frac{\Delta G}{T}\right)_p = \Delta H.$$

4.57. If we assume that ΔH remains relatively constant as the temperature changes, the change in $\Delta G/T$ at constant pressure can be estimated from $\Delta\left(\frac{\Delta G}{T}\right) = \Delta H\cdot\Delta\left(\frac{1}{T}\right)$ (see Example 4.12). We want the Gibbs energy to change from -228.61 kJ at 25 °C (298 K) to zero at some other temperature T_0, which means $\Delta\left(\frac{\Delta G}{T}\right) = \frac{0\text{ kJ}}{T_0} - \frac{-228.61\text{ kJ}}{298\text{ K}} = +0.767\text{ kJ/K}$.

Dividing this by ΔH gives $\frac{0.767\text{ kJ/K}}{-241.8\text{ kJ}} = \frac{-0.00317}{\text{K}}$, which is approximately equal to $\Delta\left(\frac{1}{T}\right) = \frac{1}{T_0} - \frac{1}{298\text{ K}}$. Solving this for $1/T_0$ gives $\frac{1}{T_0} = \frac{-0.00317}{\text{K}} + \frac{1}{298\text{ K}} = \frac{0.000183}{\text{K}}$, or $T_0 = 5.47\times10^3$ K (or about 5470 K).

4.59. The molar volume of 18.02 cm^3/mol indicates that we are dealing with liquid water. Therefore, we cannot use equation 4.46, but need to use equation 4.45. For 1.00 mol of water:

$$V = 18.0\text{ cm}^3 = 0.0180\text{ L, so } \Delta G = (0.0180\text{ L})(100.0\text{ atm} - 1.00\text{ atm})\left(\frac{101.32\text{ J}}{\text{L}\cdot\text{atm}}\right) = 181\text{ J}.$$

4.61. For the derivation of the Gibbs-Helmholtz equation, we started with $G = H - TS$, while we would start with $A = U - TS$ to derive an equation for A. Because $\left(\frac{\partial G}{\partial T}\right)_p = \left(\frac{\partial A}{\partial T}\right)_V = -S$, we can simply substitute G by A, H by U, and the subscript p by V to obtain $\frac{\partial}{\partial T}\left(\frac{\Delta A}{T}\right)_V = -\frac{\Delta U}{T^2}$.

If we want to use this equation, we must make sure that the volume is held constant, as indicated by the subscript.

4.63. Equation 4.55 refers to a *molar* quantity, and therefore n is inherently assumed to be 1 mol. This is apparent in the units as well: while ΔG has units of J, $\Delta\mu$ has units of J/mol.

4.65. The chemical potential of a pure, single-phase substance is equal to its molar Gibbs energy. As a molar quantity, it is also an intensive property: it does not change if the total amount of substance changes, because it is inherently referenced to 1 mol. The total Gibbs energy changes (because it is an extensive property), but μ or $\bar{G}$ do not.

4.67. $dG = -S\,dT + V\,dp + \mu_{N_2}dn_{N_2} + \mu_{O_2}dn_{O_2}$, where μ_{N_2} is the chemical potential of nitrogen in a mixture of 1.0 mol N_2 and 1.0 mol O_2, and μ_{O_2} is that of oxygen in the same mixture. The equation expresses how the total Gibbs energy of the mixture changes if the temperature and/or pressure and/or amount of N_2 and/or amount of O_2 are changed.

4.69. 1 atm (exactly) equals 1.01325 bar. Therefore, changing the pressure from 1.00 atm to 1.00 bar: $\Delta\mu = RT\ln\frac{p_f}{p_i} = \left(8.314\frac{\text{J}}{\text{mol}\cdot\text{K}}\right)(273.15\text{ K})\left(\ln\frac{1.00\text{ bar}}{1.01325\text{ bar}}\right) = -29.9\text{ J/mol}.$

(Note that although 1.00 atm are 1.01 bar, you should use the exact value of 1.01325 bar in the calculation to avoid a quite significant truncation error.) This may seem like a lot, but most chemical processes occur with energy changes in the thousands of joules per mole, so the difference is relatively trivial.

4.71. **(a)** 10.0 g of Fe at 25 °C has the greater chemical potential. Remember that for a pure system (single component in a single phase), the chemical potential equals the molar Gibbs energy: $\mu = \bar{G}$. According to $\left(\frac{\partial G}{\partial T}\right)_p = -S$, we then have $\left(\frac{\partial \mu}{\partial T}\right)_p = -\bar{S} < 0$ (entropies are always positive), which means that the chemical potential decreases as temperature increases (at constant pressure).

(b) Compressed air has the greater chemical potential. Since the composition, amount, or temperature of the air do not change, we can use $\Delta\mu = RT\ln\frac{p_f}{p_i}$: if $p_f > p_i$, then $\Delta\mu > 0$.

4.73. To use equation 4.62 to determine ϕ, first we need to rewrite the shortened van der Waals expression in terms of $Z = \frac{pV}{nRT}$: $p(V - nb) = nRT$ is multiplied out and rewritten to $pV = nRT + pnb$. Dividing both sides by nRT leads to $Z = 1 + \frac{pb}{RT}$. The argument of the integral in equation 4.62 is $\frac{Z-1}{p} = \frac{\left(1+\frac{pb}{RT}\right)-1}{p} = \frac{\frac{pb}{RT}}{p} = \frac{b}{RT}$, so

$$\ln\phi = \int_0^p \frac{b}{RT}dp = \frac{b}{RT}\int_0^p dp = \frac{b}{RT}(p-0) = \frac{bp}{RT}.$$

Therefore, taking the inverse logarithm on both sides, we find that $\phi = e^{bp/(RT)}$.

CHAPTER 5

Introduction to Chemical Equilibrium

5.1. A battery that has a voltage is not at equilibrium because there still exists a driving force for a reaction to occur. On the other hand, a completely dead battery is at equilibrium because there is no driving force for a chemical reaction to occur.

5.3. **(a)** Rb^+ and OH^- and H_2 are the prevalent equilibrium species.

(b) NaCl (crystal) is the prevalent equilibrium species.

(c) H^+ (aq) and Cl^- (aq) are the prevalent equilibrium species.

(d) C (graphite) is the thermodynamically stable species under standard conditions, and would be expected to be the prevalent equilibrium species when compared to C (diamond). However, from experience we know that diamond is chemically very stable once formed, because establishing the chemical equilibrium between diamond and graphite takes an immeasurable amount of time.

5.5. The minimum possible value of ξ is 0 (as it is for any reaction); the maximum possible value can be found by determining which reagent is the limiting reagent. The following amounts of substance are initially present:

$$\text{Zn: } (100.0\text{ g})\left(\frac{1\text{ mol}}{65.38\text{ g}}\right) = 1.530\text{ mol; HCl: } (0.1500\text{ L})\left(\frac{2.25\text{ mol}}{\text{L}}\right) = 0.338\text{ mol}$$

The reaction requires a 1:2 molar ratio of Zn to HCl, so 1.530 mol of Zn would require 3.059 mol of HCl. But we only have 0.338 mol, so HCl will be the limiting reagent. Therefore, the maximum possible value for ξ is obtained when the entire amount of HCl has reacted:

$$\xi = \frac{(0-0.338)\text{ mol}}{-2} = 0.169\text{ mol}$$

5.7. **(a)** The amount of Al decreases from 5.0 mol to (5.0 – 3.5) mol = 1.5 mol, so $\xi = \frac{(1.5-5.0)\text{ mol}}{-2} = 1.8\text{ mol}$. (Note that this would require $\frac{3}{2}\cdot 3.5\text{ mol} = 5.3\text{ mol } Cl_2$, and is therefore not limited by the amount of available Cl_2.)

(b) $\xi = 5$ is not possible because we only have 5.0 mol of Al available, so the maximum possible value would be $\xi = \frac{(0-5.0)\text{ mol}}{-2} = 2.5\text{ mol}$.

(This would require $\frac{3}{2}\cdot 5.0\text{ mol} = 7.5\text{ mol } Cl_2$, which is within the available limits.)

5.9. The first reaction is $C_{12}H_{22}O_{11}$ (s) → $C_{12}H_{22}O_{11}$ (aq) and is a complete reaction for 1.00 g sucrose and 100.0 mL water because the sucrose dissolves completely and the solution has not yet reached its saturation limit.

However, the second process is better written as an equilibrium between solid and dissolved, $C_{12}H_{22}O_{11}$ (s) $\rightleftharpoons$ $C_{12}H_{22}O_{11}$ (aq), because 200.0 g – 164.0 g = 36.0 g of sucrose remain undissolved and in equilibrium with the saturated sucrose solution. (Note the use of the double arrows, indicating that a dynamic equilibrium is established.)

5.11. Using 1 atm as standard pressure, the reaction quotient is given by

$$Q = \frac{\left(p_{NH_3}/p°\right)^2}{\left(p_{N_2}/p°\right)\left(p_{H_2}/p°\right)^3} = \frac{\left(\frac{0.26\text{ atm}}{1\text{ atm}}\right)^2}{\left(\frac{1.4\text{ atm}}{1\text{ atm}}\right)\left(\frac{0.044\text{ atm}}{1\text{ atm}}\right)^3} = 5.7\times10^2.$$

Q is (always) unitless, because only ratios of pressures appear (so the units cancel out).

5.13. We start by constructing the following table:

Amount of	**SO_3 (g)**	**SO_2 (g)**	**O_2 (g)**
Initial	2.00 mol	0	0
At equilibrium	(2.00 – $2x$) mol	$2x$ mol	x mol

The ratio of SO_2:SO_3 at equilibrium is then $\frac{n_{SO_2}}{n_{SO_3}} = \frac{2x\text{ mol}}{(2.00-2x)\text{ mol}} = 0.663$, which we can solve for x: $x = \frac{0.663\cdot2.00}{2+2\cdot0.663} = 0.399$. This leads to:

Amount of	**SO_3 (g)**	**SO_2 (g)**	**O_2 (g)**
At equilibrium	1.20 mol	0.797 mol	0.399 mol

The corresponding partial pressures can be calculated from the ideal gas law $p = nRT/V$, which, with R = 0.08314 L bar/(mol K), T = 350 K, and V = 10.0 L, yields the following table:

Partial pressure of	**SO_3 (g)**	**SO_2 (g)**	**O_2 (g)**
At equilibrium	3.50 atm	2.32 atm	1.16 atm

Using 1 atm as standard pressure, the equilibrium constant is then

$$K = \frac{\left(p_{SO_2}/p°\right)^2\left(p_{O_2}/p°\right)}{\left(p_{SO_3}/p°\right)^2} = \frac{\left(\frac{2.32\text{ atm}}{1\text{ atm}}\right)^2\left(\frac{1.16\text{ atm}}{1\text{ atm}}\right)}{\left(\frac{3.50\text{ atm}}{1\text{ atm}}\right)^2} = 0.510$$. The value of $\Delta G°$ at 350 K is then

$$\Delta G° = -RT\ln K = -\left(8.314\frac{\text{J}}{\text{K}\cdot\text{mol}}\right)(350\text{ K})\ln(0.510) = +1.96\times10^3\text{ J/mol} = +1.96\text{ kJ/mol}.$$

5.15. ΔG = 0.00 kJ indicates that equilibrium has been reached, which means that we can use $\Delta G° = -RT\ln K$ to calculate the equilibrium constant. $\Delta G°$ at 298 K and standard pressure is obtained using $\Delta_f G°$ data from Appendix 2 using Hess's law:

$$\Delta G° = \left[(1\cdot97.79)-(2\cdot51.30)\right]\text{kJ/mol} = -4.81\text{ kJ/mol}.$$

This results in $\ln K = -\frac{\Delta G^\circ}{RT} = -\frac{-4.81\times 10^3 \text{ J/mol}}{\left(8.314\frac{\text{J}}{\text{mol}\cdot\text{K}}\right)(298 \text{ K})} = 1.94$ and $K = 6.97$.

Using 1 atm as standard pressure: $K = \frac{\left(p_{N_2O_4}/p^\circ\right)}{\left(p_{NO_2}/p^\circ\right)^2} = \frac{\left(\frac{0.077 \text{ atm}}{1 \text{ atm}}\right)}{\left(\frac{p_{NO_2}}{1 \text{ atm}}\right)^2}$, so

$$\left(\frac{p_{NO_2}}{1 \text{ atm}}\right)^2 = \frac{0.077}{6.97} = 0.011$$

and $p_{NO_2} = \sqrt{0.011}\cdot(1 \text{ atm}) = 0.11 \text{ atm}$.

5.17. **(a)** $K = \frac{\left(p_{NO_2}/p^\circ\right)^2}{\left(p_{NO}/p^\circ\right)^2\left(p_{O_2}/p^\circ\right)}$

(b) $\Delta G^\circ = \left[(2\cdot 51.30)-(2\cdot 86.60+0)\right] \text{kJ/mol} = -70.60 \text{ kJ/mol at 298 K}$

(c) $\ln K = -\frac{\Delta G^\circ}{RT} = -\frac{-70.60\times 10^3 \text{ J/mol}}{\left(8.314\frac{\text{J}}{\text{mol}\cdot\text{K}}\right)(298 \text{ K})} = +28.5$ and $K = 2.4\times 10^{12}$

(d) We calculate the value of Q and compare it to K to figure out in which direction (more product or more reactant) the reaction would move:

$$Q = \frac{\left(p_{NO_2}/p^\circ\right)^2}{\left(p_{NO}/p^\circ\right)^2\left(p_{O_2}/p^\circ\right)} = \frac{\left(\frac{0.250 \text{ bar}}{1 \text{ bar}}\right)^2}{\left(\frac{1.00 \text{ bar}}{1 \text{ bar}}\right)^2\left(\frac{1.00 \text{ bar}}{1 \text{ bar}}\right)} = 0.0625.$$

This means that the reaction will move to the right (products) in order to reach the equilibrium.

5.19. The system would not necessarily be at equilibrium, because the p_i or p_j values in equation 5.9 now have different values. Only if there were the same number of moles on either side of the chemical reaction would these partial pressures cancel mathematically and the equilibrium constants have the same value.

5.21. For example, let's consider the following gas–phase reaction:

$2\,H_2\,(g) + O_2\,(g) \rightleftharpoons 2\,H_2O\,(g)$

Its equilibrium constant expression is $K = \frac{\left(p_{H_2O}/p^\circ\right)^2}{\left(p_{H_2}/p^\circ\right)^2\left(p_{O_2}/p^\circ\right)}$.

Now, divide all coefficients of the chemical reaction by 2:

$H_2\,(g) + \frac{1}{2}\,O_2\,(g) \rightleftharpoons H_2O\,(g)$

For this reaction, the equilibrium constant expression is $K' = \dfrac{(p_{H_2O}/p^\circ)}{(p_{H_2}/p^\circ)(p_{O_2}/p^\circ)^{1/2}}$.

The exponents on all of the partial pressures are half of what they were in K, meaning that all of the terms in K' are the square root of the terms in K. Therefore, $K' = K^{1/2}$ when the reaction itself is halved.

5.23. **(a)** $\Delta_{rxn}G^\circ = [(2\cdot(-16.4) - (3\cdot 0 + 1\cdot 0)]$ kJ/mol $= -32.8$ kJ/mol.

(b) $Q = \dfrac{(p_{NH_3}/p^\circ)^2}{(p_{N_2}/p^\circ)(p_{H_2}/p^\circ)^3} = \dfrac{\left(\frac{0.500\text{ bar}}{1\text{ bar}}\right)^2}{\left(\frac{0.500\text{ bar}}{1\text{ bar}}\right)\left(\frac{0.500\text{ bar}}{1\text{ bar}}\right)^3} = 4.00$ (using a standard pressure of 1 bar).

$$\Delta_{rxn}G = \Delta_{rxn}G^\circ + RT\ln Q = \left[-32.8\frac{\text{kJ}}{\text{mol}} \times \frac{1000\text{ J}}{1\text{ kJ}}\right] + \left[\left(8.314\frac{\text{J}}{\text{mol}\cdot\text{K}}\right)(298\text{ K})(\ln 4.00)\right]$$

$$\Delta_{rxn}G = -32.8\times 10^3\frac{\text{J}}{\text{mol}} + 3.44\times 10^3\frac{\text{J}}{\text{mol}} = -29.4\times 10^3\text{ J/mol} = -29.4\text{ kJ/mol}$$

5.25. $K = \dfrac{(p_{HD}/p^\circ)^2}{(p_{H_2}/p^\circ)(p_{D_2}/p^\circ)} = \dfrac{\left(\frac{2x\text{ atm}}{1\text{ atm}}\right)^2}{\left(\frac{(0.50-x)\text{ atm}}{1\text{ atm}}\right)\left(\frac{(0.10-x)\text{ atm}}{1\text{ atm}}\right)} = 4.00$. The units cancel out; multiplying both sides by the denominator leads to $4x^2 = 4.00\,(0.50 - x)\,(0.10 - x) = 0.20 - 2.4x + 4.00x^2$.

The quadratic terms on both sides cancel out, so we can solve directly for x:
$x = 0.20/2.4 = 0.083$.

Therefore, the final pressures are: $p(H_2) = 0.42$ atm, $p(D_2) = 0.02$ atm, and $p(HD) = 0.17$ atm.

To calculate the extent of reaction, we need to convert these pressures into moles. For HD:

$$n_{HD} = \frac{p_{HD}V}{RT} = \frac{(0.17\text{ atm})(20.0\text{ L})}{\left(0.08205\frac{\text{L}\cdot\text{atm}}{\text{mol}\cdot\text{K}}\right)(488\text{ K})} = 0.083\text{ mol, so}$$

$$\xi = \frac{(0.083-0)\text{ mol}}{+2} = 0.042\text{ mol.}$$

5.27. **(a)** $\Delta_{rxn}G^\circ$ at 298 K and standard pressure is obtained using $\Delta_f G^\circ$ data from Appendix 2: $\Delta_{rxn}G^\circ = [(1\cdot 97.79) - (2\cdot 51.30)]$ kJ/mol $= -4.81$ kJ/mol. From $\Delta_{rxn}G^\circ = -RT\ln K$:

$$\ln K = -\frac{\Delta G^\circ}{RT} = -\frac{-4.81\times 10^3\text{ J/mol}}{\left(8.314\frac{\text{J}}{\text{mol}\cdot\text{K}}\right)(298\text{ K})} = 1.94 \text{ and } K = 6.97.$$

Amount of	NO_2 (g)	N_2O_4 (g)
Initial	1.00 mol	0
At equilibrium	$(1.00 - 2x)$ mol	x mol

(b) We first construct the following table:

Converting moles to partial pressures:

$$p_{NO_2} = \frac{n_{NO_2}RT}{V} = \frac{\left[(1.00-2x)\text{ mol}\right]\left(0.08314\frac{\text{L}\cdot\text{bar}}{\text{mol}\cdot\text{K}}\right)(298\text{ K})}{20.0\text{ L}} = (1.24-2.48x)\text{ bar}$$

$$p_{N_2O_4} = \frac{n_{N_2O_4}RT}{V} = \frac{(x\text{ mol})\left(0.08314\frac{\text{L}\cdot\text{bar}}{\text{mol}\cdot\text{K}}\right)(298\text{ K})}{20.0\text{ L}} = 1.24x\text{ bar}$$

We substitute this into the equilibrium constant, using 1 bar as standard pressure:

$K = \dfrac{\left(p_{N_2O_4}/p^\circ\right)}{\left(p_{NO_2}/p^\circ\right)^2} = \dfrac{\left(\frac{1.24x\text{ bar}}{1\text{ bar}}\right)}{\left(\frac{(1.24-2.48x)\text{ bar}}{1\text{ bar}}\right)^2} = \dfrac{1.24x}{(1.24-2.48x)^2} = 6.97$. Rearranging terms leads to the following equation: $0.178x = 1.53 - 6.14x + 6.14x^2$ or $6.14x^2 - 6.32x + 1.53 = 0$.

Solving the quadratic equation for x: $x_{1/2} = \dfrac{-(-6.32)\pm\sqrt{(-6.32)^2 - 4(6.14)(1.53)}}{2\cdot 6.14}$,

which gives

$x = 0.393$ or $x = 0.636$. If $x = 0.636$, we would have a negative final pressure for NO_2, so we reject this answer. Therefore, $x = 0.393$ and we can now calculate the equilibrium pressures:

$p(NO_2) = (1.24 - 2.48x)$ bar $= 0.26$ bar ; $p(N_2O_4) = 1.24x$ bar $= 0.487$ bar.

To determine ξ, we need to convert one of the pressures (for example of NO_2) back to moles:

$n_{NO_2} = \dfrac{p_{NO_2}V}{RT} = \dfrac{(0.26\text{ bar})(20.0\text{ L})}{\left(0.08314\frac{\text{L}\cdot\text{bar}}{\text{mol}\cdot\text{K}}\right)(298\text{ K})} = 0.21\text{ mol}$. Finally, using the equation for ξ:

$$\xi = \frac{(0.21-1.00)\text{ mol}}{-2} = 0.39\text{ mol}.$$

5.29. Because the number of moles of gas are the same for products and reactants in $CH_3COOC_2H_5$ (g) + H_2O (g) ⇌ CH_3COOH (g) + C_2H_5OH (g), changing the volume (and accordingly the partial pressures) would not change the equilibrium constant of the system. Therefore, the amounts at equilibrium would be the same, and ξ at equilibrium would be the same as well.

5.31. (a) $K=\left(\frac{\gamma_{Pb^{2+}}m_{Pb^{2+}}}{m^\circ}\right)\left(\frac{\gamma_{Cl^-}m_{Cl^-}}{m^\circ}\right)^2$. As a solid, $PbCl_2$ does not appear in the expression.

(b) $K=\frac{\left(\gamma_{H^+}m_{H^+}/m^\circ\right)\left(\gamma_{NO_2^-}m_{NO_2^-}/m^\circ\right)}{\left(\gamma_{HNO_2}m_{HNO_2}/m^\circ\right)}$; (c) $K=\frac{\left(p_{CO_2}/p^\circ\right)}{\left(\gamma_{H_2C_2O_4}m_{H_2C_2O_4}/m^\circ\right)}$. As condensed phases,

$CaCO_3$ (s), CaC_2O_4 (s) and H_2O (ℓ) do not appear in the expression.

5.33. First, we calculate ΔG° at 298 K and standard pressure using $\Delta_f G^\circ$ data from Appendix: $\Delta G^\circ = \left(\left[1\cdot(-553.54)+1\cdot(-528.1)\right]-\left[1\cdot(-1127.8)\right]\right)$ kJ/mol = +46.2 kJ/mol. K is then obtained from $\Delta G^\circ = -RT\ln K$: $\ln K = -\frac{\Delta G^\circ}{RT} = -\frac{+46.2\times 10^3\text{ J/mol}}{\left(8.314\frac{\text{J}}{\text{mol}\cdot\text{K}}\right)(298\text{ K})} = -18.6$ and $K = 8.10\times10^{-9}$.

5.35. Using $0=\Delta_{rxn}G^\circ + RT\ln\frac{a_{dia}}{a_{gra}}$ and $\Delta_f G^\circ$(dia) = $\Delta_{rxn}G^\circ$ = +2.90 kJ/mol from Appendix 2, and substituting from equation 5.14:

$$0 = 2.90\text{ kJ/mol} + RT\left[\frac{\bar{V}_{dia}}{RT}(p-1\text{ atm}) - \frac{\bar{V}_{gra}}{RT}(p-1\text{ atm})\right].$$

The RT terms cancel, leaving us with $0 = 2.90\times10^3\frac{\text{J}}{\text{mol}} + \left(\bar{V}_{dia}-\bar{V}_{gra}\right)(p-1\text{ atm})$.

Now we need to substitute the molar volumes of diamond and graphite. 1 mol of carbon has a mass of 12.011 g, so

$\left(12.011\frac{\text{g}}{\text{mol}}\right)\left(\frac{1\text{ cm}^3}{2.25\text{ g}}\right)\left(\frac{1\text{ L}}{1000\text{ cm}^3}\right) = 5.34\times10^{-3}\frac{\text{L}}{\text{mol}}$ for the molar volume of graphite,

and

$\left(12.011\frac{\text{g}}{\text{mol}}\right)\left(\frac{1\text{ cm}^3}{3.51\text{ g}}\right)\left(\frac{1\text{ L}}{1000\text{ cm}^3}\right) = 3.42\times10^{-3}\frac{\text{L}}{\text{mol}}$ for the molar volume of diamond.

Substituting:

$$0 = 2.90\times10^3\frac{\text{J}}{\text{mol}} + \left(3.42\times10^{-3}\frac{\text{L}}{\text{mol}} - 5.34\times10^{-3}\frac{\text{L}}{\text{mol}}\right)(p-1\text{ atm})\left(\frac{101.32\text{ J}}{1\text{ L}\cdot\text{atm}}\right)$$

and solving for p: $(p - 1\text{ atm}) = 1.49\times10^4$ atm, so $p = 1.49\times10^4$ atm.

$\Delta_{rxn}G$ decreases from +2.90 kJ/mol at p = 1 atm to 0 at 1.49×10^4 atm because $\bar{V}_{dia} - \bar{V}_{gra} < 0$. Above that pressure, $\Delta_{rxn}G$ becomes negative, so diamond becomes the stable form of carbon.

5.37. 1 mol of C_{60} has a mass of 720.66 g, so its molar volume is

$$\left(720.66\frac{\text{g}}{\text{mol}}\right)\left(\frac{1\text{ cm}^3}{1.65\text{ g}}\right)\left(\frac{1\text{ L}}{1000\text{ cm}^3}\right) = 0.437\frac{\text{L}}{\text{mol}}.$$

Using equation 5.14 and solving for $a(C_{60})$:

$$\ln a_{C_{60}} = \frac{\bar{V}_{C_{60}}}{RT}(p - 1\text{ bar}) = \frac{0.437\text{ L/mol}}{\left(0.08314\frac{\text{L}\cdot\text{bar}}{\text{mol}\cdot\text{K}}\right)(373\text{ K})}(1500\text{ bar} - 1\text{ bar}) = 21.1;$$

$a_{C_{60}} = 1.46\times10^9$.

5.39. **(a)** $\Delta G° = -RT\ln K = -\left(8.314\frac{\text{J}}{\text{mol}\cdot\text{K}}\right)\left(\frac{1\text{ kJ}}{1000\text{ J}}\right)(298.2\text{ K})\ln\left(1.2\times10^{-2}\right) = +11\text{ kJ/mol}$

(b) $K = \dfrac{\left(\gamma_{H^+}m_{H^+}/m°\right)\left(\gamma_{SO_4^{2-}}m_{SO_4^{2-}}/m°\right)}{\left(\gamma_{HSO_4^-}m_{HSO_4^-}/m°\right)} \approx \dfrac{\left(m_{H^+}/m°\right)\left(m_{SO_4^{2-}}/m°\right)}{\left(m_{HSO_4^-}/m°\right)}$ at low concentrations.

To determine the molalities of HSO_4^-, H^+, and SO_4^{2-}, we first assume that HSO_4^- starts at a molality of 0.010 $m°$ (with $m°$ = 1 mol/kg), and that some amount dissociates, leaving
$(0.010 - x)\ m°$. The molalities of SO_4^{2-} and H^+ are therefore $+x\ m°$:

$K = 1.2\times10^{-2} = \dfrac{(x\,m°/m°)(x\,m°/m°)}{(0.010 - x)m°/m°} = \dfrac{x^2}{0.010 - x}$. Solving for x using the quadratic formula gives either $x = -0.018$ or 0.0065. We reject the negative root since the molalities of H^+ and SO_4^{2-} must be positive, so x = 0.0065. Therefore, the final molalities are $m_{H^+} = m_{SO_4^{2-}} = 6.5\times10^{-3}$ mol/kg and $m_{HSO_4^-} = (0.010 - 0.0065)\text{ mol/kg} = 0.004\text{ mol/kg}$.

5.41. **(a)** $K = \dfrac{a_{CO_2}}{a_C\,a_{O_2}} \approx \dfrac{a_{CO_2}}{a_{O_2}}$ (activities of condensed phases can be approximated as 1).

(b) $K = \dfrac{a_{P_4O_{10}}}{a_{P_4}\left(a_{O_2}\right)^5} \approx \dfrac{1}{\left(a_{O_2}\right)^5}$; **(c)** $K = \dfrac{\left(a_{NCl_3}\right)^2 a_{H_2}\left(a_{O_2}\right)^2}{\left(a_{HNO_2}\right)^2\left(a_{Cl_2}\right)^3}$

5.43. We use $\ln\frac{K_2}{K_1}=\frac{\Delta H°}{R}\left(\frac{1}{T_1}-\frac{1}{T_2}\right)$ to solve for $\Delta H°$: $\Delta H°=R\ln\frac{K_2}{K_1}\Big/\left(\frac{1}{T_1}-\frac{1}{T_2}\right)$

$$\Delta H°=\left(8.314\frac{\text{J}}{\text{mol}\cdot\text{K}}\right)\ln\frac{1.86\times10^{-1}}{3.76\times10^{-2}}\Big/\left(\frac{1}{350\text{ K}}-\frac{1}{450\text{ K}}\right)=20.9\text{ kJ/mol}$$

The reaction is endothermic, which explains why K increases with increasing temperature.

5.45. **(a)** 25.0 °C is 298.2 K, and 100.0 °C is 373.2 K, so from $\ln\frac{K_2}{K_1}=\frac{\Delta H°}{R}\left(\frac{1}{T_1}-\frac{1}{T_2}\right)$:

$$\Delta H°=R\ln\frac{K_2}{K_1}\Big/\left(\frac{1}{T_1}-\frac{1}{T_2}\right)=\left(8.314\frac{\text{J}}{\text{mol}\cdot\text{K}}\right)\ln\frac{5.60\times10^{-13}}{1.01\times10^{-14}}\Big/\left(\frac{1}{298.2\text{ K}}-\frac{1}{373.2\text{ K}}\right)$$

$\Delta H°=49.5$ kJ/mol

(b) $\Delta H°=\left(8.314\frac{\text{J}}{\text{mol}\cdot\text{K}}\right)\ln\frac{7.67\times10^{-14}}{1.10\times10^{-15}}\Big/\left(\frac{1}{298.2\text{ K}}-\frac{1}{373.2\text{ K}}\right)=52.3\text{ kJ/mol}$

Both autoionizations are endothermic, consistent with the fact that K increases with increasing temperature.

5.47. 0 °C is 273 K, and 100 °C is 373 K, so from $\ln\frac{K_2}{K_1}=\frac{\Delta_{\text{rxn}}H°}{R}\left(\frac{1}{T_1}-\frac{1}{T_2}\right)$:

$$\Delta_{\text{rxn}}H°=R\ln\frac{K_2}{K_1}\Big/\left(\frac{1}{T_1}-\frac{1}{T_2}\right)=\left(8.314\frac{\text{J}}{\text{mol}\cdot\text{K}}\right)\ln\frac{0.065}{58}\Big/\left(\frac{1}{273\text{K}}-\frac{1}{373\text{K}}\right)=-58\text{ kJ/mol}$$

The reaction is exothermic, which explains why K decreases with increasing temperature.

5.49. **(a)** If the pressure is increased, the equilibrium shifts toward the product side, which has fewer gas molecules (two on the product side versus three on the reactant side).

(b) We need to know the ΔH of this reaction. Using data from the appendix (at 298 K):

$\Delta H=(2\cdot(-395.77)-[2\cdot(-296.81)+0])$ kJ $=-197.92$ kJ. Thus, as an exothermic reaction, if the temperature is decreased, the reaction shifts toward products.

(c) Same as (a); the origin of the pressure increase is not important. (Note: increasing the pressure by adding nitrogen (at constant volume) is not that different from decreasing the volume, as the added nitrogen will take up some of the volume, which is then inaccessible to the reaction partners).

5.51. **(a)** $\Delta_{\text{rxn}}G°=[(-1048.01-394.35-228.61)-2\cdot(-851.0)]$ kJ/mol $=31.0$ kJ/mol

$\Delta_{\text{rxn}}H°=[(-1130.77-393.51-241.8)-2\cdot(-950.81)]$ kJ/mol $=135.5$ kJ/mol

(b) $\Delta_{rxn}G° = -RT\ln K$, so $\ln K = -\frac{\Delta_{rxn}G°}{RT} = -\frac{31.0\times10^3 \text{ J/mol}}{\left(8.314\frac{\text{J}}{\text{mol}\cdot\text{K}}\right)(298\text{ K})} = -12.5$;

$K = 3.64\times10^{-6}$

(c) $K = \left(p_{CO_2}/p°\right)\left(p_{H_2O}/p°\right)$. Using 1 bar as standard pressure, and given that $p_{CO_2} = p_{H_2O}$:

$$p_{CO_2} = p_{H_2O} = \sqrt{K}\,p° = \left(\sqrt{3.64\times10^{-6}}\right)(1\text{ atm}) = 1.91\times10^{-3}\text{ atm}.$$

(d) We first calculate the new equilibrium constant at 1150 °C, which is 1423 K:

$$\ln K_2 = \ln K_1 + \frac{\Delta_{rxn}H°}{R}\left(\frac{1}{T_1}-\frac{1}{T_2}\right) = \ln\left(3.64\times10^{-6}\right) + \frac{135.5\times10^3\text{ J/mol}}{8.314\frac{\text{J}}{\text{mol}\cdot\text{K}}}\left(\frac{1}{298\text{ K}}-\frac{1}{1423\text{ K}}\right)$$

This gives ln K_2 = 30.7 and $K_2 = 2.21\times10^{13}$, so $p_{CO_2} = p_{H_2O} = \sqrt{K_2}\,p° = 4.70\times10^6$ atm.

5.53. The easiest way to show that equations 5.18 and 5.19 are equivalent is to differentiate $1/T$ with respect to T: $\frac{d(1/T)}{dT} = -\frac{1}{T^2}$, or $dT = -T^2 d(1/T)$. Substitute this into equation 5.18:

$\frac{d\ln K}{-T^2 d(1/T)} = \frac{\Delta_{rxn}H°}{RT^2}$ and multiply both sides by $-T^2$: $\frac{d\ln K}{d(1/T)} = -\frac{\Delta_{rxn}H°}{R}$ (equation 5.19).

5.55. **(a)** For an exothermic process, $\Delta H° < 0$ and therefore the right–hand side of $\frac{d\ln K}{dT} = \frac{\Delta_{rxn}H°}{RT^2}$ is negative. ln K (and K) decreases with increasing temperature; the equilibrium shifts toward the reactants side.

(b) For an endothermic process, $\Delta H° > 0$, meaning that the right–hand side is positive, and K increases with increasing temperature (shift toward the products).

These results are consistent with Le Chatelier's principle in that increasing the temperature will shift the equilibrium in the direction where heat is taken up by the reaction.

5.57. At low concentrations, activity coefficients are approximately 1 and the activity of a dissolved solute equals its molality. Furthermore, it is common to use the molarity c instead of molality m, since mol/kg and mol/L are approximately equal for dilute solutions (1.00 L of water has a mass of 1 kg). Therefore, the two equilibrium constants are ($c° = 1$ mol/L):

$$K_1 = 10^{-2.34} = \frac{\left(\gamma_{gly}m_{gly}/m°\right)\left(\gamma_{H^+}m_{H^+}/m°\right)}{\left(\gamma_{glyH^+}m_{glyH^+}/m°\right)} \approx \frac{\left(m_{gly}/m°\right)\left(m_{H^+}/m°\right)}{\left(m_{glyH^+}/m°\right)} \approx \frac{\left(c_{gly}/c°\right)\left(c_{H^+}/c°\right)}{\left(c_{glyH^+}/c°\right)}$$

$$K_2 = 10^{-9.60} = 2.51 \times 10^{-10} \approx \frac{\left(c_{\text{gly}^-}/c^\circ\right)\left(c_{\text{H}^+}/c^\circ\right)}{\left(c_{\text{gly}}/c^\circ\right)}$$

The two main reactions that will occur (assuming that glycine is fully dissolved) are:

(i) gly + $H_2O \rightleftharpoons$ glyH^+ + OH^-

(ii) gly $\rightleftharpoons$ gly$^-$ + H^+

Furthermore, we have to take into account the autoionizaton of water:

$H_2O \rightleftharpoons H^+ + OH^-$; $K_w = 1 \times 10^{-14} \approx \left(c_{\text{H}^+}/c^\circ\right)\left(c_{\text{OH}^-}/c^\circ\right)$

Equilibrium (ii) is characterized by K_2, while equilibrium (i) is given by $K_w/K_1 = K_1$':

$$K_1' = \frac{K_w}{K_1} = \frac{1.00 \times 10^{-14}}{10^{-2.34}} = 2.19 \times 10^{-12} = \frac{\left(c_{\text{H}^+}/c^\circ\right)\left(c_{\text{OH}^-}/c^\circ\right)\left(c_{\text{glyH}^+}/c^\circ\right)}{\left(c_{\text{gly}}/c^\circ\right)\left(c_{\text{H}^+}/c^\circ\right)} = \frac{\left(c_{\text{OH}^-}/c^\circ\right)\left(c_{\text{glyH}^+}/c^\circ\right)}{\left(c_{\text{gly}}/c^\circ\right)}$$

The initial amount of glycin is distributed between glyH^+, gly, and gly$^-$, so that $1.0\text{ mol/L} = c_{\text{glyH}^+} + c_{\text{gly}} + c_{\text{gly}^-}$. Also, because of electroneutrality, we have

$c_{\text{glyH}^+} + c_{\text{H}^+} = c_{\text{gly}^-} + c_{\text{OH}^-}$. Altogether, we now have 5 independent equations with 5 unknown concentrations. This set of equations, however, is difficult to solve, and we will have to make some additional approximations.

Realizing that reaction (i) will produce OH^- and reaction (ii) will produce about the same amount of H^+, we can assume, to first approximation, that the pH of the solution remains close to neutral ($c_{H+} = 1.00 \times 10^{-7}$ mol/L). [This is not quite true because $K_1' < K_2$, so more H^+ than OH^- will be generated, and the final solution will be slightly acidic.] Furthermore, we will assume that very little glycine will actually dissociate, so $c_{\text{gly}} \approx 1.0$ mol/L. This leads to

$$c_{\text{glyH}^+} = \frac{c_{\text{gly}}c_{\text{H}^+}}{K_1 c^\circ} = \frac{(1.0\text{ mol/L})(1.00 \times 10^{-7}\text{ mol/L})}{(10^{-2.34})(1\text{ mol/L})} = 2.2 \times 10^{-5}\text{ mol/L and}$$

$$c_{\text{gly}^-} = K_2 c^\circ \frac{c_{\text{gly}}}{c_{\text{H}^+}} = (10^{-9.60})(1\text{ mol/L})\frac{1.0\text{ mol/L}}{1.00 \times 10^{-7}\text{ mol/L}} = 2.5 \times 10^{-3}\text{ mol/L}.$$

5.59. With activities equal to molalities, and with K_1 for glutamic acid (glu) from Table 5.1:

$$K_1 = 10^{-2.19} = \frac{\left(\gamma_{\text{glu}^-} m_{\text{glu}^-}/m^\circ\right)\left(\gamma_{\text{H}^+} m_{\text{H}^+}/m^\circ\right)}{\left(\gamma_{\text{glu}} m_{\text{glu}}/m^\circ\right)} \approx \frac{\left(m_{\text{glu}^-}/m^\circ\right)\left(m_{\text{H}^+}/m^\circ\right)}{\left(m_{\text{glu}}/m^\circ\right)}$$

Assuming that MSG (Na^+glu$^-$) is fully dissolved, the main reaction occuring in solution will be:

glu$^-$ + $H_2O \rightleftharpoons$ glu + OH^-

The equilibrium constant for this reaction (K_1') can be calculated using K_1 and the equilibrium constant for the autoionization of water,

$$H_2O \rightleftharpoons H^+ + OH^- ;\ K_w = 1.00 \times 10^{-14} \approx \left(m_{H^+}/m°\right)\left(m_{OH^-}/m°\right),$$

$$\text{Namely } K_1' = \frac{K_w}{K_1} = \frac{1.00 \times 10^{-14}}{10^{-2.19}} = 1.55 \times 10^{-12} = \frac{\left(m_{glu}/m°\right)\left(m_{OH^-}/m°\right)}{\left(m_{glu^-}/m°\right)}.$$

At equilibrium, $m_{glu^-} = (0.010 - x)$ mol/kg and $m_{glu} = m_{OH^-} = x$ mol/kg, which leads to a quadratic equation. However, K_1' is so small that we can assume $m_{glu^-} \approx 0.010$ mol/kg:

$$x = \sqrt{K_1' m° m_{glu^-}} == \sqrt{\left(1.55 \times 10^{-12}\right)(1\ \text{mol/kg})(0.010\ \text{mol/kg})} = 1.24 \times 10^{-7}\ \text{mol/kg}$$

The zwitterionic species is glu, with m_{glu} of about 1.2×10^{-7} mol/kg.

CHAPTER 6

Equilibria in Single-Component Systems

6.1. **(a)** 1 component **(b)** 2 components **(c)** 4 components **(d)** 2 components **(e)** 2 components.

6.3. $FeCl_2$ and $FeCl_3$ are the only chemically-stable, single-component materials that can be made from iron and chlorine. Note that we are identifying a single component as a chemical compound, not the elements that make up that compound.

6.5. The water is boiling because the vapor pressure of the water equals the ambient pressure inside the syringe. By drawing back the plunger of the syringe enough, we can reduce the pressure above the water surface inside the syringe sufficiently so that the water starts to boil. (The physical requirement for boiling is that the vapor pressure and ambient pressure are equal.)

6.7. By definition, any pure substance has only one normal boiling point. The normal boiling point is the temperature at which the substance is boiling when the ambient pressure is exactly 1 atm.

6.9. **(a)** $\Delta_{trans}H$ should be positive because energy has to put into a solid in order to sublime it.

(b) $\Delta_{trans}H$ should be negative because energy has to be removed from a gas to condense it.

6.11. Let us break this process into five steps. Each step is endothermic (heat needs to be put into the system), so the overall process is endothermic as well:

Step 1: Warming the ice from −15.0 ºC to 0.0 ºC; Step 2: Melting the ice to water at 0.0 ºC;

Step 3: Warming the water from 0.0 ºC to 100.0 ºC; Step 4: Boiling the water at 100.0 ºC; and

Step 5: Warming the steam from 100.0 ºC to 110 ºC.

The constants we need are the specific heat capacities of ice, water, and steam (Table 2.1):

c(ice) = 2.06 J/(g·K); c(water) = 4.184 J/(g·K); c(steam) = 2.04 J/(g·K) at 100 ºC. For the calculatuions, we will assume that these values are constant for the temperature range covered by each individual step. Table 2.3 lists the enthalpies of fusion and vaporization for water:

$\Delta_{fus}H$ = 333.5 J/g; $\Delta_{vap}H$ = 2,260 J/g. Recall that temperature *changes* are the same in ºC and K.

Step 1: $q = m \cdot c \cdot \Delta T = (100.0\ \text{g})\ [2.06\ \text{J/(g·K)}]\ (+15.0\ \text{K}) = 3.09\times10^3\ \text{J} = 3.09\ \text{kJ}$

Step 2: $q = m \cdot \Delta_{fus}H = (100.0\ \text{g})\ (333.5\ \text{J/g}) = 33.35\times10^3\ \text{J} = 33.35\ \text{kJ}$

Step 3: $q = m{\cdot}c{\cdot}\Delta T$ = (100.0 g) [4.184 J/(g·K)] (+100.0 K) = 41.84×10³ J = 41.84 kJ

Step 4: $q = m{\cdot}\Delta_{vap}H$ = (100.0 g) (2,260 J/g) = 226.0×10³ J = 226.0 kJ

Step 5: $q = m{\cdot}c{\cdot}\Delta T$ = (100.0 g) [2.04 J/(g·K)] (+10 K) = 2.0×10³ J = 2.0 kJ

The total energy is the sum of the five parts:

q = (3.09 +33.35 + 41.84 + 226.0 + 2.0) kJ = 306.3 kJ.

6.13. **(a)** The liquid is taking up the energy $\Delta_{vap}H$ during the evaporation process, which means that it loses the energy as heat q.

(b) 1.00 g of water requires

$$\Delta_{vap}H = n\cdot\Delta_{vap}\bar{H} = (1.00\text{ g})\left(\frac{1\text{ mol}}{18.0152\text{ g}}\right)\left(43.99\frac{\text{kJ}}{\text{mol}}\right) = +2.44\text{ kJ}$$ to evaporate at

25.0 °C. Therefore, the remaining 99.0 g of water lose $q = -2.44$ kJ as heat:

$$\Delta T = \frac{q}{m\cdot c} = \frac{-2.44\times10^3\text{ J}}{(99.0\text{ g})\left(4.18\frac{\text{J}}{\text{g}\cdot\text{K}}\right)} = -5.90\text{ K}$$

This means that the final temperature of the water is (25.0 – 5.90) °C = 19.1 °C. (Recall that *changes* in temperature have the same magnitude in kelvins as they have in degrees Celsius.)

6.15. For a pure substance, the chemical potential is simply the molar Gibbs energy:

$\mu = \left(\frac{\partial G}{\partial n}\right)_{T,p} = \bar{G}$. Therefore, equation 4.40, $\left(\frac{\partial G}{\partial T}\right)_p = -S$, can be revised to

$\left(\frac{\partial \mu}{\partial T}\right)_p = -\bar{S}$, so we need the molar entropy of carbon dioxide gas at 298 K, which is

listed in Appendix 2 as $S°$ = 213.785 J/(mol·K). Therefore, $\left(\frac{\partial \mu}{\partial T}\right)_p = -213.785\ \frac{\text{J}}{\text{mol}\cdot\text{K}}$.

6.17. We solve equation 6.7 $\Delta_{fus}S = \frac{\Delta_{fus}H}{T_{fus}}$ for T_{fus}: $T_{fus} = \frac{\Delta_{fus}H}{\Delta_{fus}S} = \frac{17.61\times10^3\text{ J/mol}}{10.21\frac{\text{J}}{\text{mol}\cdot\text{K}}} = 1725\text{ K}$

Nickel melts at 1725 K or 1452 °C, close to the experimental value.

6.19. The principle involved here is the variation of G (or μ) with change in pressure:

$$\left(\frac{\partial G}{\partial p}\right)_T = V.$$

Thus, changes in μ are related to the differences in molar volume, $\bar{V}$. In order to do a rough calculation, we would have to know the pressure exerted by the skate's blades and the change in molar volume going from ice to water. Because ice is less dense than liquid water, the molar volume *decreases* when going from ice to water (that is, the change in $\bar{V}$ is negative), so increasing the pressure should decrease the chemical potential, indicating that liquid water becomes the more stable phase as the pressure is increased. However, without knowing the exact pressure involved, it is difficult to say if this is the mechanism. If it is the mechanism, then skating on a frozen solid may be virtually unique to water, as it is one of very few substances (Bi metal being another) whose solid phase is less dense than its liquid phase. Skating on bismuth, anyone?

6.21. The derivation of equation 6.12 from 6.11 assumes that the molar enthalpy change, $\Delta\bar{H}$, and the molar volume change, $\Delta\bar{V}$, do not vary with temperature.

6.23. First, we will need to calculate the molar volumes of rhombic and monoclinic S_8:

$$\left(256.53\frac{\text{g}}{\text{mol}}\right)\left(\frac{1\text{ cm}^3}{2.07\text{ g}}\right)\left(\frac{1\text{ L}}{1000\text{ cm}^3}\right)=0.124\text{ L/mol for the molar volume for rhombic S}_8$$

$$\left(256.53\frac{\text{g}}{\text{mol}}\right)\left(\frac{1\text{ cm}^3}{1.96\text{ g}}\right)\left(\frac{1\text{ L}}{1000\text{ cm}^3}\right)=0.131\text{ L/mol for the molar volume for monoclinic S}_8$$

In going from monoclinic to rhombic sulfur, the change in molar volume is (0.124 – 0.131) L/mol = −0.007 L/mol. In order to make rhombic sulfur the stable phase at

100 °C, the phase transition temperature needs to shift from 95.5 °C to above 100.0 °C, that is by *at least* 4.5 °C (or 4.5 K). We use the Clapeyron equation $\frac{\Delta p}{\Delta T}\approx\frac{\Delta\bar{S}}{\Delta\bar{V}}$ to calculate the required pressure change Δp:

$$\Delta p\approx\frac{\Delta\bar{S}}{\Delta\bar{V}}\Delta T=\frac{\left(-1.00\ \frac{\text{J}}{\text{mol}\cdot\text{K}}\right)}{(-0.007\text{ L/mol})}(+4.5\text{ K})\left(\frac{1\text{ L}\cdot\text{atm}}{101.32\text{ J}}\right)=+6\text{ atm}.$$

Assuming that the phase transition at 95.5 ºC occurs at 1 atm of pressure, this means that increasing the pressure well above 7 atm should be enough to make rhombic sulfur the stable form at 100 ºC.

6.25. We first calculate the molar volumes for the two allotropes of P_4:

$$\left(123.895\frac{\text{g}}{\text{mol}}\right)\left(\frac{1\text{ cm}^3}{1.823\text{ g}}\right)\left(\frac{1\text{ L}}{1000\text{ cm}^3}\right)=0.06796\text{ L/mol for white P}_4$$

$$\left(123.895\frac{\text{g}}{\text{mol}}\right)\left(\frac{1\text{ cm}^3}{2.270\text{ g}}\right)\left(\frac{1\text{ L}}{1000\text{ cm}^3}\right)=0.05458\text{ L/mol for red P}_4$$

The change in molar volume for converting white P to red P is then

(0.05458 – 0.06796) L/mol = −0.01338 L/mol. We solve $\Delta p=\frac{\Delta\bar{H}}{\Delta\bar{V}}\ln\frac{T_f}{T_i}$ for $\ln(T_f/T_i)$, assuming that the phase transition temperature for the white–to–red phorphorus conversion at atmospheric pressure occurs at 250 °C (523 K):

$$\ln\frac{T_f}{523\text{ K}} = \frac{\Delta\bar{V}}{\Delta\bar{H}}\Delta p = \left(\frac{-0.01338\text{ L/mol}}{-18\times10^3\text{ J/mol}}\right)\left[(625-1)\text{ atm}\right]\left(\frac{101.32\text{ J}}{1\text{ L}\cdot\text{atm}}\right) = 4.7\times10^{-2},$$

so T_f = 548 K or 275 °C. Here we have used that fact that 1 L·atm = 101.32 J to convert L·atm into J.

6.27. Evaporation of a liquid is always accompanied by an increase in molar entropy and molar volume, so that, according to the Clapeyron equation, $dp/dT > 0$. This means that an increase in pressure leads to an increase in the boiling point temperature. Because 1 atm = 1.013 bar, 1 atm is the higher pressure and has the higher boiling point (normal boiling point).

6.29. Equation 6.9 allows us to calculate how the pressure involved in the phase–change equilibrium changes with temperature; or, alternately, how the temperature of the phase–change equilibrium changes with pressure.

6.31. 122 °C are 395 K, and 222 °C are 495 K. We solve $\Delta p = \frac{\Delta\bar{H}}{\Delta\bar{V}}\ln\frac{T_f}{T_i}$ for $\Delta\bar{H}$:

$$\Delta\bar{H} = \Delta p\Delta\bar{V}\Big/\ln\frac{T_f}{T_i} = \left(1.334\times10^6\text{ bar}\right)\left(-3.22\frac{\text{cm}^3}{\text{mol}}\right)\left(\frac{1\text{ L}}{1000\text{ cm}^3}\right)\Big/\ln\frac{395\text{ K}}{495\text{ K}} = 19.0\times10^3\ \frac{\text{L}\cdot\text{bar}}{\text{mol}}.$$

Using 1 L·bar = 100 J, this is $\Delta\bar{H} = \left(19.0\times10^3\ \frac{\text{L}\cdot\text{bar}}{\text{mol}}\right)\left(\frac{100\text{ J}}{1\text{ L}\cdot\text{bar}}\right) = 1.90\times10^3$ kJ/mol.

6.33. Use the form of the Clausius–Clapeyron equation in equation 6.14 to determine the vapor pressure at 25 °C (298 K), with the understanding that at the normal boiling points of each alcohol, the vapor pressure is (exactly) 1 atm:

For 1–butanol:

$$\ln\frac{p_2}{1\text{ atm}} = -\frac{45.90\times10^3\text{ J/mol}}{8.314\ \frac{\text{J}}{\text{mol}\cdot\text{K}}}\left(\frac{1}{298\text{ K}} - \frac{1}{(117.2+273.15)\text{ K}}\right) = -4.4;\quad p_2 = 0.013\text{ atm}$$

(The difference in inverse temperatures is 0.00335 K^{-1} – 0.00256 K^{-1} = 0.00079 K^{-1}, so the results from this calculation can only be given to two significant digits.)

For 2–butanol:

$$\ln\frac{p_2}{1\text{ atm}} = -\frac{44.82\times10^3\text{ J/mol}}{8.314\ \frac{\text{J}}{\text{mol}\cdot\text{K}}}\left(\frac{1}{298\text{ K}} - \frac{1}{(99.5+273.15)\text{ K}}\right) = -3.6;\quad p_2 = 0.027\text{ atm}$$

For isobutanol:

$$\ln\frac{p_2}{1\text{ atm}} = -\frac{45.76\times10^3\text{ J/mol}}{8.314\ \frac{\text{J}}{\text{mol}\cdot\text{K}}}\left(\frac{1}{298\text{ K}} - \frac{1}{(108.1+273.15)\text{ K}}\right) = -4.0;\quad p_2 = 0.018\text{ atm}$$

For *tert*–butanol:

$$\ln\frac{p_2}{1\text{ atm}} = -\frac{43.57\times10^3\text{ J/mol}}{8.314\dfrac{\text{J}}{\text{mol}\cdot\text{K}}}\left(\frac{1}{298\text{ K}}-\frac{1}{(82.3+273.15)\text{ K}}\right) = -2.8;\quad p_2 = 0.059\text{ atm}$$

From highest to lowest vapor pressures: *tert*–butanol > 2–butanol > isobutanol > 1–butanol.

The lower the vapor pressure, the higher the temperature needs to be to reach ambient pressure, so we expect the normal boiling points to increase in the same order (from *tert*-butanol to 1-butanol).

The lower the vapor pressure, the more energy is required to overcome the intermolecular forces in the liquid, so we also expect $\Delta_{vap}H$ to increase in the same order (from *tert*-butanol to 1-butanol).

6.35. We will use $\ln\frac{p_2}{p_1} = -\frac{\Delta\bar{H}}{R}\left(\frac{1}{T_2}-\frac{1}{T_1}\right)$ and solve for T_2:

$$\left(\frac{1}{T_2}-\frac{1}{293.2\text{ K}}\right) = -\frac{R}{\Delta\bar{H}}\ln\frac{p_2}{p_1} = -\frac{8.314\dfrac{\text{J}}{\text{mol}\cdot\text{K}}}{38.6\times10^3\text{ J/mol}}\ln\frac{250.0\text{ mmHg}}{43.7\text{ mmHg}} = -3.76\times10^{-4}/\text{K};$$

$T_2 = 329.4$ K (or 56.3 ºC).

6.37. At 22.0 ºC (or 295.2 K), the vapor pressure of naphthalene is 7.9×10^{-5} bar, so at 100 ºC (373 K): $\ln\frac{p_2}{7.9\times10^{-5}\text{ bar}} = -\frac{71.40\times10^3\text{ J/mol}}{8.314\dfrac{\text{J}}{\text{mol}\cdot\text{K}}}\left(\frac{1}{373\text{ K}}-\frac{1}{295\text{ K}}\right) = 6.1;\ p_2 = 0.035$ bar.

6.39. We can use equation 6.18: $p^* = p\exp\left(\frac{\bar{V}(\ell)\Delta P}{RT}\right)$.

We need to be careful that the units in the argument to the exponential function cancel out and convert 103.9 cm^3/mol into 0.1039 L/mol:.

$$\frac{\bar{V}(\ell)\Delta P}{RT} = \frac{(0.1039\text{ L/mol})(3.30\text{ atm})}{\left(0.08205\dfrac{\text{L}\cdot\text{atm}}{\text{mol}\cdot\text{K}}\right)(298.2\text{ K})} = 0.0140$$

and $p^* = (18.3\text{ torr})\exp(0.0140) = 18.6\text{ torr}$.

6.41. According to equation 6.16 and the discussion in the text, if the temperature is increased linearly, then the vapor pressure will increase exponentially. Thus, at high temperatures nearing the boiling point, even small changes in temperature can lead to large changes in vapor pressure.

6.43. The normal boiling point of water (at 1 atm or 760 torr) is exactly 100 °C or 373.15 K. We rearrange the Clausius–Clapeyron equation $\ln\frac{p_2}{p_1} = -\frac{\Delta\bar{H}}{R}\left(\frac{1}{T_2} - \frac{1}{T_1}\right)$ and solve for temperature:

$$\left(\frac{1}{T_2} - \frac{1}{373.15\text{ K}}\right) = -\frac{R}{\Delta\bar{H}}\ln\frac{p_2}{p_1} = -\frac{8.314\ \frac{\text{J}}{\text{mol}\cdot\text{K}}}{40.7\times10^3\text{ J/mol}}\ln\frac{582\text{ torr}}{760\text{ torr}} = -5.45\times10^{-5}\text{ /K};$$

$T_2 = 365.71$ K or 92.56 °C.

6.45. The normal boiling point of water (at 1 atm) is exactly 100 °C or 373.15 K. We rearrange the Clausius–Clapeyron equation $\ln\frac{p_2}{p_1} = -\frac{\Delta\bar{H}}{R}\left(\frac{1}{T_2} - \frac{1}{T_1}\right)$ using $\Delta_{vap}\bar{H} = 40.66$ kJ/mol from Table 6.2:

$$\left(\frac{1}{T_2} - \frac{1}{373.15\text{ K}}\right) = -\frac{R}{\Delta\bar{H}}\ln\frac{p_2}{p_1} = -\frac{8.314\ \frac{\text{J}}{\text{mol}\cdot\text{K}}}{40.66\times10^3\text{ J/mol}}\ln\frac{2.02\text{ atm}}{1\text{ atm}} = -1.44\times10^{-4}\text{ /K};$$

$T_2 = 394.3$ K or 121.2 °C.

6.47. We use the Clausius–Clapeyron equation in the form $\ln\frac{f_{den,2}}{f_{den,1}} = -\frac{\Delta_{den}\bar{H}}{R}\left(\frac{1}{T_2} - \frac{1}{T_1}\right)$ with $T_1 = (273.2 + 45.0)$ K = 318.2 K and $T_2 = (273.2 + 75.0)$ K = 348.2 K:

$$\ln\frac{f_{den,2}}{0.113} = -\frac{160.8\times10^3\text{ J/mol}}{8.314\ \frac{\text{J}}{\text{mol}\cdot\text{K}}}\left(\frac{1}{348.2\text{ K}} - \frac{1}{318.2\text{ K}}\right) = 5.24;\ f_{den,2} = 21.3.$$

6.49. According to Figure 6.5, a decrease in air pressure at higher altitudes lowers the boiling point of water (we move down and left on the liquid–gas equilibrium line). Therefore, boiling water is not as hot as at sea level, and more cooking time is needed.

6.51. The slope of the solid–liquid equilibrium line is equal to dp/dT, which is $\Delta\bar{S}/\Delta\bar{V}$ according to the Clapeyron equation (equation 6.9). For any solid–to–liquid transition the change in molar entropy is always positive, so the sign of $\Delta\bar{V}$ determines whether dp/dT is positive or negative. In the case of water, $\Delta\bar{V} < 0$ because solid H_2O is less dense than liquid H_2O, which means $dp/dT < 0$: the solid–liquid equilibrium line is slanted towards the left as the pressure is increased. From Figure 6.6, we can identify Ice I ("normal" ice) as having such a negative slope; however, Ice III, Ice V, and Ice VI (the other solid phases bordering the liquid phase) all have positive slopes, suggesting that these solids are all denser than liquid water.

6.53. P can never be greater than three because degrees of freedom cannot be negative.

6.55. If we had four phases at equilibrium, their chemical potentials would all be equal:

$\mu_1 = \mu_2 = \mu_3 = \mu_4 \rightarrow F_1(p,T) = F_2(p,T) = F_3(p,T) = F_4(p,T)$. We would have three independent equations relating only two variables; this is a situation that is mathematically overdetermined and does not allow for a definite mathematical solution. Therefore, having four phases in equilibrium is not possible.

6.57. One can argue that at high enough pressures, only solid phases exist, while at high enough temperatures, only gases exist (or supercritical fluids, depending on the conditions). Thus, an argument can be made that liquids are only 'metastable' because they exist only under certain temperature and pressure conditions.

6.59. The negatively sloped line for part of the equilibrium implies that in a certain pressure range, increasing the temperature of ^{3}He promotes a liquid–to–solid phase change, rather than a solid–to–liquid phase change (as is normally the case when increasing the temperature).

6.61. Using the given data:

(a)

$$T_{tp} = -\frac{15.52\times10^3\ \text{J/mol}}{\left(8.314\dfrac{\text{J}}{\text{mol}\cdot\text{K}}\right)\ln\dfrac{200.0\ \text{torr}}{10.00\ \text{torr}} + \dfrac{41.57\times10^3\ \text{J/mol}}{410\ \text{K}} - \dfrac{57.09\times10^3\ \text{J/mol}}{364\ \text{K}}} = 5.1\times10^2\ \text{K}$$

(Note that with the transition enthalpies from Table 6.2, one obtains $T_{tp} = 3.5\times10^2$ K, or about 350 K, which is in much better agreement with the experimental value of about 387 K.)

(b) $\ln\frac{p_2}{p_1} = -\frac{\Delta\bar{H}}{R}\left(\frac{1}{T_2} - \frac{1}{T_1}\right) = -\frac{41.57\times10^3\ \text{J/mol}}{8.314\frac{\text{J}}{\text{mol}\cdot\text{K}}}\left(\frac{1}{5.1\times10^2\ \text{K}} - \frac{1}{457.4\ \text{K}}\right) = +1.1$, and with

$p_1 = 1$ atm (or 760 torr) we obtain $p_2 = 3$ atm (or 2×10^3 torr).

(Using the more accurate value of $T_{tp} = 3.5\times10^2$ K from part (a), we obtain $p_2 = 0.04$ atm or about 30 torr, which is again in better agreement with the experimental value of about 90 torr.)

6.63. The critical point represents the point in the phase diagram beyond which only a single fluid phase, a supercritical fluid, exists. Since $P = 1$, according to the Gibbs phase rule, degrees of freedom = 3 – 1 = 2. This means that two variables need to be specified (both pressure and temperature) to fully characterize a supercritical fluid.

6.65. Line on bottom left: dp/dT for the solid–gas transition (sublimation); line passing through point B: dp/dT for the solid-liquid transition (fusion); line passing through point D: dp/dT for the liquid-gas transition (evaporation).

6.67. **(a)** The allotropes are the two solid phases shown, namely rhombic and monoclinic sulfur.

(b) Under normal conditions, for example at 298 K and 1 atm of pressure as indicated in the diagram, the stable form of sulfur is rhombic.

(c) Starting at this point and increasing the temperature at a constant pressure of 1 atm (horizontal line), we first turn rhombic sulfur into the monoclinic form of the solid before melting it into a liquid; then, at higher temperatures, it will vaporize into a gas.

6.69. A change of −3.00 °C corresponds to −3.00 K, and because we will assume constant pressure, equation 6.20 simplifies to $d\mu = -\bar{S}\,dT$. Integration leads to the macroscopic change of $\Delta\mu = -\bar{S}\Delta T = -\left(69.54\,\frac{\text{J}}{\text{mol}\cdot\text{K}}\right)(-3.00\text{ K}) = +209\text{ J/mol}$.

CHAPTER 7

Equilibria in Multiple-Component Systems

7.1. The addition of the olive would add one more solid phase and one more chemical component to the system (to make the problem tractable, we will assume that the olive has only a single chemical component; this is, of course, not true). Thus, $C = 3$ and $P = 3$, so the number of degrees of freedom is $F = C - P + 2 = 3 - 3 + 2 = 2$, which could be, as in Example 7.1, the temperature and mole fraction of water in the liquid phase.

With temperature, we know the pressure because water in the liquid and solid phases is in equilibrium. And because we know the total amounts of water, ethanol, and "olive" in the system, we also know the mole fraction of ethanol in the liquid phase (which can be determined by subtraction), the amount of water in the form of ice cubes, and the amount of "olive" in the olive.

7.3. $C = 3$; we want $F = 0 = C - P + 2 = 3 - P + 2$, so $P = 5$. We would need 5 separate phases in equilibrium with each other to have zero degrees of freedom (for example, three solid phases, one liquid phase and the gas phase).

7.5. We have four chemical components in the system ($NaHCO_3$, Na_2CO_3, H_2O, and CO_2), but because the are linked by a chemical equilibrium, only three of them are independent components (the amount of the fourth can be determined from the equilibrium constant). We also have 4 phases (solid $NaHCO_3$, solid Na_2CO_3, liquid water and CO_2 gas), so there are

$F = 3 - 4 + 2 = 1$ degrees of freedom.

7.7. We start with equation 7.4: $x_i = \dfrac{n_i}{\sum_{\text{all } i} n_i}$. Take the sum over all x_i: $\sum_{\text{all } i} x_i = \sum_{\text{all } i} \left(n_i \Big/ \sum_{\text{all } i} n_i \right)$.

The denominator can be factored out of the sum, so that $\sum_{\text{all } i} x_i = \left(\dfrac{1}{\sum_{\text{all } i} n_i} \right) \left(\sum_{\text{all } i} n_i \right) = 1$.

7.9. If $x_{H_2O} = 0.35$, then $x_{CH_3OH} = 1 - 0.35 = 0.65$. Using Raoult's law:

$$p_{H_2O} = x_{H_2O} p_{H_2O}^* = (0.35)(23.76 \text{ torr}) = 8.3 \text{ torr} \; ; \; p_{CH_3OH} = (0.65)(125.0 \text{ torr}) = 81 \text{ torr}.$$

The total pressure is 8.3 torr + 81 torr = 90 torr. The mole fractions in the gas phase are equal to the ratio of the partial pressures to the total pressure:

$$y_{H_2O} = \frac{8.3}{90} = 0.093 \; ; \; y_{CH_3OH} = \frac{81}{90} = 0.91.$$

7.11. Starting again with $p_{tot} = x_1 p_1^* + x_2 p_2^*$, but now substituting $x_1 = 1 - x_2$, we obtain $p_{tot} = (1 - x_2)p_1^* + x_2 p_2^* = p_1^* - x_2 p_1^* + x_2 p_2^* = p_1^* + (p_2^* - p_1^*)x_2$. This has again the form of a straight line $y = mx + b$, with a slope of $m = p_2^* - p_1^*$ and a y intercept of $b = p_1^*$.

7.13. We start with the expressions similar to those in equation 7.18, but in terms of y_2:

$y_2 = \frac{p_2}{p_{tot}} = \frac{p_2}{p_2 + p_1} = \frac{x_2 p_2^*}{x_2 p_2^* + x_1 p_1^*}$. Next, we recognize that $x_1 = 1 - x_2$, so we substitute in the denominator: $y_2 = \frac{x_2 p_2^*}{x_2 p_2^* + (1 - x_2)p_1^*}$. The denominator can be rearranged:

$x_2 p_2^* + (1 - x_2)p_1^* = x_2 p_2^* + p_1^* - x_2 p_1^* = p_1^* + (p_2^* - p_1^*)x_2$, thus $y_2 = \frac{x_2 p_2^*}{p_1^* + (p_2^* - p_1^*)x_2}$.

7.15. A 2:1 molar ratio of hexane and cyclohexane means $x_{hex} = 2/3$ and $x_{cyc} = 1/3$. The partial pressures are obtained from Raoult's law:

$p_{hex} = x_{hex} p_{hex}^* = (2/3)(151.4 \text{ torr}) = 100.9 \text{ torr}$; $p_{cyc} = (1/3)(97.6 \text{ torr}) = 32.5 \text{ torr}$. The total pressure is (100.9 + 32.5) torr = 133.5 torr.

7.17. The partial pressures are obtained from Raoult's law:

$p_{H_2O} = x_{H_2O} p_{H_2O}^* = (0.045)(17.5 \text{ torr}) = 0.79 \text{ torr}$;

$p_{C_2H_5OH} = (0.955)(43.7 \text{ torr}) = 41.7 \text{ torr}$.

The total pressure is (0.79 + 41.7) torr = 42.5 torr.

7.19. 50 g of C_2H_5OH equal $(50.0 \text{ g})\left(\frac{1 \text{ mol}}{46.069 \text{ g}}\right) = 1.09 \text{ mol}$, and 50 g of C_3H_7OH equal

$(50.0 \text{ g})\left(\frac{1 \text{ mol}}{60.096 \text{ g}}\right) = 0.832 \text{ mol}$. The total amount is (1.09 + 0.832) mol = 1.92 mol, and

the mole fractions are: $x_{C_2H_5OH} = \frac{1.09 \text{ mol}}{1.92 \text{ mol}} = 0.566$; $x_{C_3H_7OH} = \frac{0.83 \text{ mol}}{1.92 \text{ mol}} = 0.434$. From

Raoult's law: $p_{C_2H_5OH} = (0.566)(43.7 \text{ mmHg}) = 24.7 \text{ mmHg}$;

$p_{C_3H_7OH} = (0.434)(18.0 \text{ mmHg}) = 7.81 \text{ mmHg}$.

Thus p_{tot} = (24.7 + 7.81) mmHg = 32.5 mmHg (or 32.5 torr).

7.21. We solve equation 7.17, $p_{tot} = p_2^* + (p_1^* - p_2^*)x_1$ for x_1: $x_1 = \frac{p_{tot} - p_2^*}{p_1^* - p_2^*}$. If we associate ethanol with index 1 and n-propanol with index 2:

$$x_{EtOH} = \frac{28.6 \text{ mmHg} - 18.0 \text{ mmHg}}{43.7 \text{ mmHg} - 18.0 \text{ mmHg}} = 0.412$$

and $x_{PrOH} = 1 - 0.412 = 0.588$.

7.23. In a 1:1 molar ratio mixture of hexane to cyclohexane, the mole fraction of each is exactly 0.5. Therefore, the total pressure is p_{tot} = (0.5)(151.4 torr) + (0.5)(97.6 torr) = 75.7 torr + 48.8 torr = 124.5 torr. The mole fraction of hexane in the vapor phase is then $y_{C_6H_{14}} = \frac{p_{C_6H_{14}}}{p_{tot}} = \frac{75.7\text{ torr}}{124.5\text{ torr}} = 0.608$, while that of cyclohexane is $y_{C_6H_{12}} = \frac{48.8\text{ torr}}{124.5\text{ torr}} = 0.392$.

7.25. The mole fractions of this mixture are the same as in exercise 7.19. Using Raoult's law, the total pressure is

p_{tot} = (0.566)(43.7 mmHg) + (0.434)(18.0 mmHg) = 24.7 mmHg + 7.81 mmHg = 32.5 mmHg. The mole fractions in the vapor phase are therefore

$$y_{EtOH} = \frac{p_{EtOH}}{p_{tot}} = \frac{24.7\text{ mmHg}}{32.5\text{ mmHg}} = 0.760 \text{ and } y_{PrOH} = \frac{7.81\text{ mmHg}}{32.5\text{ mmHg}} = 0.240.$$

7.27. In problem 7.13, we found that equation 7.18, written in terms of y_2, is given by $y_2 = \frac{x_2 p_2^*}{p_1^* + (p_2^* - p_1^*)x_2}$. Solving this equation for x_2 gives $x_2 = \frac{y_2 p_1^*}{p_2^* + (p_1^* - p_2^*)y_2}$ (just multiply by the denominator, collect all terms in x_2 on one side, and isolate x_2).

Substituting x_2 into equation 7.22 (using i = 2): $p_{tot} = \frac{x_2 p_2^*}{y_2} = \frac{\frac{y_2 p_1^*}{p_2^* + (p_1^* - p_2^*)y_2} p_2^*}{y_2}$.

Simplify: $p_{tot} = \frac{y_2 p_1^* p_2^*}{\left[p_2^* + (p_1^* - p_2^*)y_2\right]y_2} = \frac{p_1^* p_2^*}{p_2^* + (p_1^* - p_2^*)y_2}$.

7.29. $\Delta_{mix}\bar{G} = \left(8.314\frac{\text{J}}{\text{mol}\cdot\text{K}}\right)(293.2\text{ K})\left[(0.500)(\ln 0.500) + (0.500)(\ln 0.500)\right]$, which gives

$\Delta_{mix}\bar{G} = -1.69 \times 10^3$ J/mol and

$$\Delta_{mix}G = n_{tot}\Delta_{mix}\bar{G} = (2.00\text{ mol})\left(-1.69 \times 10^3\frac{\text{J}}{\text{mol}}\right) = -3.38 \times 10^3\text{ J}.$$

$$\Delta_{mix}\bar{S} = -\left(8.314\frac{\text{J}}{\text{mol}\cdot\text{K}}\right)\left[(0.500)(\ln 0.500) + (0.500)(\ln 0.500)\right] = +5.76\frac{\text{J}}{\text{mol}\cdot\text{K}};$$

$$\Delta_{mix}S = n_{tot}\Delta_{mix}\bar{S} = (2.00\text{ mol})\left(+5.76\frac{\text{J}}{\text{mol}\cdot\text{K}}\right) = +11.5\text{ J/K}.$$

7.31. Acetone is fully miscible with water and can therefore be used to rinse wet glassware. Because of its low boiling point (about 56 °C), this helps in getting the glassware dry. In addition, the formation of a low-boiling azeotrope between water and acetone encourages evaporation of any remaining water in the form of the azeotrope and dries out the glassware more quickly.

7.33. If an initial composition of $x_1 = 0.4$ were used, the tie lines that you would draw in Figure 7.15 would eventually lead you to the left side of the phase diagram, at which point the final product of our distillation would be pure, low-boiling component #2 (that is, x_1 would equal 0).

7.35. Minimum-boiling azeotropes have a boiling point that is lower than that of each of the pure components: CCl_4 & HCOOH; CH_3OH & CH_3COCH_3; H_2O & C_2H_5OH; H_2O & $C_2H_5CO_2CH_3$; H_2O & pyridine. Maximum-boiling azeotropes (its boiling point is higher than that of each of the pure components): HCl & $(CH_3)_2O$, HCOOH & pyridine.

7.37. Benzene is poisonous and a suspected carcinogen. Unless just the necessary amount of benzene is added to remove all of the water in the ethanol-water azeotrope, there may be residual benzene left over in the remaining "pure" ethanol. Thus, such ethanol should not be ingested!

7.39. Use Table 1.1 for the proper conversion factors:

$$\left(1.67 \times 10^8 \text{ Pa}\right)\left(\frac{1 \text{ bar}}{100{,}000 \text{ Pa}}\right) = 1.67 \times 10^3 \text{ bar} \;;$$

$$\left(1.67 \times 10^3 \text{ bar}\right)\left(\frac{1 \text{ atm}}{1.01325 \text{ bar}}\right) = 1.65 \times 10^3 \text{ atm};$$

$$\left(1.67 \times 10^3 \text{ bar}\right)\left(\frac{1 \text{ atm}}{1.01325 \text{ bar}}\right)\left(\frac{760 \text{ torr}}{1 \text{ atm}}\right)\left(\frac{1 \text{ mmHg}}{1 \text{ torr}}\right) = 1.25 \times 10^6 \text{ mmHg}.$$

7.41. Table 7.2 lists Henry's law constant for H_2 in aqueous solution as 7.03×10^9 Pa:
$p_{H_2} = K_{H_2} x_{H_2} = \left(7.03 \times 10^9 \text{ Pa}\right)(0.185) = 1.30 \times 10^9$ Pa, or 1.28×10^4 atm.

7.43. $p_{CHCl_3} = K_{CHCl_3} x_{CHCl_3} = \left(2.40 \times 10^6 \text{ Pa}\right)(0.0010) = 2.4 \times 10^3$ Pa

7.45. **(a)** A mole fraction of 1.388×10^{-5} implies that there are 1.388×10^{-5} moles of air dissolved in 1 mol of the air/water solution. Because the number is so small, we will assume that 1 mol of the solution contains exactly 1 mol of water (the difference will only show up in the 5th decimal place) and that no volume change takes place when the air is dissolving in water. The volume of 1 mol of water is

$(1 \text{ mol})\left(18.0152 \frac{\text{g}}{\text{mol}}\right)\left(\frac{\text{cm}^3}{1.00 \text{ g}}\right) = 18.0 \text{ cm}^3$ or 0.0180 L, and the molarity of the solution is $\dfrac{1.388 \times 10^{-5} \text{ mol}}{0.0180 \text{ L}} = 7.71 \times 10^{-4}$ mol/L.

(b) Using $p_{air} = K_{air}x_{air}$ and an ambient air pressure of 1 atm = 101,325 Pa, we solve for K_{air}:

$K_{air} = \frac{p_{air}}{x_{air}} = \frac{101,325 \text{ Pa}}{1.388 \times 10^{-5}} = 7.300 \times 10^{9} \text{ Pa}$. This value is close to the values for nitrogen and oxygen from Table 7.2.

(c) The solubilities of gases decrease when temperatures are increased.

7.47. A higher Henry's law constant means that a gas is less soluble in a liquid. According to equation 7.31, for a given gas pressure p_i, the mole fraction x_i will be lower with a higher constant K_i.

7.49. As impure water freezes, a new solid phase containing pure H_2O is formed. The impurities remain in the liquid phase, and their concentration increases until they eventually precipitate as the last liquid solidifies in the core of the ice cube.

7.51. Using equation 7.39: .

$$\ln x_{solute} = -\frac{11.29 \times 10^{3} \text{ J/mol}}{8.314 \frac{\text{J}}{\text{mol} \cdot \text{K}}}\left(\frac{1}{298 \text{ K}} - \frac{1}{(40.9 + 273.15) \text{ K}}\right) = -0.23, \text{ so } x_{solute} = 0.79.$$

This means that 0.79 mol phenol (or 75 g) would "dissolve" in 0.21 mol of water (3.7 g), which is equivalent to more than 2000 g of phenol per 100 g of water and several orders of magnitude higher than the experimental value given in the previous exercise. The reason for this discrepancy is that phenol and water do not form an ideal solution.

7.53. There is no inherent reason that equation 7.39 will not work for gases, since the system definition can be altered to apply to a gas-dissolved gas equilibrium. In the derivation of equation 7.39, we would simply substitute the gas phase for the solid phase of the solute; T_{MP} would become T_{BP} (boiling point of the pure solute), and $-\Delta_{fus}\bar{G}$ would become $+\Delta_{vap}\bar{G}$. Why don't you try to derive some Henry's law constants with it?

7.55. **(a)** $\ln x_{solute} = -\frac{15.27 \times 10^{3} \text{ J/mol}}{8.314 \frac{\text{J}}{\text{mol} \cdot \text{K}}}\left[\frac{1}{298.2 \text{ K}} - \frac{1}{(112.9 + 273.15) \text{ K}}\right] = -1.40; \; x_{solute} = 0.246$

To determine the mole fraction from the weight percent, assume 100.00 g of solution. Therefore, 14.09 g is I_2 and 85.91 g is C_6H_6. Convert to moles, add together, and determine mole fraction: $(14.09 \text{ g})\left(\frac{1 \text{ mol}}{253.8090 \text{ g}}\right) = 0.05551 \text{ mol } I_2$;

$(85.91 \text{ g})\left(\frac{1 \text{ mol}}{78.113 \text{ g}}\right) = 1.100 \text{ mol } C_6H_6$;

Total: 0.05551 mol + 1.100 mol = 1.155 mol. Mole fraction of I_2:

$\frac{0.05551 \text{ mol}}{1.155 \text{ mol}} = 0.04805$.

This is only about 1/5th of the predicted solubility.

(b) $x_{solute} = 0.246$, as in (a).

We again assume 100.00 g of solution to convert weight percent into mole fraction; 2.72 g of the solution is I_2 and 97.28 g is C_6H_{12}. We proceed as before:

$$(2.72\text{ g})\left(\frac{1\text{ mol}}{253.8090\text{ g}}\right) = 0.0107\text{ mol I}_2\ ;\ (97.28\text{ g})\left(\frac{1\text{ mol}}{84.161\text{ g}}\right) = 1.156\text{ mol C}_6\text{H}_{12}.$$

Total: 0.0107 mol + 1.156 mol = 1.167. Mole fraction of I_2: $\frac{0.0107\text{ mol}}{1.167\text{ mol}} = 0.00919$.

This is only about 1/25th of the predicted solubility.

(c) $$\ln x_{solute} = -\frac{17.15\times10^3\text{ J/mol}}{8.314\frac{\text{J}}{\text{mol}\cdot\text{K}}}\left[\frac{1}{298.2\text{ K}} - \frac{1}{(52.7+273.15)\text{ K}}\right] = -0.588;\ x_{solute} = 0.555$$

To convert weight percent into mole fraction, assume 100.00 g of solution. Therefore, 20.57 g is $C_6H_4Cl_2$ and 79.43 g is C_6H_{14}. Proceeding as before:

$$(20.57\text{ g})\left(\frac{1\text{ mol}}{147.003\text{ g}}\right) = 0.1399\text{ mol C}_6\text{H}_4\text{Cl}_2\ ;$$

$$(79.43\text{ g})\left(\frac{1\text{ mol}}{86.177\text{ g}}\right) = 0.9217\text{ mol C}_6\text{H}_{14};$$

Total: 0.1399 mol + 0.9217 = 1.0616. Mole fraction of $C_6H_4Cl_2$:

$$\frac{0.1399\text{ mol}}{1.0616\text{ mol}} = 0.1318$$

This is only about 1/4th of the predicted solubility.

7.57. Using degrees of freedom $F = C - P + 2$, if there are 2 components and 3 phases (the liquid solution and the two intimately-mixed solids that make up the eutectic), then $F = 2 - 3 + 2 = 1$ degree of freedom.

7.59. If you start with an equimolar, liquid mixture of Na and K ($x_{Na} = 0.5$) and cool it down, you get to a point in the phase diagram where solid Na_2K begins to form (at about 0 °C). Solid Na_2K is richer in Na ($x_{Na} = 0.66$), so the liquid becomes more and more dilute in Na, tracing the curved line down to the left. When the mole fraction of Na in the liquid reaches about 0.29, a eutectic between solid K and solid Na_2K forms, and the entire liquid solidifies as K and Na_2K.

7.61. Carbon has a sublimation point of 3642 °C. By contrast, Si has a melting point of 1414 °C. Thus, while we might be able to generate temperatures of over 3600 °C, the carbon will escape into the gas phase instead of holding together in a liquid state that can carefully resolidify.

7.63. The stoichiometric compound is found at the vertical line that intersects the *x*-axis at x_{Na} = 0.66. This represents a composition that is 2/3 Na (by mole) and 1/3 K (by mole) or a Na-to-K-ratio of 2:1, which corresponds to a formula of Na_2K.

7.65. Molarity automatically includes the concept of partial molar volume because it is defined using the volume of the *final solution*. This volume is not simply the sum of the volumes of the components, but the sum of their partial molar volumes (multiplied by the corresponding amount of moles of the components in the final solution), all taken at the composition of the final solution.

7.67. The mole fraction of isopropanol in the final mixture is
$x = \dfrac{2.00 \text{ mol}}{2.00 \text{ mol} + 0.500 \text{ mol}} = 0.800$. According to Raoult's law, the vapor pressure of isopropanol above the solution is then $p = x p^* = (0.800)(47.0 \text{ mmHg}) = 37.6 \text{ mmHg}$. Because the other component (hexachlorobenzene) is non-volatile, this is also total vapor pressure above the solution.

7.69. We first determine the mole fraction of benzene (the solvent) according to the equation
$p_{solv} = x_{solv} p_{solv}^*$: $x_{solv} = \dfrac{p_{solv}}{p_{solv}^*} = \dfrac{73.29 \text{ torr}}{76.03 \text{ torr}} = 0.9640$ (the vapor pressure of the solution is equal to the vapor pressure of benzene because the solute is nonvolatile). The mole fraction of the solid is then $x_{solute} = 1 - 0.9640 = 0.0360$.

10.00 g of benzene are $(10.00 \text{ g})\left(\dfrac{1 \text{ mol}}{78.113 \text{ g}}\right) = 0.1280 \text{ mol } C_6H_6$, so

$0.0360 = \dfrac{n_{solute}}{n_{solute} + 0.1280 \text{ mol}}$. Solve for n_{solute}: 0.00479 mol, which gives a molar mass of 1.66 g/0.00479 mol = 347 g/mol.

7.71. Some people may think that by adding salt to water during high-altitude cooking, the boiling point of the liquid may be raised enough to compensate for the decreased boiling point of water due to the altitude. However, let's calculate the amount of NaCl needed to raise the boiling point of water by 3 °C (which would return the boiling point of water in, say, Denver, back to 100 °C): (3 K) = (2 particles) (*m*) (0.512 K/molal) gives *m* = 2.93 molal.

Thus, a 2.93 mol/kg solution of NaCl is necessary to raise the temperature of water by 3 °C. For 1 liter of water (1.00 kg), that would require 2.93 mol of NaCl, or 171 grams – more than a dozen teaspoons! Adding small amounts of salt to water has a negligible effect on its boiling point, and is added for taste more than its effect on the boiling point of water.

7.73. A 23.0% NaCl solution corresponds to 23.0 g NaCl dissolved in 77.0 g of water. The molar amounts are $(23.0\text{ g})\left(\frac{1\text{ mol}}{58.4425\text{ g}}\right) = 0.394$ mol NaCl and

$(77.0\text{ g})\left(\frac{1\text{ mol}}{18.0152\text{ g}}\right) = 4.27$ mol H_2O. The mole fraction of NaCl in solution is then

$$x_{NaCl} = \frac{0.394\text{ mol}}{0.394\text{ mol} + 4.27\text{ mol}} = 0.0843.$$

We use equation 7.57 with $N = 2$ (because NaCl dissociates into Na^+ and Cl^-), $T = (273.15 + 37.0)\text{ K} = 310.2\text{ K}$, and a molar volume of 0.0180 L/mol for water. Solve for Π:

$$\Pi = \frac{Nx_{NaCl}RT}{\overline{V}} = \frac{(2)(0.0843)\left(0.08205\frac{\text{L}\cdot\text{atm}}{\text{mol}\cdot\text{K}}\right)(310.2\text{ K})}{0.0180\text{ L/mol}} = 238\text{ atm}$$

7.75. First calculate the cryoscopic constant for Hg, with $T_{MP} = (273.15 - 39)\text{ K} = 234\text{ K}$:

$$K_f = \frac{M_{solv}RT_{MP}^2}{1000\Delta_{fus}\overline{H}} = \frac{\left(200.59\frac{\text{g}}{\text{mol}}\right)\left(8.314\frac{\text{J}}{\text{mol}\cdot\text{K}}\right)(234\text{ K})^2}{(1000\text{ g/kg})(2{,}331\text{ J/mol})} = 39.2\frac{\text{K}\cdot\text{kg}}{\text{mol}}.$$

Next, we need to determine the approximate molality of the solution. 1 mol of the solution would contain 0.0477 mol of Na and 0.9523 mol of Hg. The atomic mass of Hg is 200.59 g/mol, so 0.9523 mol are $(0.9523\text{ mol})\left(\frac{200.59\text{ g}}{1\text{ mol}}\right)\left(\frac{1\text{ kg}}{1000\text{ g}}\right) = 0.1910\text{ kg}$. Now

we can determine the molality of the solution, $m_{solute} = \frac{0.0477\text{ mol}}{0.1910\text{ kg}} = 0.250\text{ mol/kg}$, and the change in freezing point:

$\Delta T_f = K_f \cdot m_{solute} = (39.2\text{ K·kg/mol})\ (0.250\text{ mol/kg}) = 9.79\text{ K}$. Therefore, the freezing point decreases by 9.79 K, to about –49 K.

7.77. We rearrange the equation for K_f and solve for $\Delta_{fus}H$:

$$\Delta_{fus}\overline{H} = \frac{M_{solv}RT_{MP}^2}{1000K_f} = \frac{\left(84.161\frac{\text{g}}{\text{mol}}\right)\left(8.314\frac{\text{J}}{\text{mol}\cdot\text{K}}\right)(279.6\text{ K})^2}{(1000\text{ g/kg})(20.3\text{ K/molal})} = 2.69\times10^3\frac{\text{J}\cdot\text{kg}\cdot\text{molal}}{\text{mol}^2}.$$

Because molal = mol/kg, the units are simply J/mol:

$$\Delta_{fus}\overline{H} == 2.69\times10^3\frac{\text{J}}{\text{mol}} = 2.69\text{ kJ/mol}.$$

7.79. Problem 7.62 results in $x_{Na} = 0.739$ at 0 °C (see "Answers to Selected Exercises"), so $x_{Hg} = 1 - 0.739 = 0.261$. Raoult's law: $p_{Hg} = x_{Hg} \cdot p_{Hg}{}^* = (0.261)\ (0.000185\text{ torr}) = 4.83\times10^{-5}$ torr, which is a vapor pressure decrease of 0.000137 torr.

7.81. Both constants are calculated in a similar fashion, and involve $T_{MP}^2 / \Delta_{fus}\overline{H}$ and $T_{BP}^2 / \Delta_{vap}\overline{H}$ for the cryoscopic and ebullioscopic constants, respectively. Because the enthalpy of fusion is usually much smaller than the enthalpy of evaporation, the cryoscopic constant is usually larger (the temperatures in K are much less different and, even squared, have a lesser impact).

7.83. We rearrange $\Pi\overline{V} = x_{solute}RT$ and solve for x (being careful to convert pressure units so that they cancel out):

$$x_{solute} = \frac{\Pi\overline{V}}{RT} = \frac{(30\text{ Pa})(0.0180\text{ L/mol})}{\left(0.08314\frac{\text{L}\cdot\text{bar}}{\text{mol}\cdot\text{K}}\right)\left(\frac{100{,}000\text{ Pa}}{1\text{ bar}}\right)(310\text{ K})} = 2.09\times10^{-7}.$$

Therefore, we need 2.09×10^{-7} mol of polymer per mole of solution; given the small mole fraction, this is per mole of solvent (18.0152 g = 0.0180152 kg of water). This corresponds to a molality of 2.10×10^{-7} mol / 0.0180152 kg = 1.16×10^{-5} mol/kg. In grams of polymer, this represents $(1.16\times10^{-5}\text{ mol})\left(\frac{185{,}000\text{ g}}{\text{mol}}\right) = 2.15$ g of polymer per kg of water.

7.85. **(a)** For NaCl: $N = 2$ (Na^+ and Cl^-); $x_{NaCl} = \frac{0.100\text{ mol}}{0.100\text{ mol} + 0.900\text{ mol}} = 0.100$. The molar volume of water is 0.0180 L/mol, and with T = (273.15 + 30.0) K = 303.2 K:

$$\Pi = \frac{Nx_{NaCl}RT}{\overline{V}} = \frac{(2)(0.100)\left(0.08205\frac{\text{L}\cdot\text{atm}}{\text{mol}\cdot\text{K}}\right)(303.2\text{ K})}{0.0180\text{ L/mol}} = 276\text{ atm}$$

(b) For $Ca(NO_3)_2$: N = 3 (Ca^{2+} and two NO_3^{2-}), all other values are the same as in (a).

$$\Pi = \frac{Nx_{Ca(NO_3)_2}RT}{\overline{V}} = \frac{(3)(0.100)\left(0.08205\frac{\text{L}\cdot\text{atm}}{\text{mol}\cdot\text{K}}\right)(303.2\text{ K})}{0.0180\text{ L/mol}} = 415\text{ atm}$$

(c) For $Al(NO_3)_3$: $N = 4$ (Al^{3+} and three NO_3^{2-}), all other values are the same as in (a).

$$\Pi = \frac{Nx_{Al(NO_3)_3}RT}{\overline{V}} = \frac{(4)(0.100)\left(0.08205\frac{\text{L}\cdot\text{atm}}{\text{mol}\cdot\text{K}}\right)(303.2\text{ K})}{0.01801\text{ L/mol}} = 553\text{ atm}$$

CHAPTER 8

Electrochemistry and Ionic Solutions

8.1. Solve $F = \dfrac{q_1 q_2}{4\pi\varepsilon_0 r^2}$ for q_1:

$$q_1 = \frac{4\pi\varepsilon_0 r^2 F}{q_2} = \frac{4\pi\left(8.854\times10^{-12}\,\dfrac{\mathrm{C}^2}{\mathrm{J\cdot m}}\right)(100.0\ \mathrm{m})^2(0.0225\ \mathrm{N})}{1.00\ \mathrm{C}} = 2.50\times10^{-8}\ \mathrm{C}.$$

8.3. **(a)** $F = \dfrac{q_1 q_2}{4\pi\varepsilon_0\varepsilon_r r^2}$ with the negative charge $q_1 = -2q_2$ gives $F = \dfrac{-2q_2{}^2}{4\pi\varepsilon_0\varepsilon_r r^2}$. Solving for q_2 (remember that F is attractive and therefore negative, and don't forget to convert cm into m):

$$q_2{}^2 = -2\pi\varepsilon_0\varepsilon_r r^2 F = -2\pi\left(8.854\times10^{-12}\,\frac{\mathrm{C}^2}{\mathrm{J\cdot m}}\right)(78)\left(6.075\times10^{-2}\ \mathrm{m}\right)^2\left(-1.55\times10^{-6}\ \mathrm{N}\right)$$

$= +2.5\times10^{-17}\ \mathrm{C}^2$, which gives $q_2 = +5.0\times10^{-9}$ C and $q_1 = -10.0\times10^{-9}$ C.

(b) The electric fields for the two particles are equal to the force divided by their charges. Therefore, the electric field on the first particle is $\dfrac{-1.55\times10^{-6}\ \mathrm{N}}{+5.0\times10^{-9}\ \mathrm{C}} = -3.1\times10^{2}\,\dfrac{\mathrm{J}}{\mathrm{C\cdot m}}$, while the electric field on the other particle is $\dfrac{-1.55\times10^{-6}\ \mathrm{N}}{-10.0\times10^{-9}\ \mathrm{C}} = +156\,\dfrac{\mathrm{J}}{\mathrm{C\cdot m}}$. (Because 1 V = 1 J/C, the units can also be written as V/m.)

The magnitudes of the electric fields are simply about 310 V/m and 156 V/m, respectively.

8.5. The charge on the electron is -1.602×10^{-19} C, that on the proton $+1.602\times10^{-19}$ C:

$$F = \frac{q_1 q_2}{4\pi\varepsilon_0 r^2} = \frac{\left(-1.602\times10^{-19}\ \mathrm{C}\right)\left(+1.602\times10^{-19}\ \mathrm{C}\right)}{4\pi\left(8.854\times10^{-12}\,\dfrac{\mathrm{C}^2}{\mathrm{J\cdot m}}\right)\left(0.529\times10^{-10}\ \mathrm{m}\right)^2} = -8.24\times10^{-8}\ \mathrm{N}.$$

The force is negative and therefore attractive.

8.7. Similar to equation 8.8, $dw = \phi\cdot dQ$, the infinitesimal amount of electric work required to move a charge Q to a new location where the potential has changed by an infinitesimal amount $d\phi$ is given by $dw = Q\cdot d\phi$; integration for constant Q gives $w = Q\cdot\Delta\phi$. In this exercise, the potential difference is given as 1.00 V and the charge is the charge on one electron, or -1.602×10^{-19} C. Therefore, the work is (1.00 V) (-1.602×10^{-19} C) = -1.60×10^{-19} J. This amount of work is defined as an electron volt (sometimes written electronvolt, and abbreviated eV). As the negative sign in the result indicates, this work is done by the electron when the potential increases during translocation ($\Delta\phi = \phi_f - \phi_I = +1.00$ V), and it is required when the potential decreases ($\Delta\phi = \phi_f - \phi_I = -1.00$ V).

8.9. $E°$ values are intensive variables, meaning that they are independent of the amount of substance. In some cases, $E°$ values can be combined directly, but only under certain special circumstances (like the canceling out of all electrons from an overall reaction). However, in circumstances in which all electrons do not cancel, $E°$ values cannot be combined directly but must be converted to some extensive variable, like ΔG, so proper combinations of the extensive variables can be made and reconverted to $E°$ values.

8.11. **(a)** The two half reactions are:

$MnO_2 + 2\,H_2O \rightarrow MnO_4^- + 4\,H^+ + 3\,e^-$ $\quad E° = -1.679$ V

$4\,e^- + O_2 + 2\,H_2O \rightarrow 4\,OH^-$ $\quad E° = 0.401$ V

To balance the electrons, we multiply the first reaction by 4 and the second reaction by 3. Combining the H^+ and OH^- ions to H_2O and consolidating the water molecules on both sides:

$4\,MnO_2 + 3\,O_2 + 2\,H_2O \rightarrow 4\,MnO_4^- + 4\,H^+$ $\quad E° = -1.278$ V

To calculate $\Delta G°$:

$$\Delta G° = -nFE° = -(12\text{ mol})(96{,}485\text{ C/mol})(-1.278\text{ V}) = +1.480 \times 10^6\text{ J} = 1480\text{ kJ}$$

(b) The two half reactions are:

$Cu^+ \rightarrow Cu^{2+} + e^-$ $\quad E° = -0.153$ V

$e^- + Cu^+ \rightarrow Cu$ $\quad E° = 0.521$ V

Because both half reactions have one electron, they can be combined without multiplication:

$2\,Cu^+ \rightarrow Cu + Cu^{2+}$ $\quad E° = 0.368$ V

To calculate $\Delta G°$:

$$\Delta G° = -nFE° = -(1\text{ mol})(96{,}485\text{ C/mol})(0.368\text{ V}) = -3.55 \times 10^4\text{ J} = -35.5\text{ kJ}$$

(c) The two half reactions are:

$Br_2 + 2\,e^- \rightarrow 2\,Br^-$ $\quad E° = 1.087$ V

$2\,F^- \rightarrow F_2 + 2\,e^-$ $\quad E° = -2.866$ V

Because both half reactions have the same number of electrons, they can be combined without multiplication:

$Br_2 + 2\,F^- \rightarrow F_2 + 2\,Br^-$ $\quad E° = -1.779$ V

To calculate $\Delta G°$:

$$\Delta G° = -nFE° = -(2\text{ mol})(96{,}485\text{ C/mol})(-1.779\text{ V}) = +3.433 \times 10^5\text{ J} = 343.3\text{ kJ}$$

(d) The two half reactions are:

$H_2O_2 + 2\,H^+ + 2\,e^- \rightarrow 2\,H_2O$ $\quad E° = 1.776$ V

$2\,Cl^- \rightarrow Cl_2 + 2\,e^-$ $\quad E° = -1.358$ V

Because both half reactions have two electrons, they can be combined without multiplication:

$2\,H_2O_2 + 2\,H^+ + 2\,Cl^- \rightarrow 2\,H_2O + Cl_2$ $\quad E° = 0.418$ V

To calculate $\Delta G°$:

$$\Delta G° = -nFE° = -(2\text{ mol})(96{,}485\text{ C/mol})(0.418\text{ V}) = -8.07 \times 10^4\text{ J} = -80.7\text{ kJ}$$

8.13. The two half reactions are:

$Fe^{2+} \rightarrow Fe^{3+} + e^-$ $\quad E° = -0.771$ V

$Fe^{2+} + 2\,e^- \rightarrow Fe$ $\quad E° = -0.447$ V

To balance the electrons, we multiply the first reaction by 2. For the overall reaction, we get

$3\,Fe^{2+} \rightarrow 2\,Fe^{3+} + Fe$ $\quad E° = -1.218$ V

Since the overall voltage is negative, the reaction is not spontaneous. To calculate $\Delta G°$:

$$\Delta G° = -nFE = -(2\text{ mol})(96{,}485\text{ C/mol})(-1.218\text{ V}) = +2.350 \times 10^5\text{ J} = 235.0\text{ kJ}.$$

8.15. $\Delta G°$ is the maximum amount of usable, non-pV work that a reaction can provide, so we need to find the reactions for which $\Delta G° \leq -196$ kJ (note the negative sign: the work is *done* by the reaction). Because $\Delta G° = -nFE°$, we therefore need to identify the reactions for which

$nE° \geq -\Delta G°/F$, which evaluates to $-(-196\times10^3\text{ J}) / (96{,}485\text{ C/mol}) = +2.03$ V·mol.

(a) The half–reactions are

$Cu^{2+} + 2\,e^- \rightarrow Cu$ $\quad E° = 0.3419$ V

$H_2 \rightarrow 2\,H^+ + 2\,e^-$ $\quad E° = 0.0000$ V

Two electrons are exchanged in the balanced reaction

$Cu^{2+} + H_2 \rightarrow Cu + 2\,H^+$ $\quad E° = 0.3419$ V

and therefore $nE° = (2\text{ mol})\ (0.3419\text{ V} + 0.0000\text{ V}) = 0.6838$ V·mol. This reaction will not provide the required amount of work.

(b) The half–reactions are

$Fe \rightarrow Fe^{3+} + 3\,e^-$ $\quad E° = +0.037$ V

$e^- + Ag^+ \rightarrow Ag$ $\quad E° = 0.7996$ V

Three electrons are exchanged in the balanced reaction

$Fe + 3\,Ag^+ \rightarrow Fe^{3+} + 3\,Ag$ $\quad E° = 0.837$ V and therefore $nE° = (3 \text{ mol})(0.837 \text{ V}) = 2.51 \text{ V·mol}$. This reaction will therefore be able to provide the required work.

(c) The half reactions are

$Co \rightarrow Co^{2+} + 2\,e^-$ $\quad E° = +0.28 \text{ V}$

$2\,e^- + Ni^{2+} \rightarrow Ni$ $\quad E° = -0.257 \text{ V}$

Two electrons are exchanged in the balanced reaction

$Co + Ni^{2+} \rightarrow Co^{2+} + Ni \quad E° = 0.02 \text{ V}$

and therefore $nE° = (2 \text{ mol})(0.02 \text{ V}) = 0.05$ V□mol. This reaction will not provide the required amount of work.

(d) The half reactions are

$Au^+ + e^- \rightarrow Au$ $\quad E° = 1.692 \text{ V}$

$Zn \rightarrow Zn^{2+} + 2\,e^-$ $\quad E° = +0.7618 \text{ V}$

Two electrons are exchanged in the balanced reaction

$Au^+ + Zn \rightarrow Au + Zn^{2+} \quad E° = 2.454 \text{ V}$

and therefore $nE° = (2 \text{ mol})(2.454 \text{ V}) = 4.908 \text{ V·mol}$. This reaction will therefore be able to provide the required work.

8.17. SCE, which is the half–reaction with a higher potential (+0.2682 V versus SHE) would be the reduction reaction. The half reactions are then

$Hg_2Cl_2 + 2\,e^- \rightarrow 2\,Hg + 2\,Cl^-$

$H_2 \rightarrow 2\,H^+ + 2\,e^-$

The balanced reaction is obtained by simple addition:

$Hg_2Cl_2 + H_2 \rightarrow 2\,Hg + 2\,Cl^- + 2\,H^+$

8.19. **(a)** The half reactions are

$I_2 + 2\,e^- \rightarrow 2\,I^-$ $\quad E° = 0.5355 \text{ V}$

$3\,I^- \rightarrow I_3^- + 2\,e^-$ $\quad E° = -0.5360 \text{ V}$

and the balanced reaction is

$I_2 + I^- \rightarrow I_3^-$ $\quad E° = -0.0005 \text{ V}$

$$\Delta G° = -nFE° = -(2 \text{ mol})(96{,}485 \text{ C/mol})(-0.0005 \text{ V}) = +1\times10^2 \text{ J}$$

(b) To construct this reaction from two half reactions, we choose

$Cr^{2+} \rightarrow Cr^{3+} + e^-$ $\quad E° = +0.407 \text{ V}$

$Cr^{3+} + 3\,e^- \rightarrow Cr$ $\quad E° = -0.744 \text{ V}$

Combining them directly gives the desired reaction: $Cr^{2+} + 2\ e^- \rightarrow Cr$. Because this reaction is not balanced, however, we cannot add the two $E°$, but must first calculate the $\Delta G°$ values:

For $Cr^{2+} \rightarrow Cr^{3+} + e^-$: $\Delta G° = -nFE° = -(1\ \text{mol})(96{,}485\ \text{C/mol})(0.407\ \text{V}) = -39.3\ \text{kJ}$

For $Cr^{3+} + 3\ e^- \rightarrow Cr$: $\Delta G° = -nFE° = -(3\ \text{mol})(96{,}485\ \text{C/mol})(-0.744\ \text{V}) = 215\ \text{kJ}$

The $\Delta G°$ value for the combined reaction is then $-39.3\ \text{kJ} + 215\ \text{kJ} = 176\ \text{kJ}$, and $E°$ is

$$E° = -\frac{\Delta G°}{nF} = -\frac{176 \times 10^3\ \text{J}}{(2\ \text{mol})(96{,}485\ \text{C/mol})} = -0.913\ \text{V}.$$

8.21. The thermodynamically most favorable process (under isobaric and isothermal conditions) is the one with the most negative ΔG. For redox reactions, $\Delta G = -nFE$, which means that the redox reaction with the most positive E (or $E°$ at standard conditions) will be favored.

$E°$ for a balanced chemical reaction is $E°_{red} - E°_{ox}$ (reduction half–reaction minus oxidation half-reaction), and for a given reduction half-reaction $E°$ will be most positive for the lowest (most negative) $E°_{ox}$. Starting from the bottom of Table 8.2, we find that $Fe^{2+} + 2\ e^- \rightarrow Fe$ has the lowest $E°$, lower than $Fe^{3+} + 3\ e^- \rightarrow Fe$, $Cu^{2+} + 2\ e^- \rightarrow Cu$, and $Cu^+ + e^- \rightarrow Cu$. Therefore, the iron pipe will corrode first.

8.23. Appendix 2 lists $\Delta_f G°$ for Fe_2O_3 as -743.5 kJ/mol, Al_2O_3 as -1582.3 kJ/mol, and Al and Fe as 0 kJ/mol.

$$\Delta G° = \Delta_f G°(Al_2O_3) - \Delta_f G°(Fe_2O_3)$$

$$\Delta G° = [(1\ \text{mol})(-1582.3\ \text{kJ/mol})] + [(1\ \text{mol})(743.5\ \text{kJ/mol})] = -838.8\ \text{kJ}$$

$$E° = \frac{-\Delta G°}{nF} = \frac{-(-838.8 \times 10^3\ \text{J})}{(6\ \text{mol})(96{,}485\ \text{C/mol})} = +1.449\ \text{V}$$

8.25. According to the text, biochemical standard states include a pH of 7, meaning that the concentration of the hydrogen ion is 1×10^{-7} mol/L.

8.27. In equation 8.25, we can identify the following characteristic parameters of a straight line: dependent variable: E; independent variable: $\ln Q$; slope: $-RT/nF$; y–intercept: $E°$.

8.29. The half reactions are

$Au^+ + e^- \rightarrow Au$ $\quad E° = 1.692\ \text{V}$

$Fe \rightarrow Fe^{3+} + 3\ e^-$ $\quad E° = +0.037\ \text{V}$

To balance the electrons, we multiply the first reaction by 3 and and combine both:

$3\ Au^+ + Fe \rightarrow 3\ Au + Fe^{3+}$ $\quad E° = 1.729\ \text{V}$

Using concentrations ($c° = 1$ mol/L) instead of molalities for the activities in Q:

$$Q = \frac{c_{Fe^{3+}}/c°}{\left(c_{Au^+}/c°\right)^3} = \frac{0.219}{(0.00446)^3} = 2.47 \times 10^6$$

$$E = 1.729\text{ V} - \frac{(8.314\,\frac{\text{J}}{\text{mol·K}})(298\text{ K})}{(3)(96,485\,\frac{\text{C}}{\text{mol}})}\ln\left(2.47\times10^6\right) = 1.60\text{ V}$$

8.31. Because all of the involved species are solids, we do not expect that changing the pressure has any influence on E.

8.33. **(a)** $E° = 0$ V for any concentration cell, since both half reactions are the same but opposite.

(b) Using concentrations ($c° = 1$ mol/L) instead of molalities for the activities in Q:

$$Q = \frac{c_{Fe^{3+}}/c°}{c_{Fe^{3+}}/c°} = \frac{0.001}{0.08} = 0.01.$$

(c) $E = 0\text{ V} - \dfrac{(8.314\,\frac{\text{J}}{\text{mol·K}})(298\text{ K})}{(3)(96,485\,\frac{\text{C}}{\text{mol}})}\ln 0.01 = 0.04\text{ V}$

(d) Concentration cells involving certain metal ions require the same kind of metal as electrode materials. Also, the number of electrons involved in the redox reaction enters into the Nernst equation, so that the voltage characteristics will depend on the chemical identity of the metal ion, and not just on its concentration.

8.35. For a concentration cell involving M^{z+} ions, $E° = 0$ V and therefore

$0.050\text{ V} = 0\text{ V} - \dfrac{(8.314\,\frac{\text{J}}{\text{mol·K}})(298\text{ K})}{z(96,485\,\frac{\text{C}}{\text{mol}})}\ln Q$. Solving for $Q = \dfrac{c_{i,\text{right}}/c°}{c_{i,\text{left}}/c°} = \dfrac{c_{i,\text{right}}}{c_{i,\text{left}}}$ (where we use concentrations instead of molalities for activities) leads to:

(a) $z = 2$; $\ln Q = -3.9$; $Q = 0.020$.

(b) $z = 3$; $\ln Q = -5.8$; $Q = 0.0029$.

(c) $z = 2$ as in (a). Therefore, the result is the same as well: $Q = 0.020$.

8.37. The half reactions are

$I_2 + 2\,e^- \rightarrow 2\,I^-$ $\quad E° = 0.5355$ V

$H_2 \rightarrow 2\,H^+ + 2\,e^-$ $\quad E° = 0.0000$ V

We combine H^+ and I^- on the right−hand side to HI:

$I_2 + H_2 \rightarrow 2\,HI$ $\quad E° = 0.5355$ V

We use equation 8.27 to estimate the change in $E°$. From Appendix 2:

$\Delta S° = [2(114.7) - (130.68 + 116.14)]\text{ J/ K}) = -17.4\text{ J/K}.$

$$\Delta E^\circ \approx \frac{-17.4\ \text{J/K}}{(2)\left(96{,}485\dfrac{\text{C}}{\text{mol}}\right)}(450\ \text{K} - 298\ \text{K}) = -0.0137\ \text{V and}$$

$E^\circ(450\ \text{K}) = 0.5355\ \text{V} - 0.0137\ \text{V} = 0.5218\ \text{V}.$

8.39. The entropy change for the half reaction AgCl (s) + e^- → Ag + Cl^- can be calculated from its temperature coefficient: $\left(\dfrac{\partial E^\circ}{\partial T}\right)_p = \dfrac{\Delta S^\circ}{nF}$. Solve for ΔS°:

$$\Delta S^\circ = (1\ \text{mol})\left(96{,}485\frac{\text{C}}{\text{mol}}\right)\left(-0.73\times10^{-3}\frac{\text{V}}{\text{K}}\right) = -70\ \text{J/K}.$$

(This value is quite different from the value obtained directly from Appendix 2, namely $\Delta S^\circ = [(42.55 + 56.4) - (96.25)]\ \text{J/K} = +2.7\ \text{J/K}$.)

The half reaction appears two times in the overall reaction

$H_2\ (g) + 2\ AgCl\ (s) \rightarrow 2\ Ag\ (s) + 2\ H^+\ (aq) + 2\ Cl^-\ (aq)$

so that the overall entropy change is (using data for H^+ (aq) and H_2 (g) from Appendix 2)

$\Delta S^\circ = [2(-70) + 2(0) - 130.68)]\ \text{J/K} = -272\ \text{J/K}.$

Using only the data from Appendix 2: $\Delta S^\circ = [2(2.7) + 2(0) - 130.68)]\ \text{J/K} = -125.3\ \text{J/K}.$

8.41. We can use the equation $\Delta S = C \ln\dfrac{T_f}{T_i}$ to correct the entropies of H_2, O_2, and H_2O gases from their values at 298. From Table 2.1, we can get heat capacities in units of J/(g·K): 14.304, 0.918, and 1.864 J/g·K, respectively. For 1 mol of gas, we then have heat capacities C of $(2.0158\ \text{g})\left(14.304\dfrac{\text{J}}{\text{g}\cdot\text{K}}\right) = 28.834\ \text{J/K}$ for H_2,

$(31.9988\ \text{g})\left(0.918\dfrac{\text{J}}{\text{g}\cdot\text{K}}\right) = 29.4\ \text{J/K}$ for O_2, and $(18.0152\ \text{g})\left(1.864\dfrac{\text{J}}{\text{g}\cdot\text{K}}\right) = 33.58\ \text{J/K}$ for H_2O, which leads to the following entropies at 500 K:

H_2: $\Delta S = \left(28.834\dfrac{\text{J}}{\text{K}}\right)\left(\ln\dfrac{500\ \text{K}}{298\ \text{K}}\right) = 14.9\ \text{J/K}$ and $S^\circ_{500\ \text{K}} = (130.68 + 14.9)\ \text{J/K} =$ 145.6 J/K

O_2: $\Delta S = \left(29.4\dfrac{\text{J}}{\text{K}}\right)\left(\ln\dfrac{500\ \text{K}}{298\ \text{K}}\right) = 15.2\ \text{J/K}$ and $S^\circ_{500\ \text{K}} = (205.14 + 15.2)\ \text{J/K} = 220.3\ \text{J/K}$

H_2O: $\Delta S = \left(33.58\dfrac{\text{J}}{\text{K}}\right)\left(\ln\dfrac{500\ \text{K}}{298\ \text{K}}\right) = 17.4\ \text{J/K}$ and $S^\circ_{500\ \text{K}} = (188.83 + 17.4)\ \text{J/K} =$ 206.2 J/K.

Determining the overall $\Delta S°$ for the reaction at 500 K:

$\Delta S°_{500\text{ K}} = 2(206.2) - [2(145.6) + 220.3]$ J/K = −99.1 J/K.

Using this value instead of $\Delta S°$ at 298 K in the calculation of $E°$ at 500 K:

$$E°_{500\text{ K}} = E° + \frac{\Delta S°_{500\text{ K}}}{nF} \cdot \Delta T = 1.229\text{ V} + \frac{(-99.1\text{ J/K})(202\text{ K})}{(4\text{ mol})\left(96{,}485\frac{\text{C}}{\text{mol}}\right)} = 1.229\text{ V} - 0.0519\text{ V} = 1.177\text{ V}$$

Because we used used the approximation of a constant $\Delta S°$ in the temperature range of interest to derive equation 8.27, these two values of E calculated with $\Delta S°$ at 298 K and $\Delta S°$ at 500 K are upper and lower limits; the true value is therefore somewhere in between 1.177 V and 1.183 V.

8.43. $\Delta S° = 2(143.80) - (130.68 + 144.96)$ J/K = +11.96 J/K. Assuming that two electrons are transferred, we have $\left(\frac{\partial E°}{\partial T}\right)_p = \frac{11.96\text{ J/K}}{(2\text{ mol})\left(96{,}485\frac{\text{C}}{\text{mol}}\right)} = 6.198 \times 10^{-5}$ V/K. With increasing temperature, the value of $E°$ increases as well, meaning that the equilibrium shifts toward HD.

8.45. Start with the expression $E = -\frac{RT}{2F}\ln\frac{p_{H_2}}{[H^+]^2}$. We note that this equation is somewhat simplified, because we only supply the *numerical values* (without units) in the numerator and denominator of the logarithm term; the hydrogen pressure needs to be given in bar, and the H^+ concentration in mol/L.

If we are working at standard pressure, the numerator is 1, and due to the properties of the logarithm ($\log\frac{a}{b} = \log a - \log b$, and log 1 = 0) we obtain $E = +\frac{RT}{2F}\ln\left([H^+]^2\right)$.

Furthermore, because $\log\left(a^2\right) = 2\log a$, this leads to $E = 2\frac{RT}{2F}\ln[H^+] = \frac{RT}{F}\ln[H^+]$. We can convert to base−10 logarithms by multiplying the expression by 2.303 to get $E = 2.303\frac{RT}{F}\log[H^+]$.

Finally, because pH = −log $[H^+]$, we substitute (thereby reintroducing the negative sign) to get $E = -2.303\frac{RT}{F}\text{pH}$, our final expression.

8.47. The half reactions are

$Sn \rightarrow Sn^{2+} + 2\,e^-$ $\quad E° = +0.1375$ V

$Pb^{2+} + 2\,e^- \rightarrow Pb$ $\quad E° = -0.1262$ V

Combining both equations leads to the balanced equation

$Sn + Pb^{2+} \rightarrow Sn^{2+} + Pb \qquad E° = 0.0113\ V$

Solving $E° = \frac{RT}{nF}\ln K$ for K: $\ln K = \frac{nFE°}{RT} = \frac{(2)(96{,}485\ C/mol)(0.0113\ V)}{\left(8.314\frac{J}{mol \cdot K}\right)(298\ K)} = 0.880$;

$K = 2.41$.

8.49. Table 8.2 provides us with the following two half–reactions:

$PbSO_4 + 2\ e^- \rightarrow Pb + SO_4^{2-} \quad E° = -0.3588\ V$

$Pb^{2+} + 2\ e^- \rightarrow Pb \qquad E° = -0.1262\ V$

Inverting the second one (and changing the sign of its $E°$) and combining them leads to

$PbSO_4 \rightarrow Pb^{2+} + SO_4^{2-} \qquad E° = (-0.3588 + 0.1262)\ V = -0.2326\ V$

Solving $E° = \frac{RT}{nF}\ln K$ for K: $\ln K = \frac{nFE°}{RT} = \frac{(2)(96{,}485\ C/mol)(-0.2326\ V)}{\left(8.314\frac{J}{mol \cdot K}\right)(298\ K)} = -18.1$;

$K = 1.36 \times 10^{-8}$. The equilibrium constant is given by $K = a_{Pb^{2+}} a_{SO_4^{2-}} \approx \frac{c_{Pb^{2+}}}{c°}\frac{c_{SO_4^{2-}}}{c°}$, which is the solubility product K_{sp} of $PbSO_4$.

8.51. First, we need to determine the overall reaction. Using Table 8.2:

$MnO_4^- + 8\ H^+ + 5\ e^- \rightarrow Mn^{2+} + 4\ H_2O \qquad E° = 1.507\ V$

$H_2 \rightarrow 2\ H^+ + 2\ e^- \qquad E° = 0.0000\ V$

The 10–electron overall reaction is:

$2\ MnO_4^- + 6\ H^+ + 5\ H_2 \rightarrow 2\ Mn^{2+} + 8\ H_2O \quad E° = 1.507\ V$

However, because some of the concentrations are not standard, equation 8.35 can't be used directly. It is easiest to use the complete Nernst equation, using molalities $m_i/m°$ for MnO_4^- and Mn^{2+}, $p_{H2}/p° = 1$ for H_2, and concentrations $c_{H+}/c°$ for H^+ in the expression for Q:

$$1.200\ V = 1.507\ V - \frac{\left(8.314\frac{J}{mol \cdot K}\right)(298\ K)}{(10)(96{,}485\ C/mol)}\ln\frac{(0.288)^2}{(0.034)^2\left(c_{H^+}/c°\right)^6(1)^5}.$$

Solving for c_{H+}: $\ln\frac{(0.288)^2}{(0.034)^2\left(c_{H^+}/c°\right)^6(1)^5} = 120$; $\frac{(0.288)^2}{(0.034)^2\left(c_{H^+}/c°\right)^6(1)^5} = 8.37 \times 10^{51}$;

$c_{H^+} = 9.7 \times 10^{-9} c° = 9.7 \times 10^{-9}$ mol/L and $pH = -\log(4.5 \times 10^{-9}) = 8.0$.

8.53. The lower the pH, the less negative the first term will be in equation 8.35. Thus, the more acidic the solution, the lower the pH, and the higher the overall E will be for the reaction between Fe and H^+. Therefore, we expect that corrosion of iron is promoted by lower pHs.

8.55. Standard conditions for biochemical reactions are 37.0 °C and a pH of 7. To calculate the effect of the temperature increase of 12.0 K (equation 8.27), we need $\Delta S°$, which can be obtained from Appendix 2: $\Delta S°$ = [130.68 – 2(0)] J/K = 130.68 J/K. Therefore, the change in $E°$ is

$$\Delta E° = \frac{\Delta S°}{nF}\Delta T = \frac{(130.68\text{ J/K})(12.0\text{ K})}{(2\text{ mol})(96{,}485\text{ C/mol})} = 0.00813\text{ V};\ E° = 0.0000\text{ V} + 0.00813\text{ V} =$$
0.0081 V.

At 25.0 °C , the reduction potential of the hydrogen electrode is $E = (-0.05916\text{ V})\cdot\text{pH}$, but the factor will change at 37.0 °C to:

$$-2.303\cdot\frac{RT}{F} = -2.303\cdot\frac{\left(8.314\frac{\text{J}}{\text{mol}\cdot\text{K}}\right)(310.2\text{ K})}{96{,}485\text{ C/mol}} = -0.06155\text{ V}.$$

At a pH of exactly 7, we then have $E = (-0.06155\text{ V})\cdot 7 = -0.4308\text{ V}$.

(Note that using the pH equation at 25.0 °C *first* and calculating the change in $E°$ due to the temperature increase to 37.0 °C *second* will not work, because the $\Delta S°$ value calculated from Appendix 2 is only valid for a H^+ concentration of 1 mol/L!)

8.57. $$\phi = -\frac{\left(8.314\frac{\text{J}}{\text{mol}\cdot\text{K}}\right)(310.2\text{ K})}{96{,}485\text{ C/mol}}\ln\frac{(4.5+139)\times10^{-3}\text{ mol/L}}{139\times10^{-3}\text{ mol/L}} = -0.000851\text{ V} = 8.51\times10^{-4}\text{ V}$$

8.59. For $CuSO_4$, we have $n_+ = 1$ and $n_- = 1$, so that $n_\pm = 2$. The ideal molality of Cu^{2+} is 0.050 m, and that of SO_4^{2-} is also 0.050 m. The mean ionic molality is therefore

$m_\pm = (0.050^1\cdot 0.050^1)^{1/2}\ m = 0.050\ m$.

With the mean activity coefficient of 0.223, the mean ionic activity is then

$$a_\pm = \left(0.223\cdot\frac{0.050\ m}{1\ m}\right)^2 = 0.00012.$$

8.61. The ionic strength of a 0.100 m $Ca_3(PO_4)_2$ solution was already determined in Example 8.10(a): $I_{Ca_3(PO_4)_2} = 1.50\ m$. The ionic strength of a NaCl solution, where $z^+ = 1$ and $z^- = -1$, is given by $I_{NaCl} = \frac{1}{2}\left[m_{Na^+}\cdot 1^2 + m_{Cl^-}\cdot(-1)^2\right] = \frac{1}{2}\left(m_{Na^+} + m_{Cl^-}\right)$. If we call the unknown molality of NaCl x, then $m_{Na^+} = m_{Cl^-} = x$, and our ionic strength condition becomes:

$$1.50\ m = \frac{1}{2}(2x) = x.$$

That's already the answer: the NaCl solution needs to be 1.50 molal.

8.63. Ammonia hydrolyzes slightly in water to form NH_4^+ and OH^- ions. Therefore, aqueous solutions of ammonia have a low ionic strength.

8.65. **(a)** The reason that an entropy of formation of –138.1 J/(mol·K) doesn't violate the third law of thermodynamics is twofold. First, an entropy of formation refers to the entropy *change* of a particular reaction (a formation reaction), not necessarily to the absolute entropy of a substance. Second, because of the fact that we're working with ions and have difficulty accounting for the electron, entropies of formation are all tied to the entropy of formation of

H^+ (aq) ions, which by convention is assigned a value of 0. Having a negative entropy of formation simply implies that there is a decrease in entropy in forming Mg^{2+} (aq) relative to the formation of H^+ (aq).

(b) A negative entropy of formation implies that when the ion is formed (by losing one or more electrons and becoming surrounded by water molecules), there are less molecular states available (that is, order is increased). At the molecular level, there is a regular arrangement of water molecules around most ions, which is a less random arrangement than that found in pure water, implying a negative entropy of formation.

8.67. **(a)** The reaction is HF (g) → H^+ (aq) + F^- (aq). Using data from Appendix 2, we have:

$\Delta H° = [(-332.63 + 0) - (-273.30)]\text{ kJ} = -59.33\text{ kJ}$

$\Delta S° = [(-13.8 + 0) - (173.779)]\text{ J/K} = -187.6\text{ J/K}$

$\Delta G° = [(-278.8 + 0) - (-274.6)]\text{ kJ} = -4.2\text{ kJ}$

(b) Using equation 5.10 with the molar Gibbs energy for the reaction:

$-4.2\times10^3\text{ J/mol} = -(8.314\text{ J/(mol·K)})\ (298\text{ K})\ \ln K$, so $\ln K = 1.7$ and $K = 5.4$.

This is rather far off from the 3.5×10^{-4} value as measured. The difference is that the equilibrium constant refers to HF in the aqueous phase being the reactant, rather than the gas phase. The predicted value for K should be much closer to the measured value if the hydration of HF were included.

8.69. The reaction for dissolving CaF_2 is CaF_2 → Ca^{2+} (aq) + 2 F^- (aq). We then have

$\Delta H° = ([-542.83 + 2(-332.63)] - [-1225.9])\text{ kJ} = 17.8\text{ kJ}$, or 17.8 kJ per mole of CaF_2.

8.71. For the reaction $NaHCO_3$ (s) → Na^+ (aq) + HCO_3^- (aq):

$\Delta H° = [(-240.12 - 691.99) - (-950.81)]\text{ kJ} = +18.70\text{ kJ}$

$\Delta S° = [(59.1 + 91.2) - 101.7]\text{ J/K} = +48.6\text{ J/K}$

$\Delta G° = [(-261.88 - 586.85) - (-851.0)]\text{ kJ} = +2.3\text{ kJ}$

For the reaction Na_2CO_3 (s) → 2 Na^+ (aq) + CO_3^{2-} (aq):

$\Delta H° = ([2(-240.12) - 676.3] - [-1130.77])\text{ kJ} = -25.8\text{ kJ}$

$\Delta S° = ([2(59.1) - 53.1] - 138.79)\text{ J/K} = -73.7\text{ J/K}$

$\Delta G° = ([2(-261.88) - 528.1] - [-1048.01])\text{ kJ} = -3.9\text{ kJ}$

8.73. Using the fact that $A = 1.171\ m^{-1/2}$ at 25 °C, and the fact that the ionic strength of the KCl solution, where $z^+ = 1$ and $z^- = -1$, is given by

$I_{KCl} = \frac{1}{2}\left[(0.0020\ m)(1)^2 + (0.0020\ m)(-1)^2\right] = 0.0020\ m$, we calculate ln $\gamma_\pm$:

$\ln \gamma_\pm = (1.171\ m^{-1/2})(+1)(-1)(0.0020\ m)^{1/2} = -0.052$; $\gamma_\pm = 0.95$, which is the same as the measured value of 0.951.

To use extended Debye–Hückel theory, first we need to calculate B for water:

$$B = \left(\frac{e^2 N_A \rho_{solv}}{\varepsilon_0 \varepsilon_r kT}\right)^{1/2} = \left(\frac{(1.602\times10^{-19}\ \text{C})^2 (6.022\times10^{23}/\text{mol})(997\ \text{kg/m}^3)}{\left(8.854\times10^{-12}\ \frac{\text{C}^2}{\text{J}\cdot\text{m}}\right)(78.54)(1.381\times10^{-23}\ \text{J/K})(298\ \text{K})}\right)^{1/2}$$

$B = 2.32\times10^9 / (\text{m}\cdot\text{molal})^{1/2}$ (that is, B has units of per meters and per square root of molality).

Note that we use "molal" instead of m here, in order to better distinguish it from meters (m). Because K^+ and Cl^- have the same value of å and the same charge, we need perform only one calculation of ln γ:

$$\ln\gamma = -\frac{(1.171\ \text{molal}^{-1/2})(1)^2(0.0020\ \text{molal})^{1/2}}{1+(2.32\times10^9\ \text{m}^{-1}\text{molal}^{-1/2})(3\times10^{-10}\ \text{m})(0.0020\ \text{molal})^{1/2}} = -0.051$$

Therefore, γ = 0.95 for each ion (a negligible change from what we calculated using the limiting law) and $\gamma_\pm = (0.95^1 \cdot 0.95^1)^{1/2} = 0.95$.

8.75. The Debye–Hückel limiting law strictly applies only to very dilute solutions ($I < 0.01\ m$). For a binary symmetric ionic salts (like NaCl or $CaSO_4$), where $n^+ = n^- = 1$ and $z^- = -z^+$, equation 8.50 for the limiting law becomes $\ln \gamma_\pm = A \cdot (z^+) \cdot (-z^+) \cdot I^{1/2} = -A \cdot (z^+)^2 \cdot I^{1/2}$.

Let's compare this to ln γ_+ or ln γ_- obtained from equation 8.52. First of all, because the mean ionic activity coefficient is defined as $\gamma_\pm = \left(\gamma_+^{n_+}\cdot\gamma_-^{n_-}\right)^{1/(n_+ + n_-)}$, taking the ln of both sides and applying the usual properties of logarithms leads to

$$\ln\gamma_\pm = \ln\left(\gamma_+^{n_+}\cdot\gamma_-^{n_-}\right)^{1/(n_+ + n_-)} = \frac{1}{n_+ + n_-}\ln\left(\gamma_+^{n_+}\cdot\gamma_-^{n_-}\right) = \frac{1}{n_+ + n_-}\left[\ln\left(\gamma_+^{n_+}\right) + \ln\left(\gamma_-^{n_-}\right)\right]$$

$$\ln\gamma_\pm = \frac{1}{n_+ + n_-}\left(n_+ \ln\gamma_+ + n_- \ln\gamma_-\right) = \frac{n_+}{n_+ + n_-}\ln\gamma_+ + \frac{n_-}{n_+ + n_-}\ln\gamma_-$$

With $n^+ = n^- = 1$, this equals $\ln \gamma_\pm = ½ \ln \gamma_+ + ½ \ln \gamma_-$, which is the average of both numbers.

The extended Debye–Hückel law gives ln γ_+ and ln γ_- individually (equation 8.52). Imagine for a moment that we can set the denominator of equation 8.52 to 1: Then $\ln \gamma_+ = -A \cdot (z^+)^2 \cdot I^{1/2}$ and also $\ln \gamma_- = -A \cdot (-z^+)^2 \cdot I^{1/2} = -A \cdot (z^+)^2 \cdot I^{1/2}$, so $\ln \gamma_\pm = ½ \ln \gamma_+ + ½ \ln \gamma_- = -A \cdot (z^+)^2 \cdot I^{1/2}$.

This is the same expression as that obtained from the Debye–Hückel limiting law!

That means that both laws become the same (at least for binary symmetric ionic salts) if the denominator of equation 8.52 can be approximated to 1, which is the case if $B \cdot a \cdot I^{1/2}$ is much smaller than 1 – let's take 0.05, for example. Given that B at 25 °C is 2.32×10^9 / (m·molal)$^{1/2}$ and that a typical value of a is $5\cdot10^{-10}$ m, the value of $B \cdot a$ is on the order of 1/molal$^{1/2}$.

To make $B \cdot a \cdot I^{1/2} = 0.05$, we therefore need $I^{1/2} = 0.05$ molal$^{1/2}$, or $I = 0.0025$ molal.

8.77. **(a)** Using the given molal concentrations, we can use the Nernst equation to determine the voltage of the cell. First, we need $E°$ for the process. From Table 8.2,

$E° = +0.7618 + 0.3419 = 1.1037$ V. Now, using the Nernst equation for this 2–electron process:

$$E = 1.1037\text{ V} - \frac{(8.314\,\frac{\text{J}}{\text{mol}\cdot\text{K}})(298\text{ K})}{(2)(96{,}485\,\frac{\text{C}}{\text{mol}})}\ln\frac{(0.1\,m)/m°}{(0.05\,m)/m°} = 1.095\text{ V}$$

(b) To get a more accurate value of E, we use the extended Debye–Hückel theory. A 0.1 molal $ZnSO_4$ solution has an ionic strength of $I = ½\,[(0.1\,m)\,(+2)^2 + (0.1\,m)\,(-2)^2] = 0.4\,m$, and a

0.05 molal $CuSO_4$ solution $I = ½\,[(0.05\,m)\,(+2)^2 + (0.05\,m)\,(-2)^2] = 0.2\,m$.

We also need to calculate B for water:

$$B = \left(\frac{e^2 N_A \rho_{solv}}{\varepsilon_0 \varepsilon_r kT}\right)^{1/2} = \left(\frac{(1.602\times10^{-19}\text{ C})^2(6.022\times10^{23}/\text{mol})(997\text{ kg/m}^3)}{\left(8.854\times10^{-12}\,\frac{\text{C}^2}{\text{J}\cdot\text{m}}\right)(78.54)(1.381\times10^{-23}\text{ J/K})(298\text{ K})}\right)^{1/2}$$

$B = 2.32\times10^9$ / (m·molal)$^{1/2}$. Now use equation 8.52 to calculate the activity coefficients (the ln γ values can normally only be given to one significant digit, but we will display two for clarity):

$$\text{Zn}^{2+}:\ \ln\gamma = -\frac{(1.171\text{ molal}^{-1/2})(2)^2(0.4\text{ molal})^{1/2}}{1+(2.32\times10^9\text{ m}^{-1}\text{molal}^{-1/2})(6\times10^{-10}\text{ m})(0.4\text{ molal})^{1/2}} = -1.6;\ \gamma = 0.2$$

$$\text{Cu}^{2+}:\ \ln\gamma = -\frac{(1.171\text{ molal}^{-1/2})(2)^2(0.2\text{ molal})^{1/2}}{1+(2.32\times10^9\text{ m}^{-1}\text{molal}^{-1/2})(6\times10^{-10}\text{ m})(0.2\text{ molal})^{1/2}} = -1.3;\ \gamma = 0.3$$

Substituting the activities $a_i = \gamma_i \cdot m_i/m°$ into the Nernst equation:

$$E = 1.1037\text{ V} - \frac{(8.314\,\frac{\text{J}}{\text{mol}\cdot\text{K}})(298\text{ K})}{(2)(96{,}485\,\frac{\text{C}}{\text{mol}})}\ln\frac{(0.2)(0.1\,m)/m°}{(0.3)(0.05\,m)/m°} = 1.098\text{ V}$$

This is only a slight difference in voltage, because the activities of both ionic species decrease by about the same factor. The slight difference is due to the higher activity of Cu^{2+} ions over Zn^{2+} ions.

8.79. The data in Table 8.3 show that the activity coefficients get closer to 1 as the concentration of the solution decreases, no matter what the charge is on the ion. Therefore, we can say that the data in Table 8.3 do support the idea that $\gamma \rightarrow 1$ as $m \rightarrow 0$.

8.81. **(a)** Using values from Table 8.4:

$\Lambda_0(NaNO_3) = \Lambda_0(NaCl) + \Lambda_0(KNO_3) - \Lambda_0(KCl) = (126.45 + 144.96 - 149.86)$ $cm^2/(mol{\cdot}ohm) = 121.55\ cm^2/(mol{\cdot}ohm)$, the same as the $121.55\ cm^2/(mol{\cdot}ohm)$ listed in the table for $\Lambda_0(NaNO_3)$.

(b) NH_4NO_3 can be thought of as $NH_4Cl + NaNO_3 - NaCl$. In terms of Λ_0's:

$\Lambda_0(NH_4NO_3) = (149.7 + 121.55 - 126.45)\ cm^2/(mol{\cdot}ohm) = 144.8\ cm^2/(mol{\cdot}ohm)$.

$CaBr_2$ can be thought of as $CaCl_2 + 2KBr - 2KCl$. Performing a similar combination of Λ_0s: $\Lambda_0(CaBr_2) = [135.84 + 2(151.9) - 2(149.86)]\ cm^2/(mol{\cdot}ohm) = 139.9$ $cm^2/(mol{\cdot}ohm)$.

8.83. According to equation 8.59, $v_i = \dfrac{e\cdot|z_i|\cdot E}{f} = \dfrac{e\cdot|z_i|\cdot E}{6\pi\eta r_i}$. We have all of the data given, although we may have to convert units so that they work out properly. Substituting:

$$v_{Cu^{2+}} = \frac{e\cdot|z_{Cu^{2+}}|\cdot E}{6\pi\eta r_i} = \frac{(1.602\times10^{-19}\ C)(2)(100.0\ V/m)}{6\pi(0.00894\ poise)(4\times10^{-10}\ m)} = 5\times10^{-7}\frac{C\cdot V}{poise\cdot m}$$

Now we need to convert units properly. 1 V = 1 J/C (see equation 8.7), and

1 poise = 1 g/(cm·s), which needs to be converted to standard SI units:

$1\ poise = 1\dfrac{g}{cm\cdot s}\times\dfrac{1\ kg}{1000\ g}\times\dfrac{100\ cm}{1\ m} = 0.1\dfrac{kg}{m\cdot s}$. Substituting these two definitions:

$5\times10^{-7}\dfrac{C\cdot(J/C)}{\left(0.1\dfrac{kg}{m\cdot s}\right)\cdot m} = 5\times10^{-6}\dfrac{J\cdot s}{kg}$. To finish the unit reduction, we must break down

the J unit into kg·m/s²: $5\times10^{-6}\dfrac{(kg\cdot m/s^2)\cdot s}{kg} = 5\times10^{-6}\dfrac{kg\cdot m\cdot s}{kg\cdot s^2} = 5\times10^{-6}\dfrac{m}{s}$.

This may not seem like a large velocity, but it corresponds to over 10,000 atomic radii per second!

CHAPTER 9

Pre-Quantum Mechanics

9.1. With the kinetic energy $K=\frac{1}{2}m\dot{z}^2$ and the potential energy $V=mgz$, the Lagrangian function is $L=\frac{1}{2}m\dot{z}^2-mgz$. The Lagrangian equation of motion follows from $\frac{d}{dt}\left(\frac{\partial L}{\partial \dot{z}}\right)=\frac{\partial L}{\partial z}$, so we first need to determine these derivatives:

left–hand side: $\frac{\partial L}{\partial \dot{z}}=\frac{\partial}{\partial \dot{z}}\left(\frac{1}{2}m\dot{z}^2-mgz\right)=m\dot{z}+0$, and therefore $\frac{d}{dt}\left(\frac{\partial L}{\partial \dot{z}}\right)=\frac{d}{dt}(m\dot{z})=m\ddot{z}$;

right–hand side: $\frac{\partial L}{\partial z}=\frac{\partial}{\partial z}\left(\frac{1}{2}m\dot{z}^2-mgz\right)=0-mg=-mg$.

Note that for the first derivative, the second term's derivative with respect to $\dot{z}$ is equal to zero because $\dot{z}$ does not appear in that term. Similarly, for the last derivative, the first term's derivative with respect to z is zero because z does not appear in that term (only its derivative does).

Combining the two results, we have for the Lagrangian equation of motion: $m\ddot{z}=-mg$.

9.3. In this case, q is z, $\dot{q}$ is $\dot{z}$, and according to equation 9.10:

$$p=\frac{\partial L}{\partial \dot{q}}=\frac{\partial}{\partial \dot{z}}\left(\frac{1}{2}m\dot{z}^2-mgz\right)=m\dot{z},\text{ so } \dot{p}=m\ddot{z}.$$

Therefore, the Hamiltonian function as defined by equation 9.11 is

$$H=p\cdot\dot{q}-L=m\dot{z}\cdot\dot{z}-\left(\frac{1}{2}m\dot{z}^2-mgz\right)=\frac{1}{2}m\dot{z}^2+mgz \text{ (see problem 9.1 for } L\text{).}$$

Hamilton's equations of motion are equations 9.14 and 9.15: $\frac{\partial H}{\partial(m\dot{z})}=\dot{z}$ and $\frac{\partial H}{\partial z}=-m\ddot{z}$.

The first derivative is easy to demonstrate:

$\frac{\partial H}{\partial(m\dot{z})}=\frac{1}{m}\frac{\partial H}{\partial \dot{z}}=\frac{1}{m}\frac{\partial}{\partial \dot{z}}\left(\frac{1}{2}m\dot{z}^2+mgz\right)=\frac{1}{m}(m\dot{z}+0)=\dot{z}$, as required (the derivative of mgz with respect to $\dot{z}$ is zero because $\dot{z}$ does not appear in that term).

The second derivative is a little tricky. If we perform the differentiation, we get

$\frac{\partial H}{\partial z}=\frac{\partial}{\partial z}\left(\frac{1}{2}m\dot{z}^2+mgz\right)=0+mg=mg$ (the derivative of the first term with respect to z is zero because z itself does not appear in that term, only its derivative), which does not look like the required expression. However, remember what g is: it is the acceleration due to gravity experienced by any mass falling down, so $g = \frac{dv}{dt} = \frac{d}{dt}\left(-\frac{dz}{dt}\right) = -\frac{d^2z}{dt^2}$ (the negative sign is necessary because while the speed v is increasing the value of z is decreasing during the fall). Therefore, we do in fact have $\frac{\partial H}{\partial z} = mg = -m\ddot{z}$, as required by equation 9.15.

9.5. **(a)** Frequency (number of oscillations per time) is is a wave property.

(b) Both waves and particles can have a velocity of propagation or change in position.

(c) Both waves and particles can have kinetic energy, associated with their velocity.

(d) Interference is a phenomenon associated with (classical) waves only.

(e) Only (classical) particles can have momentum.

9.7. The drawing is left to the student.

9.9. $\tilde{\nu} = \frac{1}{\lambda} = \frac{1}{550 \times 10^{-9}\ \text{m}} = 1.82 \times 10^{6}\ \text{m}^{-1} = 1.82 \times 10^{4}\ \text{cm}^{-1}$ (remember that 1 m = 100 cm, so the reciprocals are 1 m^{-1} = 0.01 cm^{-1}.)

9.11. As n increases, the Balmer lines ($n_2 = 2$) are becoming more and more densely spaced. In the limit of n_1 tending toward infinity, the term $1/n_1^2$ becomes zero and the "last line" is given by $\tilde{\nu} = R_{\text{H}}\left(\frac{1}{4} - 0\right) = \frac{109{,}737.315685\ \text{cm}^{-1}}{4} \approx 27{,}434\ \text{cm}^{-1}$.

9.13. We use equation 9.17 with the indicated values for n_1 and n_2:

(a) $n_1 = 5$; $n_2 = 1$: $\tilde{\nu} = 105{,}348\ \text{cm}^{-1}$;

(b) $n_1 = 8$; $n_2 = 2$: $\tilde{\nu} = 25{,}720\ \text{cm}^{-1}$;

(c) $n_1 = 4$; $n_2 = 3$: $\tilde{\nu} = 5{,}334\ \text{cm}^{-1}$;

(d) $n_1 = 8$; $n_2 = 4$: $\tilde{\nu} = 5{,}144\ \text{cm}^{-1}$;

(e) $n_1 = 6$; $n_2 = 5$: $\tilde{\nu} = 1{,}341\ \text{cm}^{-1}$.

9.15. First, we need to convert the wavelengths to wavenumbers (1 nm = 10^{-9} m = 10^{-7} cm):

$\lambda = 656.2\ \text{nm} = 656.2 \times 10^{-7}\ \text{cm}$, so $\tilde{\nu} = \frac{1}{\lambda} = \frac{1}{656.2 \times 10^{-7}\ \text{cm}} = 15{,}240\ \text{cm}^{-1}$

$\lambda = 486.1\ \text{nm} = 486.1 \times 10^{-7}\ \text{cm}$, so $\tilde{\nu} = \frac{1}{\lambda} = \frac{1}{486.1 \times 10^{-7}\ \text{cm}} = 20{,}570\ \text{cm}^{-1}$

$\lambda = 434.0\ \text{nm} = 434.0 \times 10^{-7}\ \text{cm}$, so $\tilde{\nu} = \frac{1}{\lambda} = \frac{1}{434.0 \times 10^{-7}\ \text{cm}} = 23{,}040\ \text{cm}^{-1}$

(All values are rounded to four significant figures, which is the accuracy of the given wavelengths.) Using these wavenumber values, we can calculate R for the Balmer series ($n_2 = 2$) to within four significant figures, assuming that the first three lines correspond to $n_1 = 3$, 4, and 5, respectively. Using $R = \tilde{\nu} / \left(\frac{1}{2^2} - \frac{1}{n_1^2} \right)$:

$$R = 15{,}240 \text{ cm}^{-1} / \left(\frac{1}{2^2} - \frac{1}{3^2} \right) = 109{,}700 \text{ cm}^{-1};$$

$$R = 20{,}570 \text{ cm}^{-1} / \left(\frac{1}{2^2} - \frac{1}{4^2} \right) = 109{,}700 \text{ cm}^{-1};$$

$R = 23{,}040 \text{ cm}^{-1} / \left(\frac{1}{2^2} - \frac{1}{5^2} \right) = 109{,}700 \text{ cm}^{-1}$. Therefore, the average value is 109,700 cm^{-1}.

9.17. According to the values given in the text, $e = 1.601 \times 10^{-19}$ C and $m = 9.36 \times 10^{-31}$ kg, yielding an e/m ratio of $\frac{1.601 \times 10^{-19} \text{ C}}{9.36 \times 10^{-31} \text{ kg}} = 1.71 \times 10^{11}$ C/kg.

9.19. **(a)** The total radiant energy given off in watts/m^2 is given by the Stefan–Boltzmann law. At 1000 K: $\left(5.6705 \times 10^{-8} \frac{\text{W}}{\text{m}^2 \cdot \text{K}^4} \right) (1000 \text{ K})^4 = 5.670 \times 10^4 \text{ W/m}^2$.

(b) For an area of 250 cm^2:

$$\left(250 \text{ cm}^2 \right) \left(\frac{1 \text{ m}}{100 \text{ cm}} \right)^2 \left(5.670 \times 10^4 \frac{\text{W}}{\text{m}^2} \right) \left(\frac{1 \text{ kW}}{1000 \text{ W}} \right) = 1.42 \text{ kW}.$$

9.21. We use the Wien displacement law to estimate the temperature of a blackbody emitter with a peak wavelength of 9.4 μm: 2898 μm·K / 9.4 μm = 3.1×10^2 K. Curiously (but not surprisingly), 310 K or 37 °C is the body temperature of warm–blooded animals!

9.23. The total emitted power per unit area is given by the Stefan–Boltzmann law. 37 °C is 310 K, so $\left(5.6705 \times 10^{-8} \frac{\text{W}}{\text{m}^2 \cdot \text{K}^4} \right) (310 \text{ K})^4 = 525 \text{ W/m}^2$. The total power emitted by a body surface area of 0.65 m^2 is then $\left(525 \frac{\text{W}}{\text{m}^2} \right) \left(0.65 \text{ m}^2 \right) = 3.4 \times 10^2$ W.

9.25. **(a)** Substituting into the given equation:

$$\frac{d\rho}{d\lambda} = \frac{8\pi kT}{\lambda^4} = \frac{8\pi (1.381 \times 10^{-23} \text{ J/K})(1000 \text{ K})}{\left(500 \times 10^{-9} \text{ m} \right)^4} = 5.55 \times 10^6 \text{ J/m}^4.$$

(b) $\dfrac{d\rho}{d\lambda} = \dfrac{8\pi(1.381\times10^{-23}\text{ J/K})(2000\text{ K})}{(500\times10^{-9}\text{ m})^4} = 1.11\times10^7\text{ J/m}^4$

(c) $\dfrac{d\rho}{d\lambda} = \dfrac{8\pi(1.381\times10^{-23}\text{ J/K})(2000\text{ K})}{(5000\times10^{-9}\text{ m})^4} = 1111\text{ J/m}^4$

(d) $\dfrac{d\rho}{d\lambda} = \dfrac{8\pi(1.381\times10^{-23}\text{ J/K})(2000\text{ K})}{(10{,}000\times10^{-9}\text{ m})^4} = 69.42\text{ J/m}^4$

The values indicate that as the wavelength decreases from (d) to (c) to (b), the slope increases dramatically in magnitude. Because $d\rho/d\lambda$ is directly proportional to the 4th power of the reciprocal wavelength, this trend continues without limits as wavelengths approach zero.

9.27. We combine the equations $E = h\nu$ for the energy of a light quantum and its speed $c = \lambda\nu$ to obtain $E = hc/\lambda = (6.626\times10^{-34}\text{ J·s})(3.00\times10^8\text{ m/s}) / \lambda = 1.99\times10^{-25}\text{ J·m} / \lambda$:

(a) $E = (1.99\times10^{-25}\text{ J·m}) / (572\times10^{-9}\text{ m}) = 3.48\times10^{-19}\text{ J}$;

(b) $E = (1.99\times10^{-25}\text{ J·m}) / (546\times10^{-9}\text{ m}) = 3.64\times10^{-19}\text{ J}$;

(c) $E = (1.99\times10^{-25}\text{ J·m}) / (430\times10^{-9}\text{ m}) = 4.62\times10^{-19}\text{ J}$.

9.29. $E = h\nu$ is the energy of a light quantum as a function of its frequency ν. Combining this equation with $c = \lambda\nu$ leads to $E = hc/\lambda$, which gives the energy as a function of its wavelength.

Remembering that wavenumbers $\tilde{\nu}$ are reciprocal wavelengths, we can write this as $E = hc\tilde{\nu}$.

(a) $E = \dfrac{hc}{\lambda} = \dfrac{(6.626\times10^{-34}\text{ J}\cdot\text{s})(3.00\times10^8\text{ m/s})}{5.42\times10^{-6}\text{ m}} = 3.67\times10^{-20}\text{ J}$;

(b) $E = h\nu = (6.626\times10^{-34}\text{ J}\cdot\text{s})(6.69\times10^{13}\text{ /s}) = 4.43\times10^{-20}\text{ J}$;

(c) $E = \dfrac{hc}{\lambda} = \dfrac{(6.626\times10^{-34}\text{ J}\cdot\text{s})(3.00\times10^8\text{ m/s})}{3.27\times10^{-9}\text{ m}} = 6.08\times10^{-17}\text{ J}$;

(d) $E = h\nu = (6.626\times10^{-34}\text{ J}\cdot\text{s})(106.5\times10^6\text{ /s}) = 7.057\times10^{-26}\text{ J}$;

(e) $E = hc\tilde{\nu} = (6.626\times10^{-34}\text{ J}\cdot\text{s})(3.00\times10^8\text{ m/s})\left(4321\dfrac{1}{\text{cm}}\right)\left(\dfrac{100\text{ cm}}{1\text{ m}}\right) = 8.59\times10^{-20}\text{ J}$.

9.31. Planck's law is written as equation 9.23. In order to use the given integral, we will have to substitute all λ's by $x=\dfrac{hc}{\lambda kT}$. This also means that we have to substitute $d\lambda$, which can be easily done by taking the derivative: $\dfrac{dx}{d\lambda}=-\dfrac{hc}{\lambda^2 kT}$, so $-\dfrac{\lambda^2 kT}{hc}dx=d\lambda$. Because this term results in a "λ^2" being reintroduced into the equation, we first substitute $d\lambda$:

$$dE=\frac{2\pi hc^2}{\lambda^5}\left(\frac{1}{e^{hc/(\lambda kT)}-1}\right)\left(-\frac{\lambda^2 kT}{hc}\right)dx.$$

Simplifying leads to: $dE=-\dfrac{2\pi ckT}{\lambda^3}\left(\dfrac{1}{e^{hc/(\lambda kT)}-1}\right)dx$.

We now substitute λ by $hc/(xkt)$:

$$dE=-\frac{2\pi ckT\cdot x^3k^3T^3}{h^3c^3}\left(\frac{1}{e^x-1}\right)dx=-\frac{2\pi k^4T^4}{h^3c^2}\left(\frac{x^3}{e^x-1}\right)dx.$$

Integrating λ from 0 to ∞ means integrating x from ∞ to 0 (because of their reciprocal relationship): $E=\int dE=\int_0^\infty -\dfrac{2\pi ckT\cdot x^3k^3T^3}{h^3c^3}\left(\dfrac{1}{e^x-1}\right)dx=\int_0^\infty +\dfrac{2\pi k^4T^4}{h^3c^2}\left(\dfrac{x^3}{e^x-1}\right)dx$ (note that switching the integration limits means inverting the sign inside the integral). Pulling out the constant factor and integrating using the given formula leads to the Stefan–Boltzmann law:

$$E=\int_0^\infty +\frac{2\pi k^4T^4}{h^3c^2}\left(\frac{x^3}{e^x-1}\right)dx=\frac{2\pi k^4T^4}{h^3c^2}\int_0^\infty\frac{x^3}{e^x-1}dx=\frac{2\pi k^4T^4}{h^3c^2}\cdot\frac{\pi^4}{15}=\frac{2\pi^5k^4T^4}{15h^3c^2}.$$

9.33. Substituting the values of the various constants:

$$\sigma=\frac{2\pi^5k^4}{15h^3c^2}=\frac{2\pi^5\left(1.381\times10^{-23}\text{ J/K}\right)}{15\left(6.626\times10^{-34}\text{ J}\cdot\text{s}\right)^3\left(3.00\times10^8\text{ m/s}\right)^2}=5.67\times10^{-8}\text{ J/(K}^4\text{m}^2\text{s)}.$$

Because 1 W = 1 J/s, this is 5.67×10^{-8} W/(m^2·K^4), which is the value of the Stefan–Boltzmann constant quoted in the text.

9.35. A work function of 2.16 eV is equal to $(2.16\text{ eV})\left(\dfrac{1.602\times10^{-19}\text{ J}}{1\text{ eV}}\right)=3.46\times10^{-19}$ J.

We therefore calculate the energy of the light quantum and subtract 3.46×10^{-19} J from that energy to obtain the kinetic energy of the ejected electron, ½ mv^2, and its speed v.

(a) $E=h\nu=\dfrac{hc}{\lambda}=\dfrac{\left(6.626\times10^{-34}\text{ J}\cdot\text{s}\right)\left(3.00\times10^8\text{ m/s}\right)}{550\times10^{-9}\text{ m}}=3.61\times10^{-19}$ J (remember that 1 nm = 10^{-9} m). The kinetic energy is then $(3.61\times10^{-19}-3.46\times10^{-19})$ J = 0.15×10^{-19}

J. This equals ½ mv^2, so that $v^2 = \dfrac{2(0.15\times10^{-19}\text{ J})}{9.109\times10^{-31}\text{ kg}} = 3.3\times10^{10}\dfrac{\text{J}}{\text{kg}}$. Recall that 1 J = 1 N·m = 1 kg·m^2/s^2, so the units come out as m^2/s^2. Taking the square root: $v = 1.8\times10^5$ m/s.

(b) The same calculation for a wavelength of 450 nm yields $E = 4.41\times10^{-19}$ J. The kinetic energy is then ($4.41\times10^{-19} - 3.46\times10^{-19}$) J = 0.96×10^{-19} J; $v^2 = 2.1\times10^{11}$ m^2/s^2 and $v = 4.6\times10^5$ m/s.

(c) The same calculation for a wavelength of 350 nm yields $E = 5.68\times10^{-19}$ J. The kinetic energy is then ($5.68\times10^{-19} - 3.46\times10^{-19}$) J = 2.22×10^{-19} J; $v^2 = 4.87\times10^{11}$ m^2/s^2 and $v = 6.98\times10^5$ m/s.

9.37. A work function of 2.90 eV is equal to $(2.90\text{ eV})\left(\dfrac{1.602\times10^{-19}\text{ J}}{1\text{ eV}}\right) = 4.65\times10^{-19}\text{ J}.$

The energy of a light quantum with a wavelength of 1850 Å = 1850×10^{-10} m is $E = h\nu = \dfrac{hc}{\lambda} = \dfrac{(6.626\times10^{-34}\text{ J}\cdot\text{s})(3.00\times10^{8}\text{ m/s})}{1850\times10^{-10}\text{ m}} = 1.07\times10^{-18}\text{ J}$. The kinetic energy is then ($1.07\times10^{-18} - 4.65\times10^{-19}$) J = 6.1×10^{-19} J. This equals ½ mv^2, so that with the mass of a proton of $m = 1.67\times10^{-27}$ kg: $v^2 = \dfrac{2(6.1\times10^{-19}\text{ J})}{1.67\times10^{-27}\text{ kg}} = 7.3\times10^{8}\dfrac{\text{J}}{\text{kg}}$. Recall that 1 J = 1 N·m = 1 kg·m^2/s^2, so the units come out as m^2/s^2. Taking the square root: $v = 2.7\times10^4$ m/s or about 27,000 m/s.

9.39. Cs has a very low work function, so it is easy for light to knock an electron off and produce an electrical current.

9.41. **(a)** The momentum of 435 nm photons is $p = \dfrac{h}{\lambda} = \dfrac{6.626\times10^{-34}\text{ J}\cdot\text{s}}{435\times10^{-9}\text{ m}} = 1.52\times10^{-27}\dfrac{\text{J}\cdot\text{s}}{\text{m}}$.

Recall that 1 J = 1 N·m = 1 kg·m^2/s^2, so the units are kg·m/s, which are indeed units of momentum (mass times velocity). If 1.00×10^{18} photons were striking the sail each second, the transfer of momentum would be (1.00×10^{18} / s) (1.52×10^{-27} kg·m/s) = 1.52×10^{-9} kg·m/s^2.

This momentum transfer equals a force, or mass times acceleration; therefore, we can calculate the acceleration of the sail as $a = \dfrac{1.52\times10^{-9}\text{ kg}\cdot\text{m/s}^2}{4.29\text{ kg}} = 3.55\times10^{-10}\text{ m/s}^2$.

(b) 10.0 years are 3.15×10^8 s. Using $v = at$, the acquired speed is

$$v = \left(3.55\times10^{-10}\frac{\text{m}}{\text{s}^2}\right)(3.15\times10^{8}\text{ s}) = 0.112\text{ m/s}.$$

(c) $s = \frac{1}{2}at^2 = \frac{1}{2}\left(3.55 \times 10^{-10}\,\frac{\text{m}}{\text{s}^2}\right)\left(3.15 \times 10^8\ \text{s}\right)^2 = 1.77 \times 10^7\ \text{m}.$

(d) This is about 1/20th the distance between the earth and the moon, so it does not seem like solar light sails would be very useful for space travel.

9.43. Bohr postulated that the energy of the electron remains constant as the electron remains in its orbit about the nucleus. However, Maxwell's theory of electromagnetism requires that when charged matter changes its direction (and therefore its velocity vector), it must emit radiation energy as it accelerates; therefore, the fact that the energy of the electron remains constant in its orbit is a violation of Maxwell's classical laws of electrodynamics.

9.45. According to equation 9.34, the radius of the nth energy level of the Bohr hydrogen atom is $r = \frac{\varepsilon_0 n^2 h^2}{\pi m_e e^2} = \left(\frac{\varepsilon_0 h^2}{\pi m_e e^2}\right)n^2$. We start by calculating the constant factor:

$$\frac{\varepsilon_0 h^2}{\pi m_e e^2} = \frac{\left(8.854 \times 10^{-12}\,\frac{\text{C}^2}{\text{J}\cdot\text{m}}\right)\left(6.626 \times 10^{-34}\ \text{J}\cdot\text{s}\right)^2}{\pi\left(9.109 \times 10^{-31}\ \text{kg}\right)\left(1.602 \times 10^{-19}\ \text{C}\right)^2} = 5.293 \times 10^{-11}\,\frac{\text{J}\cdot\text{s}^2}{\text{kg}\cdot\text{m}}.$$

Because $1\ \text{J} = 1\ \text{N}\cdot\text{m} = 1\ \text{kg}\cdot\text{m}^2/\text{s}^2$, so the value of this prefactor is 5.293×10^{-11} m. (You may recognize this as the value of the first Bohr radius, 0.529 Å, obtained for $n = 1$).

$n = 4$: $r = (5.293\times10^{-11}\ \text{m})\ (4)^2 = 8.469\times10^{-10}\ \text{m} = 8.469$ Å;

$n = 5$: $r = (5.293\times10^{-11}\ \text{m})\ (5)^2 = 1.323\times10^{-9}\ \text{m} = 13.23$ Å;

$n = 6$: $r = (5.293\times10^{-11}\ \text{m})\ (6)^2 = 1.905\times10^{-9}\ \text{m} = 19.05$ Å.

9.47. Equation 9.35 gives the energy of the nth energy level of the Bohr hydrogen atom as $E_{\text{tot}} = -\frac{m_e e^4}{8\varepsilon_0^2 n^2 h^2} = -\left(\frac{m_e e^4}{8\varepsilon_0^2 h^2}\right)\frac{1}{n^2}$. We save some time by first calculating the constant factor:

$$\frac{m_e e^4}{8\varepsilon_0^2 h^2} = \frac{\left(9.109 \times 10^{-31}\ \text{kg}\right)\left(1.602 \times 10^{-19}\ \text{C}\right)^4}{8\left(8.854 \times 10^{-12}\,\frac{\text{C}^2}{\text{J}\cdot\text{m}}\right)^2\left(6.626 \times 10^{-34}\ \text{J}\cdot\text{s}\right)^2} = 2.179 \times 10^{-18}\,\frac{\text{kg}\cdot\text{m}^2}{\text{s}^2} = 2.179 \times 10^{-18}\ \text{J}$$

(recall that $1\ \text{J} = 1\ \text{N}\cdot\text{m} = 1\ \text{kg}\cdot\text{m}^2/\text{s}^2$). Then we calculate the values of E_{tot} for different n's:

$n = 4$: $E_{\text{tot}} = -2.179\times10^{-18}\ \text{J} / 4^2 = -1.362\times10^{-19}$ J;

$n = 5$: $E_{\text{tot}} = -2.179\times10^{-18}\ \text{J} / 5^2 = -8.716\times10^{-20}$ J;

$n = 6$: $E_{\text{tot}} = -2.179\times10^{-18}\ \text{J} / 5^2 = -6.053\times10^{-20}$ J.

9.49. Inserting the constants in equation 9.40:

$$R_H = \frac{m_e e^4}{8\varepsilon_0^2 h^3 c} = \frac{\left(9.109\times10^{-31}\text{ kg}\right)\left(1.602\times10^{-19}\text{ C}\right)^4}{8\left(8.854\times10^{-12}\frac{\text{C}^2}{\text{J}\cdot\text{m}}\right)^2\left(6.626\times10^{-34}\text{ J}\cdot\text{s}\right)^3\left(3.00\times10^8\frac{\text{m}}{\text{s}}\right)} = 1.10\times10^7\frac{\text{kg}\cdot\text{m}}{\text{J}\cdot\text{s}^2}.$$

Because 1 J = 1 N · m = 1 kg · m^2/s^2, the value obtained is $1.10\times10^7\ m^{-1}$.

Or because 1 m = 100 cm, taking the reciprocal gives 1 m^{-1} = 0.01 cm^{-1}, so that $R_H = 1.10\times10^5\ cm^{-1}$ or about 110,000 cm^{-1}.

9.51. **(a)** Combining $m_e vr = \frac{nh}{2\pi}$ and $r = \frac{\varepsilon_0 n^2 h^2}{\pi m_e e^2}$ gives $m_e v\frac{\varepsilon_0 n^2 h^2}{\pi m_e e^2} = \frac{nh}{2\pi}$. Simplifying and solving for v leads to: $v = \frac{e^2}{2\varepsilon_0 nh}$.

(b) For $n = 1$, $v = \dfrac{\left(1.602\times10^{-19}\text{ C}\right)^2}{2\left(8.854\times10^{-12}\frac{\text{C}^2}{\text{J}\cdot\text{m}}\right)(1)\left(6.626\times10^{-34}\text{ J}\cdot\text{s}\right)} = 2.187\times10^6\text{ m/s}.$

This is about 0.73% of the speed of light.

(c) $L = m_e vr = (9.109\times10^{-31}\text{ kg})\,(2.187\times10^6\text{ m/s})\,(0.529\times10^{-10}\text{ m}) = 1.05\times10^{-34}\text{ kg}\cdot\text{m}^2/\text{s}$ (or J·s), which is indeed equal to $h/(2\pi)$.

9.53. **(a)** Mass is a particle property.

(b) The de Broglie wavelength is a wave property (ascribed to particles).

(c) Diffraction is a wave property.

(d) Velocity is a property associated with both waves and particles.

(e) Momentum is a property classically associated with particles.

9.55. According to de Broglie, $\lambda = \frac{h}{p} = \frac{h}{mv}$, so for a baseball with m = 100.0 g = 0.1000 kg and $v = \left(160\frac{\text{km}}{\text{h}}\right)\left(\frac{1000\text{ m}}{1\text{ km}}\right)\left(\frac{1\text{ h}}{3600\text{ s}}\right) = 44.4\text{ m/s}$:

$$\lambda = \frac{6.626\times10^{-34}\text{ J}\cdot\text{s}}{\left(0.1000\text{ kg}\right)\left(44.4\text{ m/s}\right)} = 1.49\times10^{-34}\frac{\text{J}\cdot\text{s}^2}{\text{kg}\cdot\text{m}}.$$

Because 1 J = 1 N · m = 1 kg · m^2/s^2, this is $\lambda = 1.49\times10^{-34}$ m – a tiny wavelength, so that the wave nature of a baseball will never be noticed.

For an electron: $\lambda = \dfrac{6.626\times10^{-34}\text{ J}\cdot\text{s}}{\left(9.109\times10^{-31}\text{ kg}\right)\left(44.4\text{ m/s}\right)} = 1.64\times10^{-5}\text{ m}$ (or 16.4 μm).

9.57. We solve the de Broglie equation for v: $\lambda = \frac{h}{p} = \frac{h}{mv}$, so $v = \frac{h}{m\lambda}$. For an electron with

$\lambda = 1.00\ \text{Å} = 1.00\times10^{-10}$ m: $v = \frac{6.626\times10^{-34}\ \text{J}\cdot\text{s}}{(9.109\times10^{-31}\ \text{kg})(1.00\times10^{-10}\ \text{m})} = 7.27\times10^{6}\ \frac{\text{J}\cdot\text{s}}{\text{kg}\cdot\text{m}}$.

Because 1 J = 1 N · m = 1 kg · m^2/s^2, the units are m/s, as we would expect for a velocity.

For a proton ($m = 1.67\times10^{-27}$ kg): $v = \frac{6.626\times10^{-34}\ \text{J}\cdot\text{s}}{(1.67\times10^{-27}\ \text{kg})(1.00\times10^{-10}\ \text{m})} = 3.97\times10^{3}\ \text{m/s}$.

CHAPTER 10

Introduction to Quantum Mechanics

10.1. From the text, the following statements were given as postulates. The state of a system can be described by an expression called a wavefunction; all possible information about the observables of the system are contained in the wavefunction. For every physical observable, there exists a corresponding operator, and the value of the observable can be determined by an eigenvalue equation involving the operator and the wavefunction. Allowable wavefunctions must satisfy the Schrödinger equation. The average value of an observable can be determined from the appropriate operator and wavefunction using the expression given by equation 10.13. Other sources may list postulates differently; your answer may vary.

10.3. **(a)** Yes, the function is an acceptable wavefunction.

(b) No, the function is not acceptable because it is not finite over the range given: for example, y approaches infinity for $x \to +\infty$. (The fact that it is complex for negative x does not disqualify it as such; wavefunctions can have complex values.)

(c) No, the function is not acceptable because it is not finite; for example, the tangent tends toward $+\infty$ for $x \to \pi/2$ (approaching from $x < \pi/2$). In addition, it is not continuous across the values $x = \pi/2$ and $x = -\pi/2$.

(d) Yes, the function is an acceptable wavefunction.

10.5. Functions that satisfy the requirements of a wavefunction:

$f(x) = x^2 + 1, \quad 0 \le x \le 2\pi$

$f(x) = \tan x, \quad -\pi/3 \le x \le +\pi/3$

$f(x) = e^{-|x|}, \quad 0 \le x \le \infty$

$f(x) = 1/x, \quad 1 \le x \le \infty$

In contrast the following variations of these functions do not satisfy the requirements:

$f(x) = x^2 + 1, \quad 0 \le x \le \infty$ (not bounded)

$f(x) = \tan x, \quad -\pi/2 \le x \le +\pi/2$ (not bounded)

$f(x) = e^{-|x|}, \quad -\infty \le x \le \infty$ (not differentiable in $x = 0$)

$f(x) = 1/x, \quad 0 \le x \le \infty$ (not bounded)

10.7. Evaluation of the operation 2×3 yields the value 6; evaluation of the operation $4 \div 5$ yields the value 0.8; evaluation of the operation $\frac{d}{dx}\left(4x^3 - 7x + \frac{7}{x}\right)$ yields the function $12x^2 - 7 - \frac{7}{x^2}$.

10.9. **(a)** $\hat{P}_x(4,5,6)=(-4,5,6)$;

(b) $\hat{P}_y\hat{P}_z(0,-4,-1)=\hat{P}_y(0,-4,1)=(0,4,1)$;

(c) $\hat{P}_x\hat{P}_x(5,0,0)=\hat{P}_x(-5,0,0)=(5,0,0)$;

(d) $\hat{P}_y\hat{P}_x(\pi,\pi/2,0)=\hat{P}_y(-\pi,\pi/2,0)=(-\pi,-\pi/2,0)$;

(e) Yes, $\hat{P}_x\hat{P}_y$ should equal $\hat{P}_y\hat{P}_x$ for any set of coordinates, because the x coordinate and the y coordinate are independent of each other.

10.11. **(a)** $\frac{d}{dx}\cos(4x)=-4\sin(4x)$ (no eigenvalue equation)

(b) $\frac{d^2}{dx^2}\cos(4\pi x)=\frac{d}{dx}\left[-4\pi\sin(4\pi x)\right]=-16\pi^2\cos(4\pi x)$. This is an eigenvalue equation with the eigenvalue $-16\pi^2$.

(c) $\hat{p}_x\left(\sin\frac{2\pi x}{3}\right)=-i\hbar\frac{d}{dx}\left(\sin\frac{2\pi x}{3}\right)=-i\hbar\frac{2\pi}{3}\cos\frac{2\pi x}{3}$ (no eigenvalue equation)

(d) $\hat{x}\left(\sqrt{\frac{2}{a}}\sin\frac{2\pi x}{a}\right)=x\cdot\left(\sqrt{\frac{2}{a}}\sin\frac{2\pi x}{a}\right)=\sqrt{\frac{2}{a}}x\sin\frac{2\pi x}{a}$ (no eigenvalue equation)

(e) $\hat{3}(4\ln x^2)=3\cdot(4\ln x^2)$. This is an eigenvalue equation with the eigenvalue 3.

(f) $\frac{d}{d\theta}(\sin\phi\cos\theta)=-\sin\phi\sin\theta$ (no eigenvalue equation)

(g) $\frac{d^2}{d\theta^2}(\sin\phi\cos\theta)=\frac{d}{d\theta}(-\sin\phi\sin\theta)=-\sin\phi\cos\theta$. This is an eigenvalue equation with the eigenvalue -1.

(h) $\frac{d}{d\phi}\tan\phi=\frac{d}{dx}\left(\frac{\sin\phi}{\cos\phi}\right)=\frac{\cos^2\phi-\sin\phi(-\sin\phi)}{\cos^2\phi}=\frac{\cos^2\phi+\sin^2\phi}{\cos^2\phi}=1+\tan^2\phi$ (no eigenvalue equation)

10.13. Having zero as an eigenvalue may be problematic because it results in zero as the overall eigenvalue/eigenfunction combination. Therefore, there is no guarantee that the *original* function is reproduced, as opposed to any other function, if the overall answer is simply zero.

10.15. Multiplication of a function by i would only be considered a Hermitian operator if the eigenvalue equation always produced a real number as an eigenvalue. A simple 'multiplication by a number' operator operating on an arbitrary function leads to an eigenvalue equation (by definition), so that *any* function is an eigenfunction. This means that a simple 'multiplication by i' operator would not be a Hermitian operator, because it does not lead to real numbers as eigenvalues when operating on non-complex eigenfunctions.

10.17. For the baseball: $\Delta x(0.250 \text{ kg})\left(4\times10^{3}\dfrac{\text{m}}{\text{hr}}\right)\left(\dfrac{1 \text{ hr}}{3600 \text{ s}}\right)\geq\dfrac{6.626\times10^{-34} \text{ J}\cdot\text{s}}{4\pi}$, which gives $\Delta x\geq2\times10^{-34}$ m. This level of precision is not achievable, so we would never notice this uncertainty. For the electron having the same uncertainty in velocity:

$$\Delta x\left(9.109\times10^{-31} \text{ kg}\right)\left(4\times10^{3}\frac{\text{m}}{\text{hr}}\right)\left(\frac{1 \text{ hr}}{3600 \text{ s}}\right)\geq\frac{6.626\times10^{-34} \text{ J}\cdot\text{s}}{4\pi} \text{ or } \Delta x\geq5\times10^{-5} \text{ m}.$$

This is an easily measured distance!

10.19. $\Delta x\cdot\Delta p\geq\dfrac{\hbar}{2}$, so $\left(74\times10^{-12} \text{ m}\right)\cdot\Delta p\geq\dfrac{6.626\times10^{-34} \text{ J}\cdot\text{s}}{4\pi}$. Rearranging results in $\Delta p\geq7.1\times10^{-25}$ kg·m/s. With the mass of the electron of $m=9.109\times10^{-31}$ kg, this results in a velocity uncertainty of $\Delta v=\dfrac{\Delta p}{m}\geq\dfrac{7.1\times10^{-25} \text{ kg}\cdot\text{m/s}}{9.109\times10^{-31} \text{ kg}}$, or $\Delta v\geq7.8\times10^{5}$ m/s.

10.21. In Bohr's theory, it is presumed that we can know the exact distance of the electron from the nucleus as well as its velocity of the electron in its orbit. However, the uncertainty principle suggests that there is a limit to how exactly we can know both position and momentum (or velocity), so these two ideas are inconsistent.

10.23. The operators are defined as $\hat{x}=x\cdot$ and $\hat{p}_x=-i\hbar\dfrac{d}{dx}$, so we obtain:

$$\hat{x}\left[\hat{p}_x\sin(\pi x)\right]=\hat{x}\left(-i\hbar\frac{d}{dx}\left[\sin(\pi x)\right]\right)=x\cdot\left[-i\hbar\pi\cos(\pi x)\right]=-i\hbar\pi x\cos(\pi x)$$

$$\hat{p}_x\left[\hat{x}\sin(\pi x)\right]=-i\hbar\frac{d}{dx}\left[x\cdot\sin(\pi x)\right]=-i\hbar\left[x\cdot\pi\cos(\pi x)+\sin(\pi x)\right]=-i\hbar\left[\pi x\cos(\pi x)+\sin(\pi x)\right]$$

10.25. **(a)** $\Psi^*=3x$; **(b)** $\Psi^*=4+3i$; **(c)** $\Psi^*=\cos 4x$; **(d)** $\Psi^*=+i\hbar\sin 4x$; **(e)** $\Psi^*=e^{3\hbar\phi}$; **(f)** $\Psi^*=e^{+2\pi i\phi/\hbar}$.

10.27. To obtain the probability of finding the particle in the interval from $x=x_1$ to $x=x_2$, we need to evaluate the integral $\int_{x_1}^{x_2}\left(\sqrt{\dfrac{2}{a}}\sin\dfrac{2\pi x}{a}\right)^2dx$. This integral is listed in Appendix 1:

$$P=\frac{2}{a}\int_{x_1}^{x_2}\left(\sin\frac{2\pi x}{a}\right)^2dx=\frac{2}{a}\left[\frac{x}{2}-\frac{1}{4\cdot\frac{2\pi}{a}}\sin\left(2\frac{2\pi}{a}x\right)\right]_{x_1}^{x_2}=\frac{2}{a}\left[\frac{x}{2}-\frac{a}{8\pi}\sin\frac{4\pi x}{a}\right]_{x_1}^{x_2}$$, or with the limits:

$$P=\frac{2}{a}\left[\left(\frac{x_2}{2}-\frac{a}{8\pi}\sin\frac{4\pi x_2}{a}\right)-\left(\frac{x_1}{2}-\frac{a}{8\pi}\sin\frac{4\pi x_1}{a}\right)\right]=\left(\frac{x_2}{a}-\frac{1}{4\pi}\sin\frac{4\pi x_2}{a}\right)-\left(\frac{x_1}{a}-\frac{1}{4\pi}\sin\frac{4\pi x_1}{a}\right)$$

(a) $x_1 = 0, x_2 = 0.02a$: $P = \left(\frac{0.02a}{a} - \frac{1}{4\pi}\sin\frac{4\pi \cdot 0.02a}{a}\right) - (0 - 0) \approx 0.00021$

(b) $x_1 = 0.24a, x_2 = 0.26a$:

$$P = \left[0.26 - \frac{1}{4\pi}\sin(4\pi \cdot 0.26)\right] - \left[0.24 - \frac{1}{4\pi}\sin(4\pi \cdot 0.24)\right] \approx 0.040$$

(c) $x_1 = 0.49a, x_2 = 0.51a$:

$$P = \left[0.51 - \frac{1}{4\pi}\sin(4\pi \cdot 0.51)\right] - \left[0.49 - \frac{1}{4\pi}\sin(4\pi \cdot 0.49)\right] \approx 5.3 \times 10^{-5}$$

(d) $x_1 = 0.74a, x_2 = 0.76a$:

$$P = \left[0.76 - \frac{1}{4\pi}\sin(4\pi \cdot 0.76)\right] - \left[0.74 - \frac{1}{4\pi}\sin(4\pi \cdot 0.74)\right] \approx 0.040$$

(e) $x_1 = 98a, x_2 = 1.00a$:

$$P = \left[1.00 - \frac{1}{4\pi}\sin(4\pi \cdot 1.00)\right] - \left[0.98 - \frac{1}{4\pi}\sin(4\pi \cdot 0.98)\right] \approx 0.00021$$

In this case, the probability in the central region is very small, but the probabilities around $0.25a$ and $0.75a$ are relatively large.

10.29. First, we need to normalize the function: $1 = \int_0^a (Nk)^* (Nk)\,dx = N^2k^2 \int_0^a dx = N^2k^2\, x\big|_0^a$

(N and k are real numbers, so taking the complex conjugate has no effect). This leads to

$1 = N^2 k^2 (a - 0)$, which we can solve for N: $N = \sqrt{\frac{1}{k^2 a}} = \frac{1}{k\sqrt{a}}$. We only need to consider the positive root for N (taking the negative root would work as well). Therefore, the normalized wavefunction is $\Psi = \frac{1}{k\sqrt{a}}k = \frac{1}{\sqrt{a}}$. Note how k cancels!

For the first third of the interval:

$$P = \int_0^{a/3} \left(\frac{1}{\sqrt{a}}\right)^* \left(\frac{1}{\sqrt{a}}\right) dx = \frac{1}{a}\int_0^{a/3} dx = \frac{1}{a}\cdot x\big|_0^{a/3} = \frac{1}{a}\left(\frac{a}{3} - 0\right) = \frac{1}{3}.$$

For the last third of the interval:

$$P = \int_{2a/3}^{a} \left(\frac{1}{\sqrt{a}}\right)^* \left(\frac{1}{\sqrt{a}}\right) dx = \frac{1}{a}\int_{2a/3}^{a} dx = \frac{1}{a}\cdot x\big|_{2a/3}^{a} = \frac{1}{a}\left(a - \frac{2a}{3}\right) = \frac{1}{3}.$$

The lesson here is that if the wavefunction is some constant over an interval, it appears that a particle is equally distributed through the interval.

10.31. To obtain the normalization constant N, we replace Ψ by $N\Psi$ and write down the normalization condition $\int (N\Psi)^* (N\Psi) d\tau = 1$ (between the proper limits, and substituting $d\tau$ by dx or the correct expression). Because the normalization constant is a real number, taking the complex conjugate has no effect, and we can write $N^2 \int \Psi^* \Psi \, d\tau = 1$.

(a) $1 = N^2 \int_0^1 (x^2)^* (x^2) dx = N^2 \int_0^1 x^4 \, dx = N^2 \left(\frac{1}{5} x^5 \right) \Big|_0^1 = N^2 \left(\frac{1}{5} - 0 \right) = \frac{N^2}{5}$. This means 5 = N^2, so $N = \sqrt{5}$. The normalized wavefunction is $\Psi = \sqrt{5} x^2$.

(b) $1 = N^2 \int_5^6 \left(\frac{1}{x} \right)^* \left(\frac{1}{x} \right) dx = N^2 \int_5^6 \frac{1}{x^2} dx = N^2 \left(-\frac{1}{x} \right) \Big|_5^6 = N^2 \left(-\frac{1}{6} + \frac{1}{5} \right) = \frac{N^2}{30}$. This means $30 = N^2$, so $N = \sqrt{30}$. The normalized wavefunction is $\Psi = \sqrt{30} / x$.

(c) $1 = N^2 \int_{-\pi/2}^{\pi/2} (\cos x)^* (\cos x) dx = N^2 \int_{-\pi/2}^{\pi/2} \cos^2 x \, dx = N^2 \left[\frac{x}{2} + \frac{1}{4} \sin(2x) \right]_{-\pi/2}^{\pi/2}$ (see Appendix 1).

We evaluate the primitive function (in square brackets) between the given limits:

$$\left[\frac{x}{2} + \frac{1}{4} \sin(2x) \right]_{-\pi/2}^{\pi/2} = \left[\frac{\pi}{4} + \frac{1}{4} \sin \pi \right] - \left[-\frac{\pi}{4} + \frac{1}{4} \sin(-\pi) \right] = \left[\frac{\pi}{4} + \frac{1}{4}(0) \right] - \left[-\frac{\pi}{4} + \frac{1}{4}(0) \right] = \frac{\pi}{2},$$

so that the normalization condition becomes $1 = N^2 \frac{\pi}{2}$. Solving for N: $N = \sqrt{\frac{2}{\pi}}$. The normalized wavefunction is $\sqrt{\frac{2}{\pi}} \cos x$.

(d) $1 = N^2 \int_0^\infty \left(e^{-r/a} \right)^* \left(e^{-r/a} \right) d\tau = N^2 \int_0^\infty \left(e^{-r/a} \right)^2 \cdot 4\pi r^2 \, dr = N^2 \int_0^\infty e^{-2r/a} \cdot 4\pi r^2 \, dr$ (recall that $(a^b)^c = a^{b \cdot c}$). Simplifying leads to $1 = 4\pi N^2 \int_0^\infty r^2 e^{-2r/a} \, dr$. The *definite* integral $\int_0^\infty x^n e^{-bx} \, dx$ is given in Appendix 1. To use it, we replace n by 2, x by r and b by $2/a$ (and confirm that the conditions $n \neq -1$ and $b > 0$ are satisfied):

$1 = 4\pi N^2 \frac{2!}{(2/a)^3} = 4\pi N^2 \frac{2 \cdot 1}{8 / a^3} = \pi N^2 a^3$. This finally leads to $N = \sqrt{\frac{1}{\pi a^3}}$. The normalized wavefunction is therefore $\Psi = \sqrt{\frac{1}{\pi a^3}} e^{-r/a}$.

(e) $1 = N^2 \int_{-\infty}^{\infty} \left(e^{-r^2/a}\right)^* \left(e^{-r^2/a}\right) d\tau = N^2 \int_{-\infty}^{\infty} \left(e^{-r^2/a}\right)^2 \cdot 4\pi r^2\, dr = 4\pi N^2 \int_{-\infty}^{\infty} r^2 e^{-2r^2/a}\, dr$. Using the integral table in Appendix 1: $1 = 4\pi N^2 \frac{1}{2}\left[\frac{\pi}{(2/a)^3}\right]^{1/2} = 2\pi N^2 \left(\frac{\pi a^3}{8}\right)^{1/2} = N^2 \left(\frac{\pi^3 a^3}{2}\right)^{1/2}$.

This leads to

$N = \left(\frac{2}{\pi^3 a^3}\right)^{1/4}$. Therefore, the normalized wavefunction is $\Psi = \left(\frac{2}{\pi^3 a^3}\right)^{1/4} e^{-r^2/a}$.

10.33. We proceed in the same manner as outlined in 10.31. The normalization condition is: $1 = N^2 \int_{-\infty}^{\infty} \left(e^{-ax^2}\right)^* \left(e^{-ax^2}\right) dx = N^2 \int_{-\infty}^{\infty} \left(e^{-ax^2}\right)^2 dx = N^2 \int_{-\infty}^{\infty} e^{-2ax^2}\, dx$. In Appendix 1, this integral is not listed; however, you can find $\int_0^{\infty} e^{-bx^2}\, dx$. Our integral can be solved by realizing that only x^2 appears inside the integral, so that the integrated function e^{-2ax^2} looks the same for positive and negative values of x. This means that the integral from -∞ to 0 is the same as that from 0 to +∞, so that $\int_{-\infty}^{\infty} e^{-2ax^2}\, dx = \int_{-\infty}^{0} e^{-2ax^2}\, dx + \int_{0}^{\infty} e^{-2ax^2}\, dx = 2\int_{0}^{\infty} e^{-2ax^2}\, dx$. Using Appendix 1, the normalization condition becomes

$1 = 2N^2 \int_0^{\infty} e^{-2ax^2}\, dx = 2N^2 \frac{1}{2}\left(\frac{\pi}{2a}\right)^{1/2} = N^2 \left(\frac{\pi}{2a}\right)^{1/2}$. Solve for N: $N = \left(\frac{2a}{\pi}\right)^{1/4}$. The normalized Gaussian wavefunction is $\Psi = \left(\frac{2a}{\pi}\right)^{1/4} e^{-ax^2}$.

10.35. The normalization constant can be assumed to be part of the constants A and B in the wavefunction $\Psi = Ae^{i(2mE)^{1/2}x/\hbar} + Be^{-i(2mE)^{1/2}x/\hbar}$.

The complex conjugate of Ψ is $\Psi^* = Ae^{-i(2mE)^{1/2}x/\hbar} + Be^{+i(2mE)^{1/2}x/\hbar}$, and the normalization condition is $\int_{-\infty}^{\infty} \Psi^* \Psi\, dx = 1$. To make the expressions more manageable, write the exponent $i(2mE)^{1/2}\, x/\hbar$ simply as iax, so $\Psi = Ae^{iax} + Be^{-iax}$ and $\Psi^* = Ae^{-iax} + Be^{iax}$. Then:

$\Psi^*\Psi = \left(Ae^{-iax} + Be^{iax}\right)\left(Ae^{iax} + Be^{-iax}\right) = A^2 e^{-iax} e^{iax} + ABe^{-iax} e^{-iax} + BAe^{iax} e^{iax} + B^2 e^{iax} e^{-iax}$, or

$\Psi^*\Psi = A^2 e^0 + ABe^{-2iax} + ABe^{2iax} + B^2 e^0 = A^2 + B^2 + 2AB\left(e^{2iax} + e^{-2iax}\right) = A^2 + B^2 + 2AB\cos(2ax)$

.

(For the last step, we have substituted $e^{\pm iy}$ using Euler's formula, $e^{i\theta} = \cos\theta + i\sin\theta$, which means that $e^{-i\theta} = \cos(-\theta) + i\sin(-\theta) = \cos\theta - i\sin\theta$). The normalization condition is then

$$1 = \int_{-\infty}^{\infty} \Psi^* \Psi \, dx = \int_{-\infty}^{\infty} \left(A^2 + B^2\right) dx + \int_{-\infty}^{\infty} 2AB\cos(2ax)\, dx = \left(A^2 + B^2\right) \int_{-\infty}^{\infty} dx + 2AB \int_{-\infty}^{\infty} \cos(2ax)\, dx.$$

Unfortunately, this means that the wavefunction is not normalizable in the range from $-\infty$ to $+\infty$: the first integral is not finite (the integral from $-\infty$ to $+\infty$ of any constant, non-zero function is infinite), and the second integral does not converge as x tends toward $\pm\infty$ (meaning it does not exist).

10.37. The kinetic energy operator is based on the momentum operator, which includes a derivative. However, the potential energy operator is based on position, and the position operator is defined as multiplication by the position variable. Thus, kinetic energy operators are derivative operators, while potential energy operators are multiplicative operators.

10.39. For $V = 0$:

$$-\frac{\hbar^2}{2m}\frac{d^2}{dx^2}\left[\sqrt{2}\sin(\pi x)\right] + 0 \cdot \left[\sqrt{2}\sin(\pi x)\right] = -\frac{\hbar^2}{2m}\frac{d}{dx}\left[\sqrt{2}\,\pi\cos(\pi x)\right] = +\frac{\hbar^2\pi^2}{2m}\left[\sqrt{2}\sin(\pi x)\right].$$

Therefore, the value of the eigenvalue energy is $\frac{\hbar^2\pi^2}{2m}$.

For $V = 0.5$: $-\frac{\hbar^2}{2m}\frac{d^2}{dx^2}\left[\sqrt{2}\sin(\pi x)\right] + 0.5 \cdot \left[\sqrt{2}\sin(\pi x)\right] = \left(\frac{\hbar^2\pi^2}{2m} + 0.5\right)\left[\sqrt{2}\sin(\pi x)\right]$

Therefore, the value of the eigenvalue energy is $\frac{\hbar^2\pi^2}{2m} + 0.5$. The difference between both values is simply the difference in (constant) potential energy, as expected.

10.41. (a) $-\frac{\hbar^2}{2m}\frac{d^2}{dx^2}\left(e^{iKx}\right) = -\frac{\hbar^2}{2m}\frac{d}{dx}\left(iKe^{iKx}\right) = \left(-\frac{\hbar^2}{2m}\right)\left(i^2K^2e^{iKx}\right) = +\frac{\hbar^2K^2}{2m}\left(e^{iKx}\right).$

(Here we have used $i^2 = -1$.) Therefore, e^{iKx} is an eigenfunction with an eigenvalue of $\frac{\hbar^2K^2}{2m}$.

(b) $-\frac{\hbar^2}{2m}\frac{d^2}{dx^2}\left(e^{iKx}\right) + k \cdot \left(e^{iKx}\right) = \frac{\hbar^2K^2}{2m}\left(e^{iKx}\right) + k\left(e^{iKx}\right) = \left(\frac{\hbar^2K^2}{2m} + k\right)\left(e^{iKx}\right).$

Therefore, e^{iKx} is an eigenfunction with an eigenvalue of $\frac{\hbar^2K^2}{2m} + k$.

(c)

$$-\frac{\hbar^2}{2m}\frac{d^2}{dx^2}\left(\sqrt{\frac{2}{a}}\sin\frac{\pi x}{a}\right)=-\frac{\hbar^2}{2m}\frac{d}{dx}\left(\frac{\pi}{a}\sqrt{\frac{2}{a}}\cos\frac{\pi x}{a}\right)=-\frac{\hbar^2}{2m}\left(-\frac{\pi^2}{a^2}\sqrt{\frac{2}{a}}\sin\frac{\pi x}{a}\right)=\frac{\hbar^2\pi^2}{2ma^2}\left(\sqrt{\frac{2}{a}}\sin\frac{\pi x}{a}\right)$$

Therefore, $\sqrt{\frac{2}{a}}\sin\frac{\pi x}{a}$ is an eigenfunction with an eigenvalue of $\frac{\hbar^2\pi^2}{2ma^2}$.

10.43. $1=\int_0^a\left(N\sin\frac{n\pi x}{a}\right)^*\left(N\sin\frac{n\pi x}{a}\right)dx=\int_0^a\left(N\sin\frac{n\pi x}{a}\right)^2dx=N^2\int_0^a\left(\sin\frac{n\pi x}{a}\right)^2dx.$

Solve the integral using Appendix 1:

$$\int_0^a\left(\sin\frac{n\pi x}{a}\right)^2dx=\left(\frac{x}{2}-\frac{a}{4n\pi}\sin\frac{2n\pi x}{a}\right)\Big|_0^a=\left(\frac{a}{2}-\frac{a}{4n\pi}\sin\frac{2n\pi a}{a}\right)-(0-0)=\frac{a}{2}-\frac{a}{4n\pi}\sin(2n\pi).$$

The sine function is zero at 0, ±π, ±2π etc., so that the last term vanishes. The normalization condition is now $N^2\frac{a}{2}=1$, from which we obtain $N=\sqrt{\frac{2}{a}}$ (the positive root is chosen for simplicity; the negative root would also work).

10.45. If $n = 0$ were allowed, the wavefunction would be $\Psi=\sqrt{2/a}\sin 0=0$ everywhere in the box, which is not allowed. (Remember that the probability of finding the particle in the box is given by $P=\int_0^a\Psi^*\Psi\,dx$, which would be zero. This would imply that the particle does not exist, which is a contradiction.)

10.47. (a) $E=h\nu=\frac{hc}{\lambda}$. Rearrange: $\lambda=\frac{hc}{E}=\frac{(6.626\times10^{-34}\text{ J}\cdot\text{s})(3.00\times10^{8}\text{ m/s})}{1.00\times10^{-32}\text{ J}}=1.99\times10^{7}\text{ m}.$

This is about 50% wider than the diameter of the earth; the smaller the energy differences, the longer the wavelengths of corresponding light quanta.

(b) For $n = 1$, we have $E_1=\frac{(1)^2h^2}{8ma^2}$. Rearrange to solve for a:

$$a^2=\frac{h^2}{8mE}=\frac{(6.626\times10^{-34}\text{ J}\cdot\text{s})^2}{8(9.109\times10^{-31}\text{ kg})(1.00\times10^{-32}\text{ J})}=6.02\times10^{-6}\text{ m}^2;\ a=0.00245\text{ m}=$$

2.45 mm.

10.49. (a) Calculate the energy for a = 2.64 nm = 2.64×10^{-10} m and n = 11 using equation 10.12:

$$E=\frac{(11)^2(6.626\times10^{-34}\text{ J}\cdot\text{s})^2}{8(9.109\times10^{-31}\text{ kg})(2.64\times10^{-9}\text{ m})^2}=1.05\times10^{-18}\text{ J}.$$

(b) $n = 12$: $E = \dfrac{(12)^2(6.626\times10^{-34}\ \text{J}\cdot\text{s})^2}{8(9.109\times10^{-31}\ \text{kg})(2.64\times10^{-9}\ \text{m})^2} = 1.24\times10^{-18}\ \text{J}$

(c) $\Delta E = (1.24\times10^{-18} - 1.05\times10^{-18})\ \text{J} = 0.19\times10^{-18}\ \text{J} = 1.9\times10^{-19}\ \text{J}$

(d) $\nu = \Delta E / h = 1.9\times10^{-19}\ \text{J} / (6.626\times10^{-34}\ \text{J·s}) = 3.0\times10^{14}\ \text{s}^{-1}$ (or Hz);

$\lambda = c / \nu = (3.00\times10^8\ \text{m/s}) / (3.0\times10^{14}\ \text{s}^{-1}) = 1.0\times10^{-6}\ \text{m} = 1.0\ \mu\text{m}$.

10.51. Nodes are locations where the wavefunction is zero. If we do not count the sides of the box, there are $n - 1$ nodes for the nth wavefunction (see Figure 10.6), so there are 4 nodes for Ψ_5, 9 nodes for Ψ_{10}, and 99 nodes for Ψ_{100}.

10.53. The normalization condition for the general particle-in-a-box wavefunction is $N^2\int_0^a\left(\sin\frac{n\pi x}{a}\right)^*\left(\sin\frac{n\pi x}{a}\right)dx = 1$. Then $1 = N^2\int_0^a\left(\sin\frac{n\pi x}{a}\right)\left(\sin\frac{n\pi x}{a}\right)dx = N^2\int_0^a\left(\sin\frac{n\pi x}{a}\right)^2 dx$,

which can be solved using Appendix 1: $N^2\int_0^a \sin^2\frac{n\pi x}{a}dx = N^2\left[\frac{x}{2} - \frac{1}{4(n\pi/a)}\sin\frac{2n\pi x}{a}\right]_0^a$

Substituting the limits, and setting the entire expression to 1 (normalization condition):

$N^2\left[\left(\frac{a}{2} - \frac{a}{4n\pi}\sin 2n\pi\right) - \left(0 - \frac{a}{4n\pi}\sin 0\right)\right] = 1$. The first sine term is zero, since the sine is zero at 0, $\pm\pi$, $\pm2\pi$ etc. The second sine term is also zero, as $\sin 0 = 0$. Thus, we have $N^2\frac{a}{2} = 1$, and solving for N we obtain $N = \sqrt{\frac{2}{a}}$ as the normalization constant for any of these wavefunctions.

N is indeed independent of the value of the quantum number n.

10.55. For $n = 1$: $E = \dfrac{(1)^2(6.626\times10^{-34}\ \text{J}\cdot\text{s})^2}{8(9.109\times10^{-31}\ \text{kg})(0.50\times10^{-9}\ \text{m})^2} = 2.4\times10^{-19}\ \text{J}$. Setting $E = ½\, mv^2$ and solving for v: $v^2 = 2E / m = 2\,(2.4\times10^{-19}\ \text{J}) / (9.109\times10^{-31}\ \text{kg}) = 5.3\times10^{11}\ \text{m}^2/\text{s}^2$ and therefore $v = 7.3\times10^5$ m/s.

10.57. Of course the uncertainty principle is consistent with our particle-in-a-box system. Despite our mathematical understanding of the system, there is nothing to preclude the uncertainty principle from applying to a particle in a one-dimensional box. The wavefunctions of equation 10.11 are *not* eigenfunctions of the position or momentum operators, so we cannot obtain specific, instantaneous values for position or momentum from it. However, we can obtain average values for position and momentum, but the uncertainty principle poses no restriction on the certainty with which the average values are known.

10.59. According to the Schrödinger equation, in regions where the potential energy is infinite we have $\left[-\frac{i\hbar^2}{2m}\frac{d^2}{dx^2}+\infty\right]\Psi = E\Psi$, which can only be fulfilled if $\Psi = 0$ in these regions.

Thus, for $x > a$, the probability of finding a particle is $P = \int_a^\infty \Psi\, dx = 0$ (the same holds for $x < 0$).

10.61. From $\Psi = \sqrt{\frac{2}{a}}\sin\frac{n\pi x}{a}$ we first calculate $|\Psi|^2 = \Psi^*\Psi$:

$$|\Psi|^2 = \left(\sqrt{\frac{2}{a}}\sin\frac{n\pi x}{a}\right)^*\left(\sqrt{\frac{2}{a}}\sin\frac{n\pi x}{a}\right) = \left(\sqrt{\frac{2}{a}}\sin\frac{n\pi x}{a}\right)^2 = \frac{2}{a}\left(\sin\frac{n\pi x}{a}\right)^2.$$

The maxima and minima of $|\Psi|^2$ are located where its first derivative (with respect to x) is zero:

$$0 = \frac{d}{dx}\left[\frac{2}{a}\left(\sin\frac{n\pi x}{a}\right)^2\right] = \frac{2}{a}\left(2\sin\frac{n\pi x}{a}\right)\left(\cos\frac{n\pi x}{a}\right)\left(\frac{n\pi}{a}\right) = \frac{4n\pi}{a}\sin\frac{n\pi x}{a}\cos\frac{n\pi x}{a}.$$

This means that maxima and minima occur when $\sin\frac{n\pi x}{a}$ or $\cos\frac{n\pi x}{a}$ become zero, which is the case for $\frac{n\pi x}{a} = 0, \pi, 2\pi, \ldots$ and $\frac{n\pi x}{a} = \frac{\pi}{2}, \frac{3\pi}{2}, \frac{5\pi}{2}, \ldots$, respectively (we leave out the negative values here because n and a are positive and x is between 0 and a). However, this includes the points where Ψ itself passes through zero (nodes); these nodes are the minima of $|\Psi|^2$ and occur when $\sin\frac{n\pi x}{a}$ becomes zero. Excluding these values, the maxima of $|\Psi|^2$ therefore occur for $\frac{n\pi x}{a} = m\frac{\pi}{2}$ where $m = 1, 3, 5, \ldots$ Solve for x:

$$x = m\frac{a}{2n}.$$

(a) $n = 1$: Possible values for x are $a/2$, $3a/2$, $5a/2$ etc., but only $a/2$ is within the box (which extends from 0 to a). Therefore, the maximum is at $x = a/2$.

(b) $n = 2$: Possible values for x are $a/4$, $3a/4$, $5a/4$ etc., but only $a/4$ and $3a/4$ are within the box.

Therefore, the maxima are at $a/4$ and $3a/4$.

(c) $n = 3$: Possible values for x are $a/6$, $3a/6$, $5a/6$, $7a/6$ etc., but only the first three are within the box. Therefore, the maxima are at $a/6$, $3a/6 = a/2$, and $5a/6$.

(d) The trend is that the maxima are at $x = m\frac{a}{2n}$, where m takes on all odd integers between 1 and $2n$.

10.63. **(a)** Normalize: $1 = N^2\int_0^1 (x)^*(x)dx = N^2\int_0^1 x^2\,dx = N^2\left.\frac{x^3}{3}\right|_0^1 = N^2\left(\frac{1}{3}-0\right) = \frac{N^2}{3}$, so $N=\sqrt{3}$.

For $\Psi=\sqrt{3}x$: $\langle x\rangle = \int_0^1 \Psi^*\hat{x}\Psi\,dx = \int_0^1\left(\sqrt{3}x\right)^* x\cdot\left(\sqrt{3}x\right)dx = \int_0^1 3x^3\,dx = \left.\frac{3x^4}{4}\right|_0^1 = \frac{3}{4}-0=\frac{3}{4}$.

(b) Normalize: $1 = N^2\int_0^2 (x)^*(x)dx = N^2\int_0^2 x^2\,dx = N^2\left.\frac{x^3}{3}\right|_0^2 = N^2\left(\frac{8}{3}-0\right) = \frac{8N^2}{3}$,

so $N=\sqrt{\frac{3}{8}}$.

For $\Psi=\sqrt{\frac{3}{8}}x$: $\langle x\rangle = \int_0^2 \Psi^*\hat{x}\Psi\,dx = \int_0^2\left(\sqrt{\frac{3}{8}}x\right)^* x\cdot\left(\sqrt{\frac{3}{8}}x\right)dx = \int_0^2 \frac{3}{8}x^3\,dx = \left.\frac{3x^4}{32}\right|_0^2 = \frac{3}{2}-0=\frac{3}{2}$.

(c) Normalize: $1 = N^2\int_0^\pi [\sin(2x)]^*[\sin(2x)]dx = N^2\int_0^\pi \sin^2(2x)dx = N^2\left[\frac{x}{2}-\frac{1}{4}\sin(4x)\right]_0^\pi$.

Substituting the limits: $1 = N^2\left[\left(\frac{\pi}{2}-\frac{1}{4}\sin(4\pi)\right)-(0-\sin 0)\right] = \frac{N^2\pi}{2}$, so $N=\sqrt{\frac{2}{\pi}}$.

For $\Psi=\sqrt{\frac{2}{\pi}}\sin(2x)$: $\langle x\rangle = \int_0^\pi \Psi^*\hat{x}\Psi\,dx = \int_0^\pi\left[\sqrt{\frac{2}{\pi}}\sin(2x)\right]^* x\cdot\left[\sqrt{\frac{2}{\pi}}\sin(2x)\right]dx = \int_0^\pi \frac{2}{\pi}x\sin^2(2x)dx$.

From Appendix 1: $\int_0^\pi \frac{2}{\pi}x\sin^2(2x)dx = \frac{2}{\pi}\left[\frac{x^2}{4}-\frac{x}{8}\sin(4x)-\frac{1}{32}\cos(4x)\right]_0^\pi$.

Substituting the limits:

$$\langle x\rangle = \frac{2}{\pi}\left(\left[\frac{\pi^2}{4}-\frac{\pi}{8}\sin(4\pi)-\frac{\cos(4\pi)}{32}\right]-\left[0-0\sin 0-\frac{\cos 0}{32}\right]\right) = \frac{2}{\pi}\left[\left(\frac{\pi^2}{4}-\frac{1}{32}\right)-\left(-\frac{1}{32}\right)\right] = \frac{\pi}{2}.$$

10.65. $\langle x\rangle = \int_0^a\left(\sqrt{\frac{2}{a}}\sin\frac{2\pi x}{a}\right)^*\cdot x\cdot\left(\sqrt{\frac{2}{a}}\sin\frac{2\pi x}{a}\right)dx = \frac{2}{a}\int_0^a x\sin^2\frac{2\pi x}{a}dx$. Using Appendix 1:

$$\langle x\rangle = \frac{2}{a}\left[\frac{x^2}{4}-\frac{x}{4(2\pi/a)}\sin\frac{4\pi x}{a}-\frac{1}{8(2\pi/a)^2}\cos\frac{4\pi x}{a}\right]_0^a = \frac{2}{a}\left[\left(\frac{a^2}{4}-0-\frac{a^2}{32\pi}\right)-\left(0-0-\frac{a^2}{32\pi}\right)\right]$$

$\langle x\rangle = \frac{2}{a}\left(\frac{a^2}{4}\right) = \frac{a}{2}$. That is, the average value of position is the center of the box. This is the same value that $\langle x\rangle$ had for the first wavefunction. (In fact, it can be shown that the average value will always equal $a/2$ for the particle-in-the-box wavefunctions.)

10.67. For $n = 1$: $\left\langle x^2 \right\rangle = \int_0^a \left(\sqrt{\frac{2}{a}} \sin \frac{\pi x}{a} \right)^* \cdot x^2 \cdot \left(\sqrt{\frac{2}{a}} \sin \frac{\pi x}{a} \right) dx = \frac{2}{a} \int_0^a x^2 \sin^2 \frac{\pi x}{a} dx$.

Using Appendix 1: $\left\langle x^2 \right\rangle = \frac{2}{a} \left[\frac{x^3}{6} - \left(\frac{x^2}{4(\pi/a)} - \frac{1}{8(\pi/a)^3} \right) \sin \frac{2\pi x}{a} - \frac{x}{4(\pi/a)^2} \cos \frac{2\pi x}{a} \right]_0^a$.

We simplify $\left\langle x^2 \right\rangle = \frac{2}{a} \left[\frac{x^3}{6} - \left(\frac{x^2 a}{4\pi} - \frac{a^3}{8\pi^3} \right) \sin \frac{2\pi x}{a} - \frac{x a^2}{4\pi^2} \cos \frac{2\pi x}{a} \right]_0^a$ and evaluate the limits:

$$\left\langle x^2 \right\rangle = \frac{2}{a} \left(\left[\frac{a^3}{6} - 0 - \frac{a^3}{4\pi^2} \right] - [0 - 0 - 0] \right) = \frac{a^2}{3} - \frac{a^2}{2\pi^2}.$$

For $n = 2$: $\left\langle x^2 \right\rangle = \int_0^a \left(\sqrt{\frac{2}{a}} \sin \frac{2\pi x}{a} \right)^* \cdot x^2 \cdot \left(\sqrt{\frac{2}{a}} \sin \frac{2\pi x}{a} \right) dx = \frac{2}{a} \int_0^a x^2 \sin^2 \frac{2\pi x}{a} dx$.

Using Appendix 1: $\left\langle x^2 \right\rangle = \frac{2}{a} \left[\frac{x^3}{6} - \left(\frac{x^2}{4(2\pi/a)} - \frac{1}{8(2\pi/a)^3} \right) \sin \frac{4\pi x}{a} - \frac{x}{4(2\pi/a)^2} \cos \frac{4\pi x}{a} \right]_0^a$

We simplify $\left\langle x^2 \right\rangle = \frac{2}{a} \left[\frac{x^3}{6} - \left(\frac{x^2 a}{8\pi} - \frac{a^3}{64\pi^3} \right) \sin \frac{4\pi x}{a} - \frac{x a^2}{16\pi^2} \cos \frac{4\pi x}{a} \right]_0^a$ and evaluate the limits:

$$\left\langle x^2 \right\rangle = \frac{2}{a} \left(\left[\frac{a^3}{6} - 0 - \frac{a^3}{16\pi^2} \right] - [0 - 0 - 0] \right) = \frac{a^2}{3} - \frac{a^2}{8\pi^2}.$$

10.69. $\langle E \rangle = \int_0^a \left(\sqrt{\frac{2}{a}} \sin \frac{\pi x}{a} \right)^* \cdot \frac{-\hbar^2}{2m} \frac{d^2}{dx^2} \left(\sqrt{\frac{2}{a}} \sin \frac{\pi x}{a} \right) dx = \frac{-\hbar^2}{2m} \int_0^a \left(\sqrt{\frac{2}{a}} \sin \frac{\pi x}{a} \right) \frac{d}{dx} \left(\frac{\pi}{a} \sqrt{\frac{2}{a}} \cos \frac{\pi x}{a} \right) dx$

$$\langle E \rangle = -\frac{\hbar^2}{2m} \int_0^a \left(\sqrt{\frac{2}{a}} \sin \frac{\pi x}{a} \right) \left(-\frac{\pi^2}{a^2} \sqrt{\frac{2}{a}} \sin \frac{\pi x}{a} \right) dx = \frac{\hbar^2 \pi^2}{2ma^2} \int_0^a \left(\sqrt{\frac{2}{a}} \sin \frac{\pi x}{a} \right) \left(\sqrt{\frac{2}{a}} \sin \frac{\pi x}{a} \right) dx.$$

Rather than evaluating this integral, we remind ourselves that the wavefunction is normalized, so the value of the remaining integral is simply 1. Therefore:
$\langle E \rangle = \frac{\hbar^2 \pi^2}{2ma^2} = \frac{h^2 \pi^2}{(2\pi)^2 2ma^2} = \frac{h^2}{8ma^2}$, which is the value it should have. We justify this by recalling that the wavefunction is an eigenfunction of the Hamiltonian operator, so the average value is simply equal to the eigenvalue of the wavefunction.

10.71. The wavefunction is Ψ_1 of the particle-in-a-box. We already know from Example 10.12 that $\langle x\rangle = \frac{a}{2}$ and from 10.67 that $\langle x^2\rangle = \frac{a^2}{3} - \frac{a^2}{2\pi^2}$. We still need to evaluate $\langle p_x\rangle$ and $\langle p_x^{\,2}\rangle$:

$$\langle p_x\rangle = \int_0^a \left(\sqrt{\frac{2}{a}}\sin\frac{\pi x}{a}\right)^* \left[-i\hbar\frac{d}{dx}\left(\sqrt{\frac{2}{a}}\sin\frac{\pi x}{a}\right)\right]dx = \int_0^a \left(\sqrt{\frac{2}{a}}\sin\frac{\pi x}{a}\right)^* \left(-i\hbar\frac{\pi}{a}\sqrt{\frac{2}{a}}\cos\frac{\pi x}{a}\right)dx$$

$\langle p_x\rangle = -i\hbar\frac{2\pi}{a^2}\int_0^a \sin\frac{\pi x}{a}\cos\frac{\pi x}{a}dx$. From Appendix 1:

$$\langle p_x\rangle = -i\hbar\frac{2\pi}{a^2}\left[\frac{a}{\pi}\sin^2\frac{\pi x}{a}\right]_0^a = -i\hbar\frac{2}{a}\left(\sin^2\pi - \sin^2 0\right) = -i\hbar\frac{2}{a}(0-0) = 0.$$

$$\langle p_x^{\,2}\rangle = \int_0^a \left(\sqrt{\frac{2}{a}}\sin\frac{\pi x}{a}\right)^* \left[-\hbar^2\frac{d^2}{dx^2}\left(\sqrt{\frac{2}{a}}\sin\frac{\pi x}{a}\right)\right]dx = \int_0^a \left(\sqrt{\frac{2}{a}}\sin\frac{\pi x}{a}\right)\left[-\hbar^2\frac{d}{dx}\left(\frac{\pi}{a}\sqrt{\frac{2}{a}}\cos\frac{\pi x}{a}\right)\right]dx$$

$$\langle p_x^{\,2}\rangle = \int_0^a \left(\sqrt{\frac{2}{a}}\sin\frac{\pi x}{a}\right)\left[-\hbar^2\left(-\frac{\pi^2}{a^2}\sqrt{\frac{2}{a}}\sin\frac{\pi x}{a}\right)\right]dx = \frac{\hbar^2\pi^2}{a^2}\int_0^a \left(\sqrt{\frac{2}{a}}\sin\frac{\pi x}{a}\right)\left(\sqrt{\frac{2}{a}}\sin\frac{\pi x}{a}\right)dx$$

Realizing that the remaining integral is the condition for a normalized wavefunction, we do not need to evaluate it and can set it to 1: $\langle p_x^{\,2}\rangle = \frac{\hbar^2\pi^2}{a^2}$.

Therefore, we now can get Δx and Δp:

$$\Delta x = \sqrt{\langle x^2\rangle - \langle x\rangle^2} = \sqrt{\left(\frac{a^2}{3} - \frac{a^2}{2\pi^2}\right) - \left(\frac{a}{2}\right)^2} = \sqrt{\frac{a^2}{3} - \frac{a^2}{2\pi^2} - \frac{a^2}{4}} = a\sqrt{\frac{1}{12} - \frac{1}{2\pi^2}}.$$

$\Delta p_x = \sqrt{\langle p_x^{\,2}\rangle - \langle p_x\rangle^2} = \sqrt{\left(\frac{\hbar^2\pi^2}{a^2}\right) - 0} = \frac{\hbar\pi}{a}$. Now, we evaluate $\Delta x \cdot \Delta p_x$:

$$\Delta x \cdot \Delta p_x = a\sqrt{\frac{1}{12} - \frac{1}{2\pi^2}} \cdot \left(\frac{\hbar\pi}{a}\right) = \hbar\pi\sqrt{\frac{1}{12} - \frac{1}{2\pi^2}} = \frac{h}{2}\sqrt{\frac{1}{12} - \frac{1}{2\pi^2}} \approx 0.090\,h.$$

According to the uncertainty principle, $\Delta x \cdot \Delta p_x$ must be greater than $\frac{\hbar}{2}$, which is about 0.080 h. Thus, our result is in agreement with the uncertainty principle.

10.73. $X(x) = \sqrt{\frac{2}{a}}\sin\frac{n_x\pi x}{a}$ itself has units of $m^{-1/2}$ (because a has units of distance).

The first derivative, dX/dx, therefore has units of $m^{-1/2}/m = m^{-3/2}$; the second derivative, d^2X/dx^2, has units of $m^{-3/2}/m = m^{-5/2}$. The units of $\frac{1}{X}\frac{d^2}{dx^2}X$ are then $m^{-5/2}/\ m^{-1/2} = m^{-2}$

(that is, meters squared in the denominator). This means that the expression that equals $\frac{1}{X}\frac{d^2}{dx^2}X$ must also have units of m^{-2}, which is the case for $-\frac{2mE_x}{\hbar^2}$.

10.75. The energies of the electron in a three-dimensional box are given by equation 10.21:

$$E=\frac{h^2}{8m}\left(\frac{n_x^{\ 2}}{a^2}+\frac{n_y^{\ 2}}{b^2}+\frac{n_z^{\ 2}}{c^2}\right).$$

Here, $a = 2$ Å, $b = 3$ Å, and $c = 5$ Å: $E=\frac{h^2}{8m\text{Å}^2}\left(\frac{n_x^{\ 2}}{2^2}+\frac{n_y^{\ 2}}{3^2}+\frac{n_z^{\ 2}}{5^2}\right)$.

To find the lowest energies, we need to find the five triplets (n_x, n_y, n_z) that minimize the expression in parentheses. In order from lowest to highest, they are:

(1, 1, 1): $E\approx 0.401\frac{h^2}{8m\text{Å}^2}$; (1, 1, 2): $E\approx 0.5211\frac{h^2}{8m\text{Å}^2}$; (1, 1, 3): $E\approx 0.721\frac{h^2}{8m\text{Å}^2}$;

(1, 2, 1): $E\approx 0.734\frac{h^2}{8m\text{Å}^2}$; (1, 2, 2): $E\approx 0.854\frac{h^2}{8m\text{Å}^2}$. All other sets lead to higher energies.

The corresponding wavefunctions are:

$$\Psi_{111}=\sqrt{\frac{8}{30\ \text{Å}^3}}\sin\frac{1\cdot\pi x}{2\ \text{Å}}\sin\frac{1\cdot\pi y}{3\ \text{Å}}\sin\frac{1\cdot\pi z}{5\ \text{Å}};\ \Psi_{112}=\sqrt{\frac{8}{30\ \text{Å}^3}}\sin\frac{1\cdot\pi x}{2\ \text{Å}}\sin\frac{1\cdot\pi y}{3\ \text{Å}}\sin\frac{2\cdot\pi z}{5\ \text{Å}};$$

$$\Psi_{113}=\sqrt{\frac{8}{30\ \text{Å}^3}}\sin\frac{1\cdot\pi x}{2\ \text{Å}}\sin\frac{1\cdot\pi y}{3\ \text{Å}}\sin\frac{3\cdot\pi z}{5\ \text{Å}};\ \Psi_{121}=\sqrt{\frac{8}{30\ \text{Å}^3}}\sin\frac{1\cdot\pi x}{2\ \text{Å}}\sin\frac{2\cdot\pi y}{3\ \text{Å}}\sin\frac{1\cdot\pi z}{5\ \text{Å}};$$

$$\Psi_{122}=\sqrt{\frac{8}{30\ \text{Å}^3}}\sin\frac{1\cdot\pi x}{2\ \text{Å}}\sin\frac{2\cdot\pi y}{3\ \text{Å}}\sin\frac{2\cdot\pi z}{5\ \text{Å}}.$$

10.77. Energy of a particle in a one-dimensional box: $E_{1\text{D}}=\frac{n^2h^2}{8ma^2}$

Energy of a particle in a cubical three-dimensional box: $E_{3\text{D}}=\frac{h^2}{8ma^2}\left(n_x^{\ 2}+n_y^{\ 2}+n_z^{\ 2}\right)$.

(a) The lowest possible quantum numbers are $n = 1$ and $n_x = n_y = n_z = 1$, so $E_{1\text{D}}=\frac{1^2h^2}{8ma^2}$ and $E_{3\text{D}}=\frac{h^2}{8ma^2}\left(1^2+1^2+1^2\right)=\frac{3h^2}{8ma^2}$. Therefore $E_{1\text{D}}/E_{3\text{D}}=\frac{h^2/\left(8ma^2\right)}{3h^2/\left(8ma^2\right)}=\frac{1}{3}$.

(b) The ratio of energies is $E_{1D}/E_{3D} = \dfrac{n^2h^2/(8ma^2)}{h^2(n_x^{\ 2}+n_y^{\ 2}+n_z^{\ 2})/(8ma^2)} = \dfrac{n^2}{n_x^{\ 2}+n_y^{\ 2}+n_z^{\ 2}}$. This means that it varies depending on the values of the quantum numbers, but as long as $n_x = n_y = n_z = n$, the ratio will always be $E_{1D}/E_{3D} == \dfrac{n^2}{3n^2} = \dfrac{1}{3}$.

10.79. According to equation 10.25, the energy of a particle in a cubical box is a multiple of $h^2/(8ma^2)$, where the factor is given by the sum of squares of the three quantum numbers,

$n_x^{\ 2} + n_y^{\ 2} + n_z^{\ 2}$. The wavefunction with the highest energy will be:

(4,4,4): 48 $h^2/(8ma^2)$. Degeneracy: 1.

We now write out all the possible sets, starting with (1,1,1) and increasing values in a systematic way until we exceed the energy of 48 $h^2/(8ma^2)$.

(1,1,1): 3 $h^2/(8ma^2)$. Degeneracy: 1.

(1,1,2), (1,2,1), (2,1,1): 6 $h^2/(8ma^2)$. Degeneracy: 3.

(1,1,3), (1,3,1), (3,1,1): 11 $h^2/(8ma^2)$. Degeneracy: 3.

(1,1,4), (1,4,1), (4,1,1): 18 $h^2/(8ma^2)$. Degeneracy: 3.

(1,1,5), (1,5,1), (5,1,1): 27 $h^2/(8ma^2)$. Degeneracy: 3.

(1,1,6), (1,6,1), (6,1,1): 38 $h^2/(8ma^2)$. Degeneracy: 3.

The next set in this series, (1,1,7), has an energy of 51 $h^2/(8ma^2)$, which is higher than that of the last set (4,4,4).

(1,2,2), (2,1,2), (2,2,1): 9 $h^2/(8ma^2)$. Degeneracy: 3.

(1,2,3), (2,1,3), (1,3,2), (3,1,2), (2,3,1), (3,2,1): 14 $h^2/(8ma^2)$. Degeneracy: 6.

(1,2,4), (2,1,4), (1,4,2), (4,1,2), (2,4,1), (4,2,1): 21 $h^2/(8ma^2)$. Degeneracy: 6.

(1,2,5), (2,1,5), (1,5,2), (5,1,2), (2,5,1), (5,2,1): 30 $h^2/(8ma^2)$. Degeneracy: 6.

(1,2,6), (2,1,6), (1,6,2), (6,1,2), (2,6,1), (6,2,1): 41 $h^2/(8ma^2)$. Degeneracy: 6.

The next set in this series, (1,2,7), has an energy of 54 $h^2/(8ma^2)$, which is higher than that of the last set (4,4,4).

(1,3,3), (3,1,3), (3,3,1): 19 $h^2/(8ma^2)$. Degeneracy: 3.

(1,3,4), (3,1,4), (1,4,3), (4,1,3), (3,4,1), (4,3,1): 26 $h^2/(8ma^2)$. Degeneracy: 6.

(1,3,5), (3,1,5), (1,5,3), (5,1,3), (3,5,1), (5,3,1): 35 $h^2/(8ma^2)$. Degeneracy: 6.

(1,3,6), (3,1,6), (1,6,3), (6,1,3), (3,6,1), (6,3,1): 46 $h^2/(8ma^2)$. Degeneracy: 6.

The next set in this series, (1,3,7), has an energy of 59 $h^2/(8ma^2)$, which is higher than that of the last set (4,4,4).

(1,4,4), (4,1,4), (4,4,1): 33 $h^2/(8ma^2)$. Degeneracy: 3.

(1,4,5), (4,1,5), (1,5,4), (5,1,4), (4,5,1), (5,4,1): 42 h^2 / ($8ma^2$). Degeneracy: 6.

The next set in this series, (1,4,6), has an energy of 53 h^2 / ($8ma^2$), which is higher than that of the last set (4,4,4).

Also, the series starting with (1,5,5) has a minimum energy of 51 h^2 / ($8ma^2$), so we can move on to the next series.

(2,2,2): 12 h^2 / ($8ma^2$). Degeneracy: 1.

(2,2,3), (2,3,2), (3,2,2): 17 h^2 / ($8ma^2$). Degeneracy: 3.

(2,2,4), (2,4,2), (4,2,2): 24 h^2 / ($8ma^2$). Degeneracy: 3.

(2,2,5), (2,5,2), (5,2,2): 33 h^2 / ($8ma^2$). Degeneracy: 3.

(2,2,6), (2,6,2), (6,2,2): 44 h^2 / ($8ma^2$). Degeneracy: 3.

The next set in this series, (2,2,7), has an energy of 57 h^2 / ($8ma^2$), which is higher than that of the last set (4,4,4).

(2,3,3), (3,2,3), (3,3,2): 22 h^2 / ($8ma^2$). Degeneracy: 3.

(2,3,4), (3,2,4), (2,4,3), (4,2,3), (3,4,2), (4,3,2): 29 h^2 / ($8ma^2$). Degeneracy: 6.

(2,3,5), (3,2,5), (2,5,3), (5,2,3), (3,5,2), (5,3,2): 38 h^2 / ($8ma^2$). Degeneracy: 6.

The next set in this series, (2,3,6), has an energy of 49 h^2 / ($8ma^2$), which is higher than that of the last set (4,4,4).

(2,4,4), (4,2,4), (4,4,2): 36 h^2 / ($8ma^2$). Degeneracy: 3.

(2,4,5) (4,2,5), (2,5,4), (5,2,4), (4,5,2), (5,4,2): 45 h^2 / ($8ma^2$). Degeneracy: 6.

The next set in this series, (2,4,6), has an energy of 56 h^2 / ($8ma^2$), which is higher than that of the last set (4,4,4).

Also, the series starting with (2,5,5) has a minimum energy of 54 h^2 / ($8ma^2$), so we can move on to the next series.

(3,3,3): 27 h^2 / ($8ma^2$). Degeneracy: 1.

(3,3,4), (3,4,3), (4,3,3): 34 h^2 / ($8ma^2$). Degeneracy: 3.

(3,3,5), (3,5,3), (5,3,3): 43 h^2 / ($8ma^2$). Degeneracy: 3.

The next set in this series, (3,3,6), has an energy of 54 h^2 / ($8ma^2$), which is higher than that of the last set (4,4,4).

(3,4,4), (4,3,4), (4,4,3): 41 h^2 / ($8ma^2$). Degeneracy: 3.

The next set in this series, (3,4,5), has an energy of 50 h^2 / ($8ma^2$), which is higher than that of the last set (4,4,4).

The series starting with (3,5,5) has an even higher energy, so that we can move on to (4,4,4), which is the last wavefunction we are asked to evaluate. We have already done this initially. This produces all energies in Figure 10.13.

If we order these wavefunctions by energy, we can discover some accidental degeneracies and produce the following list (for each set, only the first permutation is given):

(1,1,1): Degeneracy 1

(1,1,2): Degeneracy 3

(1,2,2): Degeneracy 3

(1,1,3): Degeneracy 3

(2,2,2): Degeneracy 1

(1,2,3): Degeneracy 6

(2,2,3): Degeneracy 3

(1,1,4): Degeneracy 3

(1,3,3): Degeneracy 3

(1,2,4): Degeneracy 6

(2,3,3): Degeneracy 3

(2,2,4): Degeneracy 3

(1,3,4): Degeneracy 6

(1,1,5) and (3,3,3): Degeneracy 4

(2,3,4): Degeneracy 6

(1,2,5): Degeneracy 6

(1,4,4) and (2,2,5): Degeneracy 6

(3,3,4): Degeneracy 3

(1,3,5): Degeneracy 6

(2,4,4): Degeneracy 3

(1,1,6) and (2,3,5): Degeneracy 9

(1,2,6) and (3,4,4): Degeneracy 9

(1,4,5): Degeneracy 6

(3,3,5): Degeneracy 3

(2,2,6): Degeneracy 3

(2,4,5): Degeneracy 6

(1,3,6): Degeneracy 6

(4,4,4): Degeneracy 1

10.81. By analogy with the one- and three-dimension particle-in-a-box, we can deduce that the energies of the 2D particle-in-a-box are given by $\frac{h^2}{8m}\left(\frac{n_x^{\ 2}}{a^2}+\frac{n_y^{\ 2}}{b^2}\right)$. For a square box:

$E=\frac{h^2}{8ma^2}\left(n_x^{\ 2}+n_y^{\ 2}\right)$. The wavefunction with the highest energy will be:

(8,8): 128 h^2 / (8ma^2). Degeneracy: 1.

We now write out all the possible sets, starting with (1,1) and increasing values in a systematic way until we exceed the energy of 128 h^2 / (8ma^2).

(1,1): 2 h^2 / (8ma^2). Degeneracy: 1.

(1,2), (2,1): 5 h^2 / (8ma^2). Degeneracy: 2.

(1,3), (3,1): 10 h^2 / (8ma^2). Degeneracy: 2.

(1,4), (4,1): 17 h^2 / (8ma^2). Degeneracy: 2.

(1,5), (5,1): 26 h^2 / (8ma^2). Degeneracy: 2.

(1,6), (6,1): 37 h^2 / (8ma^2). Degeneracy: 2.

(1,7), (7,1): 50 h^2 / (8ma^2). Degeneracy: 2.

(1,8), (8,1): 65 h^2 / (8ma^2). Degeneracy: 2.

(1,9), (9,1): 82 h^2 / (8ma^2). Degeneracy: 2.

(1,10), (10,1): 101 h^2 / (8ma^2). Degeneracy: 2.

(1,11), (11,1): 122 h^2 / (8ma^2). Degeneracy: 2.

The next set in this series, (1,12), has an energy of 145 h^2 / (8ma^2), which is higher than that of the last set (8,8).

(2,2): 8 h^2 / (8ma^2). Degeneracy: 1.

(2,3), (3,2): 13 h^2 / (8ma^2). Degeneracy: 2.

(2,4), (4,2): 20 h^2 / (8ma^2). Degeneracy: 2.

(2,5), (5,2): 29 h^2 / (8ma^2). Degeneracy: 2.

(2,6), (6,2): 40 h^2 / (8ma^2). Degeneracy: 2.

(2,7), (7,2): 53 h^2 / (8ma^2). Degeneracy: 2.

(2,8), (8,2): 68 h^2 / (8ma^2). Degeneracy: 2.

(2,9), (9,2): 85 h^2 / (8ma^2). Degeneracy: 2.

(2,10), (10,2): 104 h^2 / (8ma^2). Degeneracy: 2.

(2,11), (11,2): 125 h^2 / (8ma^2). Degeneracy: 2.

The next set in this series, (2,12), has an energy of 148 h^2 / (8ma^2), which is higher than that of the last set (8,8).

(3,3): 18 h^2 / (8ma^2). Degeneracy: 1.

(3,4), (4,3): 25 h^2 / (8ma^2). Degeneracy: 2.

(3,5), (5,3): 34 h^2 / (8ma^2). Degeneracy: 2.

(3,6), (6,3): 45 h^2 / (8ma^2). Degeneracy: 2.

(3,7), (7,3): 58 h^2 / (8ma^2). Degeneracy: 2.

(3,8), (8,3): 73 h^2 / (8ma^2). Degeneracy: 2.

(3,9), (9,3): 90 h^2 / (8ma^2). Degeneracy: 2.

(3,10), (10,3): 109 h^2 / (8ma^2). Degeneracy: 2.

The next set in this series, (3,11), has an energy of 130 h^2 / (8ma^2), which is higher than that of the last set (8,8).

(4,4): 32 h^2 / (8ma^2). Degeneracy: 1.

(4,5), (4,4): 41 h^2 / (8ma^2). Degeneracy: 2.

(4,6), (6,4): 52 h^2 / (8ma^2). Degeneracy: 2.

(4,7), (7,4): 65 h^2 / (8ma^2). Degeneracy: 2.

(4,8), (8,4): 80 h^2 / (8ma^2). Degeneracy: 2.

(4,9), (9,4): 97 h^2 / (8ma^2). Degeneracy: 2.

(4,10), (10,4): 116 h^2 / (8ma^2). Degeneracy: 2.

The next set in this series, (4,11), has an energy of 137 h^2 / (8ma^2), which is higher than that of the last set (8,8).

(5,5): 50 h^2 / (8ma^2). Degeneracy: 1.

(5,6), (6,5): 61 h^2 / (8ma^2). Degeneracy: 2.

(5,7), (7,5): 74 h^2 / (8ma^2). Degeneracy: 2.

(5,8), (8,5): 89 h^2 / (8ma^2). Degeneracy: 2.

(5,9), (9,5): 106 h^2 / (8ma^2). Degeneracy: 2.

(5,10), (10,5): 125 h^2 / (8ma^2). Degeneracy: 2.

The next set in this series, (5,11), has an energy of 146 h^2 / (8ma^2), which is higher than that of the last set (8,8).

(6,6): 72 h^2 / (8ma^2). Degeneracy: 1.

(6,7), (7,6): 85 h^2 / (8ma^2). Degeneracy: 2.

(6,8), (8,6): 100 h^2 / (8ma^2). Degeneracy: 2.

(6,9), (9,6): 117 h^2 / (8ma^2). Degeneracy: 2.

The next set in this series, (6,10), has an energy of 136 h^2 / (8ma^2), which is higher than that of the last set (8,8).

(7,7): 98 h^2 / (8ma^2). Degeneracy: 1.

(7,8), (8,7): 113 h^2 / (8ma^2). Degeneracy: 2.

The next set in this series, (7,9), has an energy of 130 h^2 / ($8ma^2$), which is higher than that of the last set (8,8), which we have already calculated initially.

If we order these wavefunctions by energy, we can discover some accidental degeneracies and produce the following list (for each set, only the first permutation is given):

(1,1): Degeneracy 1, energy 2 h^2 / ($8ma^2$).

(1,2): Degeneracy 2, energy 5 h^2 / ($8ma^2$).

(2,2): Degeneracy 1, energy 8 h^2 / ($8ma^2$).

(1,3): Degeneracy 2, energy 10 h^2 / ($8ma^2$).

(2,3): Degeneracy 2, energy 13 h^2 / ($8ma^2$).

(1,4): Degeneracy 2, energy 17 h^2 / ($8ma^2$).

(3,3): Degeneracy 1, energy 18 h^2 / ($8ma^2$).

(2,4): Degeneracy 2, energy 20 h^2 / ($8ma^2$).

(3,4): Degeneracy 2, energy 25 h^2 / ($8ma^2$).

(1,5): Degeneracy 2, energy 26 h^2 / ($8ma^2$).

(2,5): Degeneracy 2, energy 29 h^2 / ($8ma^2$).

(4,4): Degeneracy 1, energy 32 h^2 / ($8ma^2$).

(3,5): Degeneracy 2, energy 34 h^2 / ($8ma^2$).

(1,6): Degeneracy 2, energy 37 h^2 / ($8ma^2$).

(2,6): Degeneracy 2, energy 40 h^2 / ($8ma^2$).

(4,5): Degeneracy 2, energy 41 h^2 / ($8ma^2$).

(3,6): Degeneracy 2, energy 45 h^2 / ($8ma^2$).

(1,7) and (5,5): Degeneracy 3, energy 50 h^2 / ($8ma^2$).

(4,6): Degeneracy 2, energy 52 h^2 / ($8ma^2$).

(2,7): Degeneracy 2, energy 53 h^2 / ($8ma^2$).

(3,7): Degeneracy 2, energy 58 h^2 / ($8ma^2$).

(5,6): Degeneracy 2, energy 61 h^2 / ($8ma^2$).

(1,8) and (4,7): Degeneracy 4, energy 65 h^2 / ($8ma^2$).

(2,8): Degeneracy 2, energy 68 h^2 / ($8ma^2$).

(6,6): Degeneracy 1, energy 72 h^2 / ($8ma^2$).

(3,8): Degeneracy 2, energy 73 h^2 / ($8ma^2$).

(5,7): Degeneracy 2, energy 74 h^2 / ($8ma^2$).

(4,8): Degeneracy 2, energy 80 h^2 / ($8ma^2$).

(1,9): Degeneracy 2, energy 82 h^2 / ($8ma^2$).

(2,9) and (6,7): Degeneracy 4, energy 85 h^2 / ($8ma^2$).

(5,8): Degeneracy 2, energy 89 h^2 / ($8ma^2$).

(3,9): Degeneracy 2, energy 90 h^2 / ($8ma^2$).

(4,9): Degeneracy 2, energy 97 h^2 / ($8ma^2$).

(7,7): Degeneracy 1, energy 98 h^2 / ($8ma^2$).

(6,8): Degeneracy 2, energy 100 h^2 / ($8ma^2$).

(1,10): Degeneracy 2, energy 101 h^2 / ($8ma^2$).

(2,10): Degeneracy 2, energy 104 h^2 / ($8ma^2$).

(5,9): Degeneracy 2, energy 106 h^2 / ($8ma^2$).

(3,10): Degeneracy 2, energy 109 h^2 / ($8ma^2$).

(7,8): Degeneracy 2, energy 113 h^2 / ($8ma^2$).

(4,10): Degeneracy 2, energy 116 h^2 / ($8ma^2$).

(6,9): Degeneracy 2, energy 117 h^2 / ($8ma^2$).

(1,11): Degeneracy 2, energy 122 h^2 / ($8ma^2$).

(2,11) and (5,10): Degeneracy 4, energy 125 h^2 / ($8ma^2$).

(8,8): Degeneracy 1, energy 128 h^2 / ($8ma^2$).

Plotting these sets in an energy diagram is left to the student.

10.83. We must evaluate the following integral:

$$\langle x^2 \rangle = \int_0^c \int_0^b \int_0^a \left(\sqrt{\frac{8}{abc}} \sin\frac{\pi x}{a} \cdot \sin\frac{\pi y}{b} \cdot \sin\frac{\pi z}{c} \right)^* \cdot x^2 \cdot \left(\sqrt{\frac{8}{abc}} \sin\frac{\pi x}{a} \cdot \sin\frac{\pi y}{b} \cdot \sin\frac{\pi z}{c} \right) dx\, dy\, dz$$

We can simplify this by separating the wavefunction (including the normalization constant) into three parts, one part for x, one part for y, and one part for z:

$$\langle x^2 \rangle = \int_0^c \int_0^b \int_0^a \left(\sqrt{\frac{2}{a}} \sin\frac{\pi x}{a} \cdot \sqrt{\frac{2}{b}} \sin\frac{\pi y}{b} \cdot \sqrt{\frac{2}{c}} \sin\frac{\pi z}{c} \right)^* x^2 \left(\sqrt{\frac{2}{a}} \sin\frac{\pi x}{a} \cdot \sqrt{\frac{2}{b}} \sin\frac{\pi y}{b} \cdot \sqrt{\frac{2}{c}} \sin\frac{\pi z}{c} \right) dx\, dy\, dz$$

For the (innermost) integral over x, only the parts that depend on x will matter, everything else can be factored out of the integral; the same is true for the integral over y and z, respectively. This leads to a product of three separate integrals:

$$\langle x^2 \rangle = \left[\int_0^a \left(\sqrt{\frac{2}{a}} \sin\frac{\pi x}{a} \right)^* x^2 \left(\sqrt{\frac{2}{a}} \sin\frac{\pi x}{a} \right) dx \right]$$

$$\times \left[\int_0^c \left(\sqrt{\frac{2}{c}} \sin\frac{\pi z}{c} \right)^* \left(\sqrt{\frac{2}{c}} \sin\frac{\pi z}{c} \right) dz \right] \times \left[\int_0^b \left(\sqrt{\frac{2}{b}} \sin\frac{\pi y}{b} \right)^* \left(\sqrt{\frac{2}{b}} \sin\frac{\pi y}{b} \right) dy \right]$$

We recognize that the second and third integrals represent normalized one-dimensional wavefunctions, so these integrals are simply 1. Therefore, this simplifies to:

$\langle x^2\rangle = \int_0^a \left(\sqrt{\frac{2}{a}}\sin\frac{\pi x}{a}\right)^* x^2 \left(\sqrt{\frac{2}{a}}\sin\frac{\pi x}{a}\right) dx = \frac{2}{a}\int_0^a x^2 \sin^2\frac{\pi x}{a}\, dx$. Using Appendix 1:

$$\langle x^2\rangle = \frac{2}{a}\left[\frac{x^3}{6} - \left(\frac{x^2 a}{4\pi} - \frac{a^3}{8\pi^3}\right)\sin\frac{2\pi x}{a} - \frac{xa^2}{4\pi^2}\cos\frac{2\pi x}{a}\right]_0^a = \frac{2}{a}\left[\left(\frac{a^3}{6} - 0 - \frac{a^3}{4\pi^2}\right) - (0-0-0)\right]$$

$\langle x^2\rangle = \frac{a^2}{3} - \frac{a^2}{2\pi^2}$. Finally, rather than repeat everything, by symmetry we can deduce that

$\langle y^2\rangle = \frac{b^2}{3} - \frac{b^2}{2\pi^2}$ and $\langle z^2\rangle = \frac{c^2}{3} - \frac{c^2}{2\pi^2}$.

10.85. We need to evaluate the integral

$$\int_0^c\int_0^b\int_0^a \left(\sqrt{\frac{8}{abc}}\sin\frac{\pi x}{a}\cdot\sin\frac{\pi y}{b}\cdot\sin\frac{\pi z}{c}\right)^* \left(\sqrt{\frac{8}{abc}}\sin\frac{\pi x}{a}\cdot\sin\frac{\pi y}{b}\cdot\sin\frac{2\pi z}{c}\right) dx\,dy\,dz$$

As in the previous exercise, we separate the wavefunction (including the normalization constant) into three parts, one part for x, one part for y, and one part for z:

$$\left[\int_0^a \left(\sqrt{\frac{2}{a}}\sin\frac{\pi x}{a}\right)^* \left(\sqrt{\frac{2}{a}}\sin\frac{\pi x}{a}\right) dx\right] \times \left[\int_0^b \left(\sqrt{\frac{2}{b}}\sin\frac{\pi y}{b}\right)^* \left(\sqrt{\frac{2}{b}}\sin\frac{\pi y}{b}\right) dy\right]$$

$$\times \left[\int_0^c \left(\sqrt{\frac{2}{c}}\sin\frac{\pi z}{c}\right)^* \left(\sqrt{\frac{2}{c}}\sin\frac{2\pi z}{c}\right) dz\right]$$

The first two integrals are for normalized one-dimensional wavefunctions; therefore, they are both equal to 1. The only remaining integral is

$\int_0^c \left(\sqrt{\frac{2}{c}}\sin\frac{\pi z}{c}\right)^* \left(\sqrt{\frac{2}{c}}\sin\frac{2\pi z}{c}\right) dz = \frac{2}{c}\int_0^c \sin\frac{\pi z}{c}\sin\frac{2\pi z}{c}\, dz$. Using the integral table in Appendix 1:

$$= \frac{2}{c}\left(\frac{\sin\left[\left(\frac{\pi}{c} - \frac{2\pi}{c}\right)z\right]}{2\left(\frac{\pi}{c} - \frac{2\pi}{c}\right)} - \frac{\sin\left[\left(\frac{\pi}{c} + \frac{2\pi}{c}\right)z\right]}{2\left(\frac{\pi}{c} + \frac{2\pi}{c}\right)}\right)\Bigg|_0^c = \left[\frac{\sin(\pi z/c)}{\pi} - \frac{\sin(3\pi z/c)}{3\pi}\right]_0^c = (0-0)-(0-0) = 0$$

Therefore, the entire wavefunctions are orthogonal to each other.

10.87. The first integral is set up as $\int_0^a \left(\sqrt{\frac{2}{a}}\sin\frac{\pi x}{a}\right)^* \left(\sqrt{\frac{2}{a}}\sin\frac{2\pi x}{a}\right) dx = \frac{2}{a}\int_0^a \sin\frac{\pi x}{a}\sin\frac{2\pi x}{a}\,dx$.

Using Appendix 1, this equals

$$\frac{2}{a}\left(\frac{\sin\left[\left(\frac{\pi}{a}-\frac{2\pi}{a}\right)x\right]}{2\left(\frac{\pi}{a}-\frac{2\pi}{a}\right)} - \frac{\sin\left[\left(\frac{\pi}{a}+\frac{2\pi}{a}\right)x\right]}{2\left(\frac{\pi}{a}+\frac{2\pi}{a}\right)}\right)\Bigg|_0^a = \left[\frac{\sin(\pi x/c)}{\pi} - \frac{\sin(3\pi x/c)}{3\pi}\right]_0^a = (0-0)-(0-0)=0$$

The reverse order of wavefunctions gives us

$\int_0^a \left(\sqrt{\frac{2}{a}}\sin\frac{2\pi x}{a}\right)^* \left(\sqrt{\frac{2}{a}}\sin\frac{\pi x}{a}\right) dx = \frac{2}{a}\int_0^a \sin\frac{2\pi x}{a}\sin\frac{\pi x}{a}\,dx$. We can immediately see that this is the same expression as before, because the order of factors in a product doesn't matter.

10.89. $\hat{H}\left[e^{-iEt/\hbar}\Psi(x)\right] = i\hbar\frac{\partial}{\partial t}\left[e^{-iEt/\hbar}\Psi(x)\right]$ is the expression we need to show is satisfied. On the left side, we recognize that the Hamiltonian operator does not include time, so the exponential function (which depends only on time) can be moved outside of the operator. On the right side, the spatial part of the wavefunction, $\Psi(x)$, does not depend on time, so it can be removed from the derivative: $e^{-iEt/\hbar}\cdot\hat{H}\Psi(x) = i\hbar\Psi(x)\cdot\frac{\partial(e^{-iEt/\hbar})}{\partial t}$. Evaluating the derivative gives $\frac{\partial\left(e^{-iEt/\hbar}\right)}{\partial t} = -\frac{iE}{\hbar}e^{-iEt/\hbar}$, which leads to

$e^{-iEt/\hbar}\cdot\hat{H}\Psi(x) = i\hbar\Psi(x)\cdot\left(-\frac{iE}{\hbar}\right)e^{-iEt/\hbar}$. The two exponential functions cancel. On the right side of the equation, the $\hbar$ terms cancel, as do the two i's with the negative sign ($-i^2 = 1$). What is left is $\hat{H}\Psi(x) = E\Psi(x)$, which is the time-independent Schrödinger equation. Since we assume that $\Psi(x)$ is an eigenfunction of the time-independent Schrödinger equation, we affirm that the original equation does in fact satisfy the time-dependent Schrödinger equation.

10.91. $|\Psi(x,t)|^2 = \Psi^*(x,t)\Psi(x,t) = \left[e^{-iEt/\hbar}\cdot\Psi(x)\right]^*\left[e^{-iEt/\hbar}\cdot\Psi(x)\right] = e^{+iEt/\hbar}\cdot\Psi^*(x)\cdot e^{-iEt/\hbar}\cdot\Psi(x)$

$= \Psi^*(x)\cdot\Psi(x) = |\Psi(x)|^2$. Thus, the square magnitudes of the time-dependent and the time-independent wavefunctions are the same.

CHAPTER 11

Quantum Mechanics: Model Systems and the Hydrogen Atom

11.1. $\left(3.558\,\frac{\text{mdyn}}{\text{Å}}\right)\left(\frac{1\text{ dyn}}{1000\text{ mdyn}}\right)\left(\frac{1\text{ N}}{10^5\text{ dyn}}\right)\left(\frac{10^{10}\text{ Å}}{1\text{ m}}\right) = 355.8\ \frac{\text{N}}{\text{m}}$

11.3. $\nu = \frac{1}{2\pi}\sqrt{\frac{1.00\text{ N/m}}{1.00\text{ kg}}} = \frac{1.00}{2\pi}\sqrt{\frac{(\text{kg}\cdot\text{m/s}^2)/\text{m}}{\text{kg}}} = 0.159\text{ s}^{-1}$

11.5. To answer this question, we need to think about the potential energy of a pendulum while it is moving back and forth. If h is its height relative to the equilibrium position, the potential energy is $V = mgh$. This does not satisfy the quadratic dependence on the position coordinate as required by equation 11.2, so at first sight it may look like the motion of a pendulum cannot be treated as a harmonic oscillator.

The relationship becomes clearer, however, when we realize that its swinging motion is symmetrical in the *lateral* rather than vertical direction. This means we need to introduce a position coordinate to describe its lateral motion: the angle φ between the pendulum's string and the vertical (the greater φ, the greater its lateral deviation from equilibrium).

The height h is related to φ and the string length ℓ by $h = \ell - \ell \cos \varphi$ (as you can easily work out with a simple drawing), so that the potential energy $V = mgh$ becomes $V = mg\ell(1 - \cos \varphi)$. The cosine can be expanded as a Taylor series, $\cos \varphi = 1 - \varphi^2/2! + \varphi^4/4! - \ldots$ (with angles given in radians), which we can cut off after the first two terms for small angles φ. Therefore,

$1 - \cos \varphi \approx \tfrac{1}{2}\varphi^2$, and the potential energy becomes $V = \tfrac{1}{2} mg\ell \cdot \varphi^2$. Thus, the pendulum indeed satisfies the fundamental mathematical requirement for a Hooke's-law type of oscillator in terms of its angular coordinate (and in the limit of small angular deviations).

11.7. The expression $\frac{2mE}{\hbar^2}$ has units of $\frac{\text{kg}\cdot\text{J}}{(\text{J}\cdot\text{s})^2} = \frac{\text{kg}}{\text{J}\cdot\text{s}^2}$. Because $\alpha = \frac{2\pi\nu m}{\hbar}$ has units of $\frac{\text{s}^{-1}\cdot\text{kg}}{\text{J}\cdot\text{s}} = \frac{\text{kg}}{\text{J}\cdot\text{s}^2}$, the expression $\alpha^2 x^2$ has units of $\left(\frac{\text{kg}}{\text{J}\cdot\text{s}^2}\right)^2(\text{m})^2 = \frac{\text{kg}^2\cdot\text{m}^2}{\text{J}^2\cdot\text{s}^4}$. Substituting $\text{kg}\cdot\text{m}^2/\text{s}^2$ for J into this last expression gives $\frac{\text{kg}\cdot\text{J}}{\text{J}^2\cdot\text{s}^2}$, which reduces to $\frac{\text{kg}}{\text{J}\cdot\text{s}^2}$, thus verifying that these two expressions have the same units.

As for equation 11.11, we have already verified that $\frac{2mE}{\hbar^2}$ and α have both units of $\frac{\text{kg}}{\text{J}\cdot\text{s}^2}$.

11.9. Taking the first derivative of $\Psi = e^{-\alpha x^2/2}\cdot f(x)$, we must use the chain rule because we have a product of two functions of x: $\Psi' = e^{-\alpha x^2/2}\cdot f'(x) + f(x)\cdot e^{-\alpha x^2/2}\cdot(-\alpha x)$, or $\Psi' = e^{-\alpha x^2/2}\cdot f'(x) - \alpha x\cdot f(x)\cdot e^{-\alpha x^2/2}$. In taking the second derivative, we note that the second term is now a product of three functions of x:

$$\Psi'' = e^{-\alpha x^2/2}\cdot f''(x) + f'(x)\cdot e^{-\alpha x^2/2}(-\alpha x) - \alpha\cdot f(x)e^{-\alpha x^2/2} - \alpha x\cdot f'(x)\cdot e^{-\alpha x^2/2} - \alpha x f(x)\cdot e^{-\alpha x^2/2}(-\alpha x)$$

The exponential function can be factored out of all five terms, and there are two terms of $-\alpha x f'(x)$ that can be combined. Doing that, we get for our final expression:

$\Psi'' = e^{-\alpha x^2/2}\left[\alpha^2 x^2 f(x) - \alpha f(x) - 2\alpha x f'(x) + f''(x)\right]$, which is equation 11.9.

11.11. Starting with $\alpha + 2\alpha n - \frac{2mE}{\hbar^2} = 0$, first let us multiply all terms by $\frac{\hbar^2}{2m}$. We get $\frac{\alpha\hbar^2}{2m} + \frac{\alpha n\hbar^2}{m} - E = 0$, which rearranges to $E = \frac{\alpha\hbar^2}{2m} + \frac{\alpha n\hbar^2}{m}$. Now we substitute for the definition of α: $E = \frac{2\pi\nu m\cdot\hbar^2}{\hbar\cdot 2m} + \frac{2\pi\nu m\cdot n\hbar^2}{\hbar\cdot m}$. Recalling that $\hbar = \frac{h}{2\pi}$, we can substitute and cancel out the 2π terms. We also cancel mass out of both fractions. We are left with $E = \frac{\nu h}{2} + n\nu h$. Factoring out the $h\nu$ from both terms, we are left with $E = \left(n + \frac{1}{2}\right)h\nu$, which is the required equation.

11.13. **(a)** $\Delta E = E_{n+1} - E_n = (n + 1 + ½)\,h\nu - (n + ½)\,h\nu = h\nu$, so

$\Delta E = (6.626\times10^{-34}\text{ J·s})\,(1.00\text{ s}^{-1}) = 6.63\times10^{-34}$ J.

(b) $\lambda = \frac{c}{\nu} = \frac{3.00\times10^8\text{ m/s}}{1.00\text{ s}^{-1}} = 3.00\times10^8\text{ m}$.

(c) This is in the (very-long-wavelength) radio wave region of the electromagnetic spectrum.

(d) Such a long, low-energy radio wave would have been undetectable by early 20th-century technology (and may still be undetectable today!).

11.15. For the harmonic oscillator, $E = ½\,h\nu$ for $n = 0$ and $E = (4 + ½)\,h\nu$ for $n = 4$, which means that $\Delta E = 4h\nu$. With $c = \lambda\nu$ and $\tilde{\nu} = \frac{1}{\lambda}$, this is equal to $\Delta E = 4h\frac{c}{\lambda} = 4hc\tilde{\nu}$. The energy required is thus $\Delta E = 4\,(6.626\times10^{-34}\text{ J·s})\,(3.00\times10^8\text{ m/s})\,(3650\text{ cm}^{-1})\,(100\text{ cm/m}) = 2.90\times10^{-19}$ J.

For one photon to have that amount of energy ($E = h\nu$), it must have a frequency of $\frac{2.90\times10^{-19}\ \text{J}}{6.626\times10^{-34}\ \text{J}\cdot\text{s}} = 4.38\times10^{14}\ \text{s}^{-1}$. The wavelength of a photon having that frequency is $\lambda = \frac{c}{\nu} = \frac{3.00\times10^{8}\ \text{m/s}}{4.38\times10^{14}\ \text{s}^{-1}} = 6.85\times10^{-7}\ \text{m} = 685\ \text{nm}$.

11.17. Using equation 11.19, we find that Ψ_1 has the form $\left(\frac{\alpha}{\pi}\right)^{1/4}\left(\frac{1}{2}\right)^{1/2}\left(2\alpha^{1/2}x\right)\cdot e^{-\alpha x^2/2}$.

Substituting into the Schrödinger equation:

$$-\frac{\hbar^2}{2m}\frac{d^2}{dx^2}\left[K\left(2\alpha^{1/2}x\right)e^{-\alpha x^2/2}\right]+\frac{1}{2}kx^2\cdot K\left(2\alpha^{1/2}x\right)e^{-\alpha x^2/2} = E\cdot K\left(2\alpha^{1/2}x\right)e^{-\alpha x^2/2}$$

(where we are using K to represent the collection of constants at the beginning of the wavefunction, for clarity). K can be moved outside the second derivative and can therefore be dropped from the equation entirely (simply divide both sides by K):

$$-\frac{\hbar^2}{2m}\frac{d^2}{dx^2}\left[\left(2\alpha^{1/2}x\right)e^{-\alpha x^2/2}\right]+\frac{1}{2}kx^2\cdot\left(2\alpha^{1/2}x\right)e^{-\alpha x^2/2} = E\cdot\left(2\alpha^{1/2}x\right)e^{-\alpha x^2/2}$$

First, let us evaluate the second derivative. The first derivative is $\frac{d}{dx}\left[\left(2\alpha^{1/2}x\right)e^{-\alpha x^2/2}\right] = 2\alpha^{1/2}xe^{-\alpha x^2/2}\cdot(-\alpha x)+2\alpha^{1/2}e^{-\alpha x^2/2}$ and therefore

$$\frac{d}{dx}\left(-2\alpha^{3/2}x^2e^{-\alpha x^2/2}+2\alpha^{1/2}e^{-\alpha x^2/2}\right) = -4\alpha^{3/2}xe^{-\alpha x^2/2}-2\alpha^{3/2}x^2e^{-\alpha x^2/2}(-\alpha x)+2\alpha^{1/2}e^{-\alpha x^2/2}(-\alpha x),$$

which simplifies to $2\alpha^{5/2}x^3e^{-\alpha x^2/2}-4\alpha^{3/2}xe^{-\alpha x^2/2}-2\alpha^{3/2}xe^{-\alpha x^2/2} = 2\alpha^{5/2}x^3e^{-\alpha x^2/2}-6\alpha^{3/2}xe^{-\alpha x^2/2}$.

Recalling the original definition of Ψ_1, we can rewrite this as $\left(\alpha^2x^2-3\alpha\right)\left(2\alpha^{1/2}x\right)e^{-\alpha x^2/2}$. Substituting back into the simplified Schrödinger equation:

$-\frac{\hbar^2}{2m}\left(\alpha^2x^2-3\alpha\right)\left(2\alpha^{1/2}x\right)e^{-\alpha x^2/2}+\frac{1}{2}kx^2\left(2\alpha^{1/2}x\right)e^{-\alpha x^2/2} = E\left(2\alpha^{1/2}x\right)e^{-\alpha x^2/2}$. Dividing by $\left(2\alpha^{1/2}x\right)e^{-\alpha x^2/2}$ gives us the expression for energy: $E = -\frac{\hbar^2}{2m}\left(\alpha^2x^2-3\alpha\right)+\frac{1}{2}kx^2$.

Now substitute for the definitions of α and k:

$$E = -\frac{\hbar^2}{2m}\left(\frac{2\pi\nu m}{\hbar}\right)^2x^2+\frac{3\hbar^2}{2m}\left(\frac{2\pi\nu m}{\hbar}\right)+\frac{1}{2}\left(4\pi^2\nu^2m\right)x^2 = -2\pi^2\nu^2mx^2+3\hbar\pi\nu+2\pi^2\nu^2mx^2$$

The first and last terms cancel, leaving $E = 3\hbar\pi\nu$. Writing in terms of $h/2\pi$:

$$E = \frac{3h\pi\nu}{2\pi} = \frac{3}{2}h\nu.$$

11.19. For Ψ_0: $\int_{-\infty}^{+\infty}\left[\left(\frac{\alpha}{\pi}\right)^{1/4} e^{-\frac{\alpha x^2}{2}}\right]^* \cdot \left(-i\hbar\frac{d}{dx}\left[\left(\frac{\alpha}{\pi}\right)^{1/4} e^{-\frac{\alpha x^2}{2}}\right]\right)dx = -i\hbar\left(\frac{\alpha}{\pi}\right)^{1/2}\int_{-\infty}^{+\infty} e^{-\frac{\alpha x^2}{2}}\cdot e^{-\frac{\alpha x^2}{2}}(-\alpha x)dx$,

which simplifies to $+\alpha i\hbar\left(\frac{\alpha}{\pi}\right)^{1/2}\int_{-\infty}^{+\infty} x\cdot e^{-\alpha x^2}\,dx$. Rather than evaluate this using an integral table, we will evaluate it by inspection. In the range 0 to -∞, this function has certain negative values; in the range 0 to +∞, the function has exactly the same values but now positive. That is, this function is an odd function over the interval -∞ to +∞. The integral of functions that are symmetrically odd over the region of interest is exactly zero. Therefore, $\langle p_x \rangle$ for Ψ_0 is 0.

For Ψ_1: $\int_{-\infty}^{+\infty}\left[\left(\frac{\alpha}{4\pi}\right)^{1/4} 2\alpha^{1/2} x\, e^{-\frac{\alpha x^2}{2}}\right]^* \cdot \left(-i\hbar\frac{d}{dx}\left[\left(\frac{\alpha}{4\pi}\right)^{1/4} 2\alpha^{1/2} x\, e^{-\frac{\alpha x^2}{2}}\right]\right)dx$. This evaluates to

$$-i\hbar\left(\frac{\alpha}{4\pi}\right)^{1/2}(4\alpha)\int_{-\infty}^{+\infty} x\, e^{-\frac{\alpha x^2}{2}}\cdot\left[e^{-\frac{\alpha x^2}{2}} + x\, e^{-\frac{\alpha x^2}{2}}(-\alpha x)\right]dx = -i\hbar\frac{2\alpha^{3/2}}{\pi^{1/2}}\left[\int_{-\infty}^{+\infty} x\, e^{-\alpha x^2}\,dx - \alpha\int_{-\infty}^{+\infty} x^3 e^{-\alpha x^2}\,dx\right]$$

As with the first wavefunction, here we have two terms that are both odd functions about the entire range of the integral. Therefore, this integral can be evaluated by inspection as exactly zero, and $\langle p_x \rangle$ for Ψ_1 is 0 as well.

These values make sense given that momentum is a vector quantity and that the mass is traveling back and forth in both directions.

11.21. In general, $\Psi_n(x) = N\cdot H_n(\alpha^{1/2}x)e^{-\alpha x^2/2} = N\cdot H_n(\xi)e^{-\xi^2/2}$, where we have substituted $\xi = \alpha^{1/2}x$. The normalization constant N is obtained from the condition $\int_{-\infty}^{\infty}\left[\Psi_n(x)\right]^*\Psi_n(x)dx = 1$, which can be rewritten using $\xi = \alpha^{1/2}x$ and $d\xi = \alpha^{1/2}\,dx$:

$\int_{-\infty}^{\infty}\left[\Psi_n(\xi)\right]^*\Psi_n(\xi)\frac{d\xi}{\alpha^{1/2}} = 1$, or more explicitly $\int_{-\infty}^{\infty}\left[NH_n(\xi)e^{-\xi^2/2}\right]^* NH_n(\xi)e^{-\xi^2/2}\frac{d\xi}{\alpha^{1/2}} = 1$.

The complex conjugate does not change the function (i does not appear anywhere) and we can factor out all constants, which leads to

$\frac{N^2}{\alpha^{1/2}}\int_{-\infty}^{\infty} H_n(\xi)\cdot H_n(\xi)\cdot e^{-\xi^2}\,d\xi = 1$. Table 11.2 lists this integral as $2^n n!\,\pi^{1/2}$ (note that $H_a(\xi)^* = H_a(\xi)$ in our case), so we obtain $\frac{N^2}{\alpha^{1/2}}2^n n!\pi^{1/2} = 1$. Solving this for N leads to

$N^2 = \left(\frac{\alpha}{\pi}\right)^{1/2}\frac{1}{2^n n!}$ and finally $N = \left(\frac{\alpha}{\pi}\right)^{1/4}\left(\frac{1}{2^n n!}\right)^{1/2}$. Substituting the normalization constant into the wavefunction gives the intended result:

$$\Psi_n(x) = \left(\frac{\alpha}{\pi}\right)^{1/4}\left(\frac{1}{2^n n!}\right)^{1/2} H_n(\alpha^{1/2}x)e^{-\alpha x^2/2}.$$

11.23. "As the vibrational quantum number increases, the extension of the vibration *increases* while the average length of the oscillator itself *stays the same*." Figure 11.4 shows that the wavefunction extends more and more to the left and right as the quantum number increases, which means that the likelihood of finding the mass in these positions increases (recall that it is given by $|\Psi^2|$). However, the average position (or "length" of the oscillator) is zero for all quantum numbers, because for any wavefunction the mass is just as likely to be found on the left as on the right.

11.25. $\langle x^2 \rangle = \int_{-\infty}^{\infty} \Psi_0^* \cdot \hat{x}^2 (\Psi_0) dx = \int_{-\infty}^{\infty} \left[\left(\frac{\alpha}{\pi}\right)^{1/4} e^{-\frac{\alpha x^2}{2}} \right]^* \cdot x^2 \cdot \left(\frac{\alpha}{\pi}\right)^{1/4} e^{-\frac{\alpha x^2}{2}} dx = \left(\frac{\alpha}{\pi}\right)^{1/2} \int_{-\infty}^{\infty} x^2 e^{-\alpha x^2} dx$

Using Appendix 1: $\langle x^2 \rangle = \left(\frac{\alpha}{\pi}\right)^{1/2} \int_{-\infty}^{\infty} x^2 e^{-\alpha x^2} dx = \left(\frac{\alpha}{\pi}\right)^{1/2} \cdot \frac{1}{2} \left(\frac{\pi}{\alpha^3}\right)^{1/2}$, or $\langle x^2 \rangle = \frac{1}{2\alpha}$.

It may seem strange at first that the expectation value for x^2 would be non-zero when the expectation value for x is zero. This makes sense, though: the mass is equally distributed between positive and negative x, which yields an average value of zero. All of these positions have a positive value for x^2, however, so that its average value will be nonzero.

11.27. The classical turning point is the value of x at which the total energy equals the potential energy. We have expressions for both of those energies, so we simply equate them and solve for x: $h\nu(n + ½) = ½kx^2$, so $x^2 = \frac{2h\nu(n+½)}{k} = \frac{(2n+1)h\nu}{k}$ and $x = \pm\sqrt{\frac{(2n+1)h\nu}{k}}$.

11.29. According to equation 11.19: $\Psi_3 = \left(\frac{\alpha}{\pi}\right)^{1/4} \left(\frac{1}{2^3\, 3!}\right)^{1/2} \left[8(\alpha^{1/2}x)^3 - 12(\alpha^{1/2}x) \right] e^{-\alpha x^2/2}$ or

$\Psi_3 = \left(\frac{\alpha^3}{9\pi}\right)^{1/4} (2\alpha x^2 - 3) x e^{-\alpha x^2/2}$. The nodes are located at $\Psi_3^* \Psi_3 = 0$, which is obtained

for

$x_1 = 0$ and $2\alpha x^2 - 3 = 0$. The latter leads to $x^2 = \frac{3}{2\alpha}$, or $x_2 = +\sqrt{\frac{3}{2\alpha}}$; $x_3 = -\sqrt{\frac{3}{2\alpha}}$.

11.31. The mass of the electron is 9.109×10^{-31} kg.

(a) Let us estimate the mass of a hydrogen atom by dividing its molar mass (99.98% of naturally occuring hydrogen is ^1H) by Avogadro's number: (0.0010079 kg mol^{-1}) / (6.022×10^{23} mol^{-1}) = 1.674×10^{-27} kg. This includes the mass of the electron, so the mass of the hydrogen nucleus (proton) alone is 1.674×10^{-27} kg – 9.109×10^{-31} kg = 1.673×10^{-27} kg. The reduced mass of a hydrogen atom is then

$$\frac{(1.673\times10^{-27}\text{ kg})(9.109\times10^{-31}\text{ kg})}{1.673\times10^{-27}\text{ kg} + 9.109\times10^{-31}\text{ kg}} = 9.104\times10^{-31}\text{ kg}.$$

(b) A deuterium atom is about twice the mass of a hydrogen atom, so let us use 2×(mass of H nucleus) for the mass of the deuterium nucleus: 3.346×10^{-27} kg. The reduced mass of a D atom is $\dfrac{(3.346\times10^{-27}\text{ kg})(9.109\times10^{-31}\text{ kg})}{3.346\times10^{-27}\text{ kg}+9.109\times10^{-31}\text{ kg}} = 9.107\times10^{-31}\text{ kg}$.

(c) Let us estimate the mass of carbon-12 by dividing the molar mass of carbon (which is 99% ^{12}C) by Avogadro's number: (0.012011 kg mol^{-1}) / (6.022×10^{23} mol^{-1}) = 1.995×10^{-26} kg.

A carbon atom has six electrons, so the mass of the carbon nucleus is 1.995×10^{-26} kg – 6 · 9.109×10^{-31} kg = 1.994×10^{-26} kg. The reduced mass of a C^{5+} carbon-12 atom is then

$$\frac{(1.994\times10^{-26}\text{ kg})(9.109\times10^{-31}\text{ kg})}{1.994\times10^{-26}\text{ kg}+9.109\times10^{-31}\text{ kg}} = 9.109\times10^{-31}\text{ kg}.$$

Note that as the mass of the nucleus increases, the reduced mass gets closer and closer to the mass of the electron. For the carbon atom, to four significant figures, the reduced mass equals the mass of the electron.

11.33. (a) First, we need to determine the reduced mass of CO, in kg units. The masses of carbon and oxygen are obtained by dividing their molar mass by Avogadro's number:

C: (0.012011 kg mol^{-1}) / (6.022×10^{23} mol^{-1}) = 1.995×10^{-26} kg

O: (0.0159994 kg mol^{-1}) / (6.022×10^{23} mol^{-1}) = 2.657×10^{-26} kg

The reduced mass of CO is then $\dfrac{(1.995\times10^{-26}\text{ kg})(2.657\times10^{-26}\text{ kg})}{1.995\times10^{-26}\text{ kg}+2.657\times10^{-26}\text{ kg}} = 1.139\times10^{-26}\text{ kg}$.

Now we can use equation 11.27 to determine the frequency of vibration:

$$\nu = \frac{1}{2\pi}\sqrt{\frac{1902\text{ N/m}}{1.139\times10^{-26}\text{ kg}}} = 6.503\times10^{13}\text{ s}^{-1}.$$

(b) ^{13}C: (0.01300 kg mol^{-1}) / (6.022×10^{23} mol^{-1}) = 2.159×10^{-26} kg

The reduced mass of ^{13}CO is then $\dfrac{(2.159\times10^{-26}\text{ kg})(2.657\times10^{-26}\text{ kg})}{2.159\times10^{-26}\text{ kg}+2.657\times10^{-26}\text{ kg}} = 1.191\times10^{-26}\text{ kg}$.

Now we can use equation 11.27 again to determine the frequency of vibration:

$$\nu = \frac{1}{2\pi}\sqrt{\frac{1902\text{ N/m}}{1.191\times10^{-26}\text{ kg}}} = 6.360\times10^{13}\text{ s}^{-1}.$$

11.35. (a) 99.98% of naturally occuring hydrogen is ^{1}H, so the mass of the ^{1}H atom can be estimated by dividing the molar mass of hydrogen by Avogadro's number:

^{1}H: (0.0010079 kg mol^{-1}) / (6.022×10^{23} mol^{-1}) = 1.674×10^{-27} kg

A deuterium atom is about twice the mass of a hydrogen atom, so let us use 2×(mass of ^{1}H) for the mass of the D: 3.347×10^{-27} kg. The reduced mass of HD is then $\dfrac{(3.347\times10^{-27}\text{ kg})(1.674\times10^{-27}\text{ kg})}{3.347\times10^{-27}\text{ kg}+1.674\times10^{-27}\text{ kg}}=1.116\times10^{-27}\text{ kg}$, and its vibrational frequency

$$\nu=\frac{1}{2\pi}\sqrt{\frac{575\text{ N/m}}{1.116\times10^{-27}\text{ kg}}}=1.14\times10^{14}\text{ s}^{-1}.$$

(b) If the D atom does not move, it is only the mass of the H atom that enters into the equation:

$\nu=\dfrac{1}{2\pi}\sqrt{\dfrac{575\text{ N/m}}{1.674\times10^{-27}\text{ kg}}}=9.33\times10^{13}\text{ s}^{-1}$. This is a significant difference (about 20%), much greater than in Example 11.10 because the masses of the two oscillating atoms are much more similar.

11.37. The normalization condition is $1=\int_0^{2\pi}\Psi^*\Psi\,d\phi=\int_0^{2\pi}\left(Ne^{im\phi}\right)^*\left(Ne^{im\phi}\right)d\phi$.

For Ψ_3: $1=\int_0^{2\pi}\left(Ne^{3i\phi}\right)^*\left(Ne^{3i\phi}\right)d\phi=N^2\int_0^{2\pi}e^{-3i\phi}\cdot e^{3i\phi}\,d\varphi=N^2\int_0^{2\pi}d\phi=N^2\,\phi\big|_0^{2\pi}=N^2(2\pi-0)=2\pi N^2$

This leads to $N^2=\dfrac{1}{2\pi}$ and therefore $N=\dfrac{1}{\sqrt{2\pi}}$ (only the positive root needs to be considered).

For Ψ_{13}: $1=\int_0^{2\pi}\left(Ne^{13i\phi}\right)^*\left(Ne^{13i\phi}\right)d\phi=N^2\int_0^{2\pi}e^{-13i\phi}\cdot e^{13i\phi}\,d\varphi=N^2\int_0^{2\pi}d\phi=N^2\,\phi\big|_0^{2\pi}=N^2(2\pi-0)=2\pi N^2$

Again, this leads to $N^2=\dfrac{1}{2\pi}$ and therefore $N=\dfrac{1}{\sqrt{2\pi}}$.

11.39. Nodes in the wavefunction are points where the probability $\Psi^*\Psi=0$:

$$\Psi^*\Psi=\left(\frac{1}{\sqrt{2\pi}}e^{im\phi}\right)^*\left(\frac{1}{\sqrt{2\pi}}e^{im\phi}\right)=\frac{1}{\sqrt{2\pi}}e^{-im\phi}\cdot\frac{1}{\sqrt{2\pi}}e^{im\phi}=\frac{1}{2\pi}$$

The probability does not depend on ϕ, and the rotational wavefunctions therefore have no nodes. (Note that Figure 11.9 only shows the *real part* of the wavefunction, which – unlike the probabilitiy function – does have nodes.)

11.41. We use equation 11.38 with the moment of inertia $I = mr^2$, where m is the mass of the proton (1.673×10^{-27} kg): $I = (1.673\times10^{-27}\text{ kg})(5.00\times10^{-11}\text{ m})^2 = 4.18\times10^{-48}\text{ kg m}^2$.

$$E_0 = \frac{0^2\hbar^2}{2I} = 0;\; E_1 = \frac{1^2\hbar^2}{2I} = \frac{1^2\left(6.626\times10^{-34}\text{ J}\cdot\text{s}\right)^2}{2\left(4.18\times10^{-48}\text{ kg}\cdot\text{m}^2\right)\left(2\pi\right)^2} = 1.33\times10^{-21}\text{ J};$$

$$E_2 = \frac{2^2\hbar^2}{2I} = \frac{2^2\left(6.626\times10^{-34}\text{ J}\cdot\text{s}\right)^2}{2\left(4.18\times10^{-48}\text{ kg}\cdot\text{m}^2\right)\left(2\pi\right)^2} = 5.32\times10^{-21}\text{ J}.$$

11.43. Using $e^{i\phi} = \cos\phi + i\sin\phi$, the rotational wavefunctions $\Psi_m = \frac{1}{\sqrt{2\pi}}e^{im\phi}$ can also be written as follows: $\Psi_0 = \sqrt{\frac{1}{2\pi}}$; $\Psi_1 = \sqrt{\frac{1}{2\pi}}(\cos\varphi + i\sin\varphi)$;

$$\Psi_2 = \sqrt{\frac{1}{2\pi}}\left[\cos(2\phi) + i\sin(2\phi)\right];\; \Psi_3 = \sqrt{\frac{1}{2\pi}}\left[\cos(3\phi) + i\sin(3\phi)\right].$$

11.45. We start with equation 11.33, $\frac{\hat{L}_z^{\,2}}{2I}\Psi = E\Psi$, or more explicitly $\frac{1}{2I}\hat{L}_z\left(\hat{L}_z\Psi\right) = E\Psi$. Now substitute $\hat{L}_z = -i\hbar\frac{\partial}{\partial\phi}$: $\frac{1}{2I}\left[-i\hbar\frac{\partial}{\partial\phi}\left(-i\hbar\frac{\partial\Psi}{\partial\phi}\right)\right] = E\Psi$. The derivative of the derivative is the second derivative, so $\frac{1}{2I}\left(+i^2\hbar^2\frac{\partial^2\Psi}{\partial\phi^2}\right) = E\Psi$ or $\frac{-\hbar^2}{2I}\frac{\partial^2\Psi}{\partial\phi^2} = E\Psi$. This is equation 11.35.

11.47. This is incorrect because the m in the numerator is the quantum number (with values 0, ±1, ±2, ±3 etc.), while the m in the expression for the moment of intertia $I = mr^2$ is the mass of the rotating particle. Therefore, an m in the numerator cannot be canceled by an m in the denominator (which would lead to the suggested expression).

11.49. From Figure 11.12, it is easy to see that the the length of the vertical dashed line is equal to the z coordinate. Now take a look at the right triangle formed by the dashed lines and the r vector (which is the hypotenuse of that triangle). The angle at the tip of the triangle (where the r vector ends) is equal to θ because the z axis and the vertical dashed line are parallel to each other. The definition of cosine of an angle is "adjacent side over hypotenuse", so we have $\cos\theta = \frac{z}{r}$, which rearranges to $z = r\cos\theta$, which is one of the equations that we set out to verify.

The definition of sine of an angle is "opposite side over hypotenuse", so the horizontal dashed line in the xy plane (the "opposite side") has a length of $r\sin\theta$.

The projection of this horizontal dashed line onto the x axis gives another right triangle, where the hypotenuse is the horizontal dashed line and the two legs are the x and y coordinates. The angle inside the triangle at the origin is ϕ, so we can again apply the definitions of sine and cosine to come up with expressions for x and y: The cosine is "adjacent side over hypotenuse", so $\cos\phi = \frac{x}{r\sin\theta}$ and therefore $x = r\sin\theta\cos\phi$; the sine is "opposite side over hypotenuse", so $\sin\phi = \frac{y}{r\sin\theta}$ and therefore $y = r\sin\theta\sin\phi$.

11.51. The Schrödinger equation with a constant potential energy V is given by $-\frac{\hbar^2}{2I}\left(\frac{\partial^2}{\partial\theta^2} + \cot\theta\frac{\partial}{\partial\theta} + \frac{1}{\sin^2\theta}\frac{\partial^2}{\partial\varphi^2}\right)\Psi + \hat{V}\Psi = E\Psi$. Because the potential energy operator is a multiplicative operator, we have $\hat{V}\Psi = V\cdot\Psi$ and the Schrödinger equation can be rearranged to:

$-\frac{\hbar^2}{2I}\left(\frac{\partial^2}{\partial\theta^2} + \cot\theta\frac{\partial}{\partial\theta} + \frac{1}{\sin^2\theta}\frac{\partial^2}{\partial\varphi^2}\right)\Psi = E\Psi - V\Psi = (E-V)\Psi = E_{\text{new}}\Psi$. So we could in fact use the same form of the Schrödinger equation if the potential energy were some constant value.

11.53. $\Psi_{1,0} = \frac{1}{\sqrt{2\pi}}e^{i\cdot 0\cdot\phi}\cdot\frac{1}{2}\sqrt{6}\cos\theta = \frac{1}{2}\sqrt{\frac{3}{\pi}}\cos\theta$; $\Psi_{1,1} = \frac{1}{\sqrt{2\pi}}e^{i\cdot 1\cdot\phi}\cdot\frac{1}{2}\sqrt{3}\sin\theta = \frac{1}{2}\sqrt{\frac{3}{2\pi}}e^{i\phi}\sin\theta$.

We need to evaluate $\int_{\theta=0}^{\pi}\int_{\phi=0}^{2\pi}\Psi_{1,0}{}^{*}\Psi_{1,1}\sin\theta\, d\phi\, d\theta = \int_{\theta=0}^{\pi}\int_{\phi=0}^{2\pi}\frac{1}{2}\sqrt{\frac{3}{\pi}}\cos\theta\cdot\frac{1}{2}\sqrt{\frac{3}{2\pi}}e^{i\phi}\sin^2\theta\, d\phi\, d\theta$.

The inner integral runs over $d\phi$, so all terms that only depend on θ can be treated as constants and factored out: $\frac{3}{4\pi}\frac{1}{\sqrt{2}}\int_{\theta=0}^{\pi}\cos\theta\sin^2\theta\left(\int_{\phi=0}^{2\pi}e^{i\phi}\,d\phi\right)d\theta$. Likewise, because the inner integral over $d\phi$ (highlighted with parentheses) is independent of θ, we can factor it out of the outer integral over $d\theta$ to obtain two separate integrals: $\frac{3}{4\pi}\frac{1}{\sqrt{2}}\int_{\phi=0}^{2\pi}e^{i\phi}\,d\phi\cdot\int_{\theta=0}^{\pi}\cos\theta\sin^2\theta\, d\theta$.

The first integral over ϕ yields $\int_{\phi=0}^{2\pi}e^{i\phi}\,d\phi = \frac{1}{i}e^{i\phi}\Big|_0^{2\pi} = \frac{1}{i}\left(e^{i\cdot 2\pi} - 1\right)$. We can use Euler's formula to evaluate the exponential: $e^{i\cdot 2\pi} = \cos(2\pi) + i\sin(2\pi) = 1 + 0 = 1$. Therefore, the integral is equal to zero, which makes the double integral equal to zero as well (and we don't even need to evaluate it): $\int_{\theta=0}^{\pi}\int_{\phi=0}^{2\pi}\Psi_{1,0}{}^{*}\Psi_{1,1}\sin\theta\, d\phi\, d\theta = 0$. We have thus confirmed that $\Psi_{1,0}$ and $\Psi_{1,1}$ are orthogonal.

11.55. **(a)** $E = \frac{\ell(\ell+1)\hbar^2}{2I} = \frac{(2)(2+1)(6.626\times10^{-34}\ \text{J}\cdot\text{s})^2}{2(4.445\times10^{-47}\ \text{kg}\cdot\text{m}^2)(2\pi)^2} = 7.506\times10^{-22}\ \text{J}$. This is the only possible value.

(b) $L^2 = \ell(\ell+1)\hbar^2 = (2)(2+1)(6.626\times10^{-34}\ \text{J}\cdot\text{s})^2/(2\pi)^2 = 6.673\times10^{-68}\ \text{J}^2\text{s}^2$. Therefore, the total angular momentum is the square root of this: $L = 2.583\times10^{-34}$ J·s. This is the only possible value.

(c) There are five possible values of the z component of the total angular momentum. They are $-2\hbar, -1\hbar, 0, 1\hbar$, and $2\hbar$. Their numerical values are 0, $\pm1.055\times10^{-34}$ J·s, and $\pm2.109\times10^{-34}$ J·s.

11.57. $E(\ell+1) - E(\ell) = \frac{(\ell+1)(\ell+1+1)\hbar^2}{2I} - \frac{\ell(\ell+1)\hbar^2}{2I} = \frac{2(\ell+1)\hbar^2}{2I} = \frac{(\ell+1)\hbar^2}{I}$. We can use I from Example 11.17: $E(6) - E(5) = \frac{(5+1)(6.626\times10^{-34}\ \text{J}\cdot\text{s})^2}{(1.12\times10^{-49}\ \text{kg}\cdot\text{m}^2)(2\pi)^2} = 5.96\times10^{-19}$ J.

Converting this to a wavelength of light using $\Delta E = h\nu = hc/\lambda$ gives:

$\lambda = \frac{hc}{E} = \frac{(3.00\times10^8\ \text{m/s})(6.626\times10^{-34}\ \text{J}\cdot\text{s})}{5.96\times10^{-19}\ \text{J}} = 3.34\times10^{-7}\ \text{m} = 334\ \text{nm}$. This value is only 6 nm off from the measured spectral feature at 328 nm.

For $E(8) - E(7)$: $E(8) - E(7) = \frac{(7+1)(6.626\times10^{-34}\ \text{J}\cdot\text{s})^2}{(1.12\times10^{-49}\ \text{kg}\cdot\text{m}^2)(2\pi)^2} = 7.94\times10^{-19}$ J. Converting this to a wavelength of light:

$\lambda = \frac{hc}{E} = \frac{(3.00\times10^8\ \text{m/s})(6.626\times10^{-34}\ \text{J}\cdot\text{s})}{7.94\times10^{-19}\ \text{J}} = 2.50\times10^{-7}\ \text{m} = 250\ \text{nm}$.

This is only 6 nm off from the measured spectral feature at 256 nm.

11.59. The drawings are left to the student (see Figure 11.15). The first four energy levels are given by ℓ = 0, 1, 2, 3. The degeneracies of each state are $2\ell + 1$, or 1, 3, 5, and 7, respectively, for the first four states.

11.61. The two wavefunctions are $\Psi_{3,2} = \frac{1}{\sqrt{2\pi}} e^{i\cdot2\phi} \cdot \frac{1}{4}\sqrt{105}\sin^2\theta\cos\theta$ and $\Psi_{3,-2} = \frac{1}{\sqrt{2\pi}} e^{i(-2)\phi} \cdot \frac{1}{4}\sqrt{105}\sin^2\theta\cos\theta$. The only difference is the change in sign in the exponential function. Using Euler's formula: $e^{2i\phi} = \cos(2\phi) + i\sin(2\phi)$, whereas $e^{-2i\phi} = \cos(-2\phi) + i\sin(-2\phi) = \cos(2\phi) - i\sin(2\phi)$. This means that the real parts of the two wavefunctions are identical, while the imaginary parts have opposite signs. Note that because the exponential function cancels out when taking $\Psi^*\Psi$, the probability functions $\Psi_{3,2}{}^*\Psi_{3,2}$ and $\Psi_{3,-2}{}^*\Psi_{3,-2}$ are exactly the same.

11.63. $V = -\dfrac{e^2}{4\pi\varepsilon_0 r} = -\dfrac{(1.602\times10^{-19}\ \text{C})^2}{4\pi\left(8.854\times10^{-12}\ \dfrac{\text{C}^2}{\text{J}\cdot\text{m}}\right)(0.529\times10^{-10}\ \text{m})} = -4.36\times10^{-18}\ \text{J}.$

11.65. Starting with equation 11.56:

$$\left\{-\frac{\hbar^2}{2\mu}\left[\frac{1}{r^2}\frac{\partial}{\partial r}\left(r^2\frac{\partial}{\partial r}\right)+\frac{1}{r^2\sin\theta}\frac{\partial}{\partial\theta}\left(\sin\theta\frac{\partial}{\partial\theta}\right)+\frac{1}{r^2\sin^2\theta}\frac{\partial^2}{\partial\phi^2}\right]+\hat{V}\right\}\Psi = E\Psi$$

If $V = 0$ and r is constant, we lose two things immediately: the potential energy operator and the first derivative term in r (since r is constant, any derivative with respect to r is zero). We have $-\dfrac{\hbar^2}{2\mu}\left[\dfrac{1}{r^2\sin\theta}\dfrac{\partial}{\partial\theta}\left(\sin\theta\dfrac{\partial}{\partial\theta}\right)+\dfrac{1}{r^2\sin^2\theta}\dfrac{\partial^2}{\partial\phi^2}\right]\Psi = E\Psi$. We can factor out the r^2 from the denominators in the two terms inside the square brackets, which combine with μ to form $\mu r^2 = I$: $-\dfrac{\hbar^2}{2I}\left[\dfrac{1}{\sin\theta}\dfrac{\partial}{\partial\theta}\left(\sin\theta\dfrac{\partial}{\partial\theta}\right)+\dfrac{1}{\sin^2\theta}\dfrac{\partial^2}{\partial\phi^2}\right]\Psi = E\Psi$.

Writing out the derivative with respect to θ leads to

$-\dfrac{\hbar^2}{2I}\left[\dfrac{1}{\sin\theta}\left(\cos\theta\dfrac{\partial}{\partial\theta}+\sin\theta\dfrac{\partial^2}{\partial\theta^2}\right)+\dfrac{1}{\sin^2\theta}\dfrac{\partial^2}{\partial\phi^2}\right]\Psi = E\Psi$, which we can simplify to

$-\dfrac{\hbar^2}{2I}\left(\dfrac{\cos\theta}{\sin\theta}\dfrac{\partial}{\partial\theta}+\dfrac{\partial^2}{\partial\theta^2}+\dfrac{1}{\sin^2\theta}\dfrac{\partial^2}{\partial\phi^2}\right)\Psi = E\Psi$, which is equation 11.46 (recall that $\cot\theta = \dfrac{\cos\theta}{\sin\theta}$).

11.67. **(a)** First, we must convert the wavelengths to wavenumber units by taking the reciprocal. To four significant figures:

$\dfrac{1}{656.5\ \text{nm}}\times\dfrac{10^7\ \text{nm}}{1\ \text{cm}} = 15{,}230\ \text{cm}^{-1}$ $\qquad$ $\dfrac{1}{486.3\ \text{nm}}\times\dfrac{10^7\ \text{nm}}{1\ \text{cm}} = 20{,}560\ \text{cm}^{-1}$

$\dfrac{1}{434.2\ \text{nm}}\times\dfrac{10^7\ \text{nm}}{1\ \text{cm}} = 23{,}030\ \text{cm}^{-1}$ $\qquad$ $\dfrac{1}{410.3\ \text{nm}}\times\dfrac{10^7\ \text{nm}}{1\ \text{cm}} = 24{,}370\ \text{cm}^{-1}$

These transitions are from quantum levels 3, 4, 5, and 6, respectively, to level 2. Therefore, we can calculate R as follows (again to four significant figures):

$15{,}230\ \text{cm}^{-1} = R\left(\dfrac{1}{2^2}-\dfrac{1}{3^2}\right);\quad R = 109{,}700\ \text{cm}^{-1};$

$20{,}560\ \text{cm}^{-1} = R\left(\dfrac{1}{2^2}-\dfrac{1}{4^2}\right);\quad R = 109{,}700\ \text{cm}^{-1};$

$23{,}030\ \text{cm}^{-1} = R\left(\dfrac{1}{2^2}-\dfrac{1}{5^2}\right);\quad R = 109{,}700\ \text{cm}^{-1};$

$$24{,}370 \text{ cm}^{-1} = R\left(\frac{1}{2^2} - \frac{1}{6^2}\right); \quad R = 109{,}700 \text{ cm}^{-1}.$$

Thus, to four significant figures, all four lines give the same value of the Rydberg constant.

(b) Since $Z = 2$ for helium, there is an extra Z^2 term in the numerator of the Rydberg constant for He^+, so R would equal $2^2 \times 109{,}700$ cm^{-1} = 438,800 cm^{-1}. Therefore, we would expect the lines to appear at (to four significant figures):

$$438{,}800 \text{ cm}^{-1}\left(\frac{1}{2^2} - \frac{1}{3^2}\right) = 60{,}940 \text{ cm}^{-1} \text{ or } 164.1 \text{ nm};$$

$$438{,}800 \text{ cm}^{-1}\left(\frac{1}{2^2} - \frac{1}{4^2}\right) = 82{,}280 \text{ cm}^{-1} \text{ or } 121.5 \text{ nm};$$

$$438{,}800 \text{ cm}^{-1}\left(\frac{1}{2^2} - \frac{1}{5^2}\right) = 92{,}150 \text{ cm}^{-1} \text{ or } 108.5 \text{ nm};$$

$$438{,}800 \text{ cm}^{-1}\left(\frac{1}{2^2} - \frac{1}{6^2}\right) = 97{,}510 \text{ cm}^{-1} \text{ or } 102.6 \text{ nm}.$$

11.69. Figure 11.17 in the text gives such a diagram.

11.71. The letter "h" labels the $\ell = 5$ subshell, the letter "n" the $\ell = 11$ subshell. Therefore, the degeneracies are $2 \cdot 5 + 1 = 11$ and $2 \cdot 11 + 1 = 23$, respectively.

11.73. We use $E = -\dfrac{Z^2 e^4 \mu}{8\varepsilon_0^{\,2} h^2 n^2}$ with the mass of the electron in place of the reduced mass; Z for fluorine is 9. The values for L and L_z are $\sqrt{\ell(\ell+1)}\hbar$ and $m_\ell \hbar$, respectively.

(a) $$E = -\frac{9^2 \cdot \left(1.602\times10^{-19} \text{ C}\right)^4 \left(9.109\times10^{-31} \text{ kg}\right)}{8\left(8.854\times10^{-12}\dfrac{\text{C}^2}{\text{J}\cdot\text{m}}\right)^2 \left(6.626\times10^{-34} \text{ J}\cdot\text{s}\right)^2 \cdot 1^2} = -1.765\times10^{-16} \text{ J}, L = 0,$$

and $L_z = 0$.

(b) $$E = -\frac{9^2 \cdot \left(1.602\times10^{-19} \text{ C}\right)^4 \left(9.109\times10^{-31} \text{ kg}\right)}{8\left(8.854\times10^{-12}\dfrac{\text{C}^2}{\text{J}\cdot\text{m}}\right)^2 \left(6.626\times10^{-34} \text{ J}\cdot\text{s}\right)^2 \cdot 3^2} = -1.961\times10^{-17} \text{ J},$$

$L = \sqrt{2(2+1)}\hbar = \sqrt{6}\hbar$, and $L_z = 2\hbar$.

(c) $E = -\dfrac{9^2 \cdot \left(1.602 \times 10^{-19}\ \mathrm{C}\right)^4 \left(9.109 \times 10^{-31}\ \mathrm{kg}\right)}{8\left(8.854 \times 10^{-12} \dfrac{\mathrm{C}^2}{\mathrm{J \cdot m}}\right)^2 \left(6.626 \times 10^{-34}\ \mathrm{J \cdot s}\right)^2 \cdot 2^2} = -4.412 \times 10^{-17}\ \mathrm{J}$,

$L = \sqrt{1(1+1)}\,\hbar = \sqrt{2}\,\hbar$, and $L_z = -\hbar$.

(d) $E = -\dfrac{9^2 \cdot \left(1.602 \times 10^{-19}\ \mathrm{C}\right)^4 \left(9.109 \times 10^{-31}\ \mathrm{kg}\right)}{8\left(8.854 \times 10^{-12} \dfrac{\mathrm{C}^2}{\mathrm{J \cdot m}}\right)^2 \left(6.626 \times 10^{-34}\ \mathrm{J \cdot s}\right)^2 \cdot 9^2} = -2.179 \times 10^{-18}\ \mathrm{J}$,

$L = \sqrt{6(6+1)}\,\hbar = \sqrt{42}\,\hbar$, and $L_z = -3\hbar$.

11.75. $\Psi_{4,4,0}$ would indicate $n = 4$, $\ell = 4$, $m_\ell = 0$. However, the quantum number ℓ can only take on integer values ℓ that are smaller than n (0, 1, 2, or 3); thus, $\ell = 4$ is not an allowed value.

A 3*f* subshell has $n = 3$, $\ell = 3$, but only ℓ values of 0, 1, or 2 are allowed.

11.77. We need to calculate $P = \int\limits_{\theta=0}^{\pi} \int\limits_{\phi=0}^{2\pi} \int\limits_{r=0}^{0.1\,\text{Å}} \left(\dfrac{Z^3}{\pi a^3}\right)^{1/2} e^{-Zr/a} \cdot \left(\dfrac{Z^3}{\pi a^3}\right)^{1/2} e^{-Zr/a} \cdot r^2 \sin\theta \, dr\, d\phi\, d\theta$, where the value of a is 0.529 Å; also, for hydrogen, we can substitute $Z = 1$. All the terms in the integral are either constant, or are functions of only one of the integration variables, so that by factorizing we obtain a product of three integrals:

$$P = \left(\frac{1}{\pi a^3}\right) \int\limits_{\theta=0}^{\pi} \sin\theta\, d\theta \cdot \int\limits_{\phi=0}^{2\pi} d\phi \cdot \int\limits_{r=0}^{0.1\,\text{Å}} r^2 e^{-2r/a}\, dr\,.$$

Because

$$\int\limits_{\theta=0}^{\pi} \sin\theta\, d\theta \cdot \int\limits_{\phi=0}^{2\pi} d\phi = \left[-\cos\phi\right]_0^{\pi} \cdot \left[\phi\right]_0^{2\pi} = (-\cos\pi + \cos 0)(2\pi - 0) = \left[-(-1)+1\right](2\pi) = 4\pi,$$

we obtain $P = \dfrac{1}{\pi a^3} \cdot 4\pi \cdot \int\limits_{r=0}^{0.1\,\text{Å}} r^2 e^{-2r/a}\, dr = \dfrac{4}{a^3} \int\limits_{r=0}^{0.1\,\text{Å}} r^2 e^{-2r/a}\, dr$.

Using the integral table in Appendix 1, we can solve the integral and evaluate:

$$\int\limits_{r=0}^{0.1\,\text{Å}} r^2 e^{-2r/a}\, dr = \left[e^{-2r/a}\left(\frac{r^2}{-2/a} - \frac{2r}{(-2/a)^2} + \frac{2}{(-2/a)^3}\right)\right]_0^{0.1\,\text{Å}} = \left[e^{-2r/a}\left(-\frac{r^2 a}{2} - \frac{ra^2}{2} - \frac{a^3}{4}\right)\right]_0^{0.1\,\text{Å}}$$

Evaluating this expression between the limits 0 and 0.1 Å gives

$$e^{-\frac{2\cdot 0.1\,\text{Å}}{0.529\,\text{Å}}}\left(-\frac{(0.1\,\text{Å})^2(0.529\,\text{Å})}{2}-\frac{(0.1\,\text{Å})(0.529\,\text{Å})^2}{2}-\frac{(0.529\,\text{Å})^3}{4}\right)-e^0\left(0-0-\frac{(0.529\,\text{Å})^3}{4}\right)$$

$= 2.5\times10^{-4}$ Å^3, and substituting this value into the expression for P results in:

$$P=\frac{4}{(0.529\,\text{Å})^3}\cdot\left(-2.5\times10^{-4}\,\text{Å}^3\right)=0.0068=0.68\%$$

11.79. Comparing the given energy to equation 11.66 leads us to the equation $\frac{Z^2}{8n^2}=\frac{9}{8}$ (remember that the quantum number n and the atomic number Z are integers). We rearrange this to $\frac{Z^2}{n^2}=9$, which gives $Z/n=3$. The first three systems which satisy this condition are:

If $n = 1$, then $Z = 3$. This corresponds to the Ψ_{1s} wavefunction of a Li^{2+} atom.

If $n = 2$, then $Z = 6$. This corresponds to the Ψ_{2s} and Ψ_{2p} wavefunctions of a C^{5+} atom.

If $n = 3$, then $Z = 18$. This corresponds to the Ψ_{3s}, Ψ_{3p}, or Ψ_{3d} wavefunctions of a Ar^{17+} atom.

11.81. Remember that for quantum numbers n and ℓ, there will be $n - \ell - 1$ radial nodes and ℓ angular nodes, or a total of $(n - \ell - 1) + \ell = n - 1$ nodes.

(a) For Ψ_{2s} ($n = 2$, $\ell = 0$): one radial node and no angular nodes, for a total of one node.

(b) For Ψ_{3s} ($n = 3$, $\ell = 0$): two radial nodes and no angular nodes, for a total of two nodes.

(c) For Ψ_{3p} ($n = 3$, $\ell = 1$): one radial node and one angular node, for a total of two nodes.

(d) For Ψ_{4f} ($n = 4$, $\ell = 3$): no radial nodes and three angular nodes, for a total of three nodes.

(e) For Ψ_{6g} ($n = 6$, $\ell = 4$): one radial node and four angular nodes, for a total of five nodes.

(f) For Ψ_{7s} ($n = 7$, $\ell = 0$): six radial nodes and no angular nodes, for a total of six nodes

11.83. We must evaluate $\int\limits_{\theta=0}^{\pi}\int\limits_{\phi=0}^{2\pi}\int\limits_{r=0}^{\infty}\left(\frac{1}{\pi a^3}\right)^{1/2}e^{-r/a}\cdot\frac{1}{8}\left(\frac{2}{\pi a^3}\right)^{1/2}\left(2-\frac{r}{a}\right)e^{-r/(2a)}\cdot r^2\sin\theta\, dr\, d\phi\, d\theta$.

All the terms in the integral are either constant, or are functions of only one of the integration variables, so that by factorizing we obtain a product of three integrals:

$$\frac{\sqrt{2}}{8\pi a^3}\int\limits_{\theta=0}^{\pi}\sin\theta\, d\theta\cdot\int\limits_{\phi=0}^{2\pi}d\phi\cdot\int\limits_{r=0}^{\infty}\left(2-\frac{r}{a}\right)r^2e^{-r/a}e^{-r/(2a)}\,dr\ .$$

We know that the angular integrals evaluate to 4π (see exercise 11.77), so we can simplify:

$\frac{\sqrt{2}}{2a^3}\int_0^\infty\left(2-\frac{r}{a}\right)r^2e^{-3r/(2a)}\,dr=\frac{\sqrt{2}}{2a^3}\left[2\int_0^\infty r^2e^{-3r/(2a)}\,dr-\frac{1}{a}\int_0^\infty r^3e^{-3r/(2a)}\,dr\right]$. Both of these definite integrals have a known form (the same general form, actually) that can be found in Appendix 1. Therefore, we substitute their solutions: $\frac{\sqrt{2}}{2a^3}\left[2\left(\frac{2!}{(3/2a)^3}\right)-\frac{1}{a}\left(\frac{3!}{(3/2a)^4}\right)\right]$ and simplify:

$$\frac{\sqrt{2}}{2a^3}\left[2\left(\frac{2\cdot 8a^3}{27}\right)-\frac{1}{a}\left(\frac{6\cdot 16a^4}{81}\right)\right]=\frac{\sqrt{2}}{2a^3}\left[\frac{32a^3}{27}-\frac{32a^3}{27}\right]=\frac{\sqrt{2}}{2a^3}\cdot 0=0.$$

The wavefunctions are indeed orthogonal.

11.85. $4\pi r^2\Psi^2$ for Ψ_{1s} is $4\pi r^2\left(\frac{1}{\pi a^3}\right)e^{-2r/a}=\frac{4r^2}{a^3}e^{-2r/a}$. The derivative of this expression with respect to r is $\frac{8r}{a^3}e^{-2r/a}+\frac{4r^2}{a^3}e^{-2r/a}\cdot\left(-\frac{2}{a}\right)=\frac{8r}{a^3}e^{-2r/a}-\frac{8r^2}{a^4}e^{-2r/a}=\frac{8r}{a^3}e^{-2r/a}\left(1-\frac{r}{a}\right)$. Setting this equal to 0: $\frac{8r}{a^3}e^{-2r/a}\left(1-\frac{r}{a}\right)=0$. The exponential is always greater than zero, so we can multiply both sides by $\frac{a^3}{8e^{-2r/a}}$ to simplify this condition to: $r\left(1-\frac{r}{a}\right)=0$. Both $r=0$ and $1-r/a=0$ satisfy this condition; however, $r=0$ would mean $4\pi r^2\Psi^2=0$, so this represents a minimum of probability. Thus, the maximum value of $4\pi r^2\Psi^2$ corresponds to a value of r where r equals a, the first Bohr radius – which was what we set out to show.

11.87. We will use $\Psi_{3p_x}=N\left(\Psi_{3p_{+1}}+\Psi_{3p_{-1}}\right)$, where N represents the normalization constant.

$$\int\left[N\left(\Psi_{3p_{+1}}+\Psi_{3p_{-1}}\right)\right]^*\hat{L}_z\left[N\left(\Psi_{3p_{+1}}+\Psi_{3p_{-1}}\right)\right]d\tau=N^2\int\left(\Psi_{3p_{+1}}+\Psi_{3p_{-1}}\right)^*\left(\hat{L}_z\Psi_{3p_{+1}}+\hat{L}_z\Psi_{3p_{-1}}\right)d\tau$$

Using the L_z eigenvalues: $N^2\int\left(\Psi_{3p_{+1}}+\Psi_{3p_{-1}}\right)^*\left[(+1\cdot\hbar)\Psi_{3p_{+1}}+(-1\cdot\hbar)\Psi_{3p_{-1}}\right]d\tau$. Now we can expand this into a sum of four terms:

$$N^2\int\left(\Psi_{3p_{+1}}+\Psi_{3p_{-1}}\right)^*\left[(+1\cdot\hbar)\Psi_{3p_{+1}}+(-1\cdot\hbar)\Psi_{3p_{-1}}\right]d\tau$$

$=N^2\int\left(\hbar\Psi^*_{3p_{+1}}\Psi_{3p_{+1}}+\hbar\Psi^*_{3p_{-1}}\Psi_{3p_{+1}}-\hbar\Psi^*_{3p_{+1}}\Psi_{3p_{-1}}-\hbar\Psi^*_{3p_{-1}}\Psi_{3p_{-1}}\right)d\tau$, which we can separate into four integrals:

$$N^2\hbar\int\Psi^*_{3p_{+1}}\Psi_{3p_{+1}}\,d\tau+N^2\hbar\int\Psi^*_{3p_{-1}}\Psi_{3p_{+1}}\,d\tau-N^2\hbar\int\Psi^*_{3p_{+1}}\Psi_{3p_{-1}}\,d\tau-N^2\hbar\int\Psi^*_{3p_{-1}}\Psi_{3p_{-1}}\,d\tau$$

The first and last integrals equal exactly 1, because the $3p_{+1}$ and $3p_{-1}$ wavefunctions are normalized. The second and third integrals are exactly 0, because the wavefunctions are orthogonal. Therefore, we have $\langle L_z \rangle = N^2\hbar + 0 - 0 - N^2\hbar = 0$.

The reason Ψ_{3p_x} equals zero is that it is a combination of two eigenfunctions ($\Psi_{3p_{+1}}$ and $\Psi_{3p_{-1}}$) of the L_z operator that have opposite eigenvalues.

11.89. Using Ψ_{1s} from Table 11.4 with the Schrödinger equation (equation 11.66):

$$\left\{-\frac{\hbar^2}{2\mu}\left[\frac{1}{r^2}\frac{\partial}{\partial r}\left(r^2\frac{\partial}{\partial r}\right)+\frac{1}{r^2\sin\theta}\frac{\partial}{\partial\theta}\left(\sin\theta\frac{\partial}{\partial\theta}\right)+\frac{1}{r^2\sin^2\theta}\frac{\partial^2}{\partial\phi^2}\right]+\frac{-Ze^2}{4\pi\varepsilon_0 r}\right\}\left(\frac{Z^3}{\pi a^3}\right)^{1/2}e^{-Zr/a}$$

Ψ_{1s} only depends on r, so all derivatives with respect to θ and ϕ are zero:

$\left\{-\frac{\hbar^2}{2\mu}\cdot\frac{1}{r^2}\frac{\partial}{\partial r}\left(r^2\frac{\partial}{\partial r}\right)-\frac{Ze^2}{4\pi\varepsilon_0 r}\right\}\left(\frac{Z^3}{\pi a^3}\right)^{1/2}e^{-Zr/a}$. The 2nd term is simply $-\frac{Ze^2}{4\pi\varepsilon_0 r}\left(\frac{Z^3}{\pi a^3}\right)^{1/2}e^{-Zr/a}$, but for the first term we need to evaluate the first and second derivatives:

$\frac{\partial}{\partial r}\left[\left(\frac{Z^3}{\pi a^3}\right)^{1/2}e^{-Zr/a}\right]=\left(\frac{Z^3}{\pi a^3}\right)^{1/2}e^{-Zr/a}\cdot\left(-\frac{Z}{a}\right)=-\frac{Z}{a}\left(\frac{Z^3}{\pi a^3}\right)^{1/2}e^{-Zr/a}$, so that

$$\frac{1}{r^2}\frac{\partial}{\partial r}\left[r^2\cdot\left(-\frac{Z}{a}\right)\left(\frac{Z^3}{\pi a^3}\right)^{1/2}e^{-Zr/a}\right]=\frac{1}{r^2}\left(-\frac{Z}{a}\right)\left(\frac{Z^3}{\pi a^3}\right)^{1/2}\left(2r\cdot e^{-Zr/a}+r^2e^{-Zr/a}\cdot\left(-\frac{Z}{a}\right)\right).$$

Factoring out the orginal wavefunction and rearranging leads to:

$$\frac{Z}{a}\left(\frac{Z}{a}-\frac{2}{r}\right)\left(\frac{Z^3}{\pi a^3}\right)^{1/2}e^{-Zr/a}.$$

Putting this all together, we obtain $-\frac{\hbar^2}{2\mu}\frac{Z}{a}\left(\frac{Z}{a}-\frac{2}{r}\right)\left(\frac{Z^3}{\pi a^3}\right)^{1/2}e^{-Zr/a}-\frac{Ze^2}{4\pi\varepsilon_0 r}\left(\frac{Z^3}{\pi a^3}\right)^{1/2}e^{-Zr/a}$,

where we can again factor out the original wavefunction:

$\left[-\frac{\hbar^2}{2\mu}\frac{Z}{a}\left(\frac{Z}{a}-\frac{2}{r}\right)-\frac{Ze^2}{4\pi\varepsilon_0 r}\right]\left(\frac{Z^3}{\pi a^3}\right)^{1/2}e^{-Zr/a}$. The expression in square brackets would be the eigenvalue, but we need to show that it is indeed a constant (we still have r that appears in there). We do this by substituting $a=\frac{4\pi\varepsilon_0\hbar^2}{\mu e^2}$:

$$E=-\frac{\hbar^2}{2\mu}\frac{Z\mu e^2}{4\pi\varepsilon_0\hbar^2}\left(\frac{Z\mu e^2}{4\pi\varepsilon_0\hbar^2}-\frac{2}{r}\right)-\frac{Ze^2}{4\pi\varepsilon_0 r}=-\frac{Ze^2}{8\pi\varepsilon_0}\cdot\frac{Z\mu e^2}{4\pi\varepsilon_0\hbar^2}+\frac{Ze^2}{8\pi\varepsilon_0}\cdot\frac{2}{r}-\frac{Ze^2}{4\pi\varepsilon_0 r}.$$

The last two terms cancel, so that we finally obtain $E = -\dfrac{Z^2 \mu e^4}{32\pi^2 \varepsilon_0{}^2 \hbar^2}$, or using

$\hbar = h/(2\pi)$: $E = -\dfrac{Z^2 \mu e^4}{8\varepsilon_0{}^2 h^2}$.

This is indeed equation 11.66 for $n = 1$.

11.91. $\langle r \rangle = \int_0^\infty \left[\left(\dfrac{1}{\pi a^3}\right)^{1/2} e^{-r/a}\right]^* \hat{r} \left[\left(\dfrac{1}{\pi a^3}\right)^{1/2} e^{-r/a}\right] 4\pi r^2 \, dr$, which simplifies to $\dfrac{4}{a^3}\int_0^\infty r^3 e^{-2r/a} dr$.

This definite integral has a known form (see Appendix 1). We evaluate:

$\langle r \rangle = \dfrac{4}{a^3}\left[\dfrac{3!}{(2/a)^4}\right] = \dfrac{4}{a^3} \cdot \dfrac{6a^4}{16} = \dfrac{6a^4}{4a^3} = \dfrac{3}{2}a$. This is the *average* value of r, which is different from the *most probable* value of r, which is $r_{max} = a = 0.529$ Å.

CHAPTER 12

Atoms and Molecules

12.1. Silver atoms have a single unpaired electron in their valence shell, the 5*s* orbital (see Table 12.1). Thus, there is an unbalanced overall spin to the atoms, making them susceptible to magnetic fields.

12.3. Referring to Table 12.1, we see that H, Li, B, Na, Al, K, Sc, Cu, Ga, Rb, Y, In, Cs, La, Lu, Au, Tl, Fr, Ac, Lr all have a single electron in a subshell with all other subshells filled, so a beam of these elements would also split into two parts in a magnetic field.

12.5. Use Table 11.4 and multiply with the spin wavefunction α or β:

$$\Psi_{3d_{-2}} = \frac{1}{162}\left(\frac{Z^3}{\pi a^3}\right)^{1/2}\frac{Z^2r^2}{a^2}e^{-Zr/(3a)}\sin^2\theta e^{-2i\varphi}\alpha\text{, or}$$

$$\Psi_{3d_{-2}} = \frac{1}{162}\left(\frac{Z^3}{\pi a^3}\right)^{1/2}\frac{Z^2r^2}{a^2}e^{-Zr/(3a)}\sin^2\theta e^{-2i\varphi}\beta.$$

12.7. Without spin, the degeneracy of each shell with principal quantum number n is n^2. When the spin is accounted for, the degeneracy is therefore $2n^2$.

12.9. **(a)** The quantum number s represents the total spin angular momentum quantum number. For all electrons, s is ½ (and is a characteristic of the particle). The quantum number m_s represents the z component of the total spin angular momentum. It has values of $-s$ to $+s$ in steps of 1, so for electrons with s = ½, it can be either – ½ or +½.

(b) If a particle has $s = 0$, then m_s can only equal 0. If a particle has $s = 2$, then m_s can be –2, -1, 0, +1, or +2. If a particle has $s = 3/2$, then m_s can equal –3/2, –1/2, +1/2, or +3/2.

12.11. The drawing is left to the student. The diagram should have two cones, with lengths of $S=\sqrt{\frac{1}{2}\left(\frac{1}{2}+1\right)}\cdot\hbar=\sqrt{3}\cdot\hbar/2$ and z components of $S_z=\pm1\cdot\hbar/2$.

12.13. With the He nucleus at the origin, we need to specify r_1, θ_1, and ϕ_1 to define the position of electron 1 and r_2, θ_2, and ϕ_2 to define the position of electron 2.

12.15. Following Example 12.3, we use $E=-\dfrac{Z^2e^4\mu}{8\varepsilon_0^2h^2n^2}$ for each electron. We will also use the electron mass as an approximation for the reduced mass of the electron–Li nucleus system.

(a) $$E_1=-\frac{(3)^2\left(1.602\times10^{-19}\text{ J}\right)^4\left(9.109\times10^{-31}\text{ kg}\right)}{8\left(8.854\times10^{-12}\ \dfrac{\text{C}^2}{\text{J}\cdot\text{m}}\right)^2\left(6.626\times10^{-34}\text{ J}\cdot\text{s}\right)^2(1)^2}=-1.961\times10^{-17}\text{ J, so}$$

assuming that all electrons have $n = 1$:

$E = 3\times E_1 = 3\times(-1.961\times10^{-17}\ \text{J}) = -5.883\times10^{-17}\ \text{J}.$

(b) $E_2 = -\dfrac{(3)^2\left(1.602\times10^{-19}\ \text{J}\right)^4\left(9.109\times10^{-31}\ \text{kg}\right)}{8\left(8.854\times10^{-12}\ \dfrac{\text{C}^2}{\text{J}\cdot\text{m}}\right)^2\left(6.626\times10^{-34}\ \text{J}\cdot\text{s}\right)^2(2)^2} = -4.903\times10^{-18}\ \text{J}$

If two electrons have $n = 1$ and the third has $n = 2$:

$E = 2\times E_1 + E_2 = 2\times(-1.961\times10^{-17}\ \text{J}) + (-4.903\times10^{-18}\ \text{J}) = -4.412\times10^{-17}\ \text{J}.$

(c) The second energy value, assuming that one electron has $n = 2$, comes closer to the experimental value for the electronic energy of lithium, but is still off by one third. Knowing that according to the Pauli exclusion principle, only two electrons can occupy the same orbital, we would indeed expect that the second estimate leads to a more realistic value.

12.17. **(a)** symmetric **(b)** antisymmetric **(c)** symmetric **(d)** symmetric **(e)** antisymmetric.

12.19. We first exchange the two electrons, meaning that we swap the indices 1 and 2:

$\Psi_{\text{He},1} = \frac{1}{\sqrt{2}}\left[(1s_2\alpha)(1s_1\beta)+(1s_2\beta)(1s_1\alpha)\right]$. Rearrange the order of the two terms inside the square brackets: $\Psi_{\text{He},1} = \frac{1}{\sqrt{2}}\left[(1s_2\beta)(1s_1\alpha)+(1s_2\alpha)(1s_1\beta)\right]$. Finally, we reorder the two factors in each term: $\Psi_{\text{He},1} = \frac{1}{\sqrt{2}}\left[(1s_1\alpha)(1s_2\beta)+(1s_1\beta)(1s_2\alpha)\right]$. This is the orginal equation, so the equation is symmetric.

12.21. Li^+ has two electrons, so the Slater determinant for the ground state is the same as for He:

$$\Psi_{Li^+} = \frac{1}{\sqrt{2}}\begin{vmatrix} 1s_1\alpha & 1s_1\beta \\ 1s_2\alpha & 1s_2\beta \end{vmatrix} = \frac{1}{\sqrt{2}}\left[(1s_1\alpha)(1s_2\beta)-(1s_2\alpha)(1s_1\beta)\right].$$

12.23. The antisymmetry issue doesn't have to be considered for H atoms because H atoms only have one electron. This does not imply, however, that the electrons in H atoms don't have spin!

12.25. $$\Psi_{He^*} = \frac{1}{\sqrt{2}}\begin{vmatrix} 1s_1\alpha & 2s_1\alpha \\ 1s_2\alpha & 2s_2\alpha \end{vmatrix} = \frac{1}{\sqrt{2}}\left[(1s_1\alpha)(2s_2\alpha)-(1s_2\alpha)(2s_1\alpha)\right];$$

$$\Psi_{He^*} = \frac{1}{\sqrt{2}}\begin{vmatrix} 1s_1\beta & 2s_1\alpha \\ 1s_2\beta & 2s_2\alpha \end{vmatrix} = \frac{1}{\sqrt{2}}\left[(1s_1\beta)(2s_2\alpha)-(1s_2\beta)(2s_1\alpha)\right];$$

$$\Psi_{He^*} = \frac{1}{\sqrt{2}} \begin{vmatrix} 1s_1\alpha & 2s_1\beta \\ 1s_2\alpha & 2s_2\beta \end{vmatrix} = \frac{1}{\sqrt{2}}\left[(1s_1\alpha)(2s_2\beta) - (1s_2\alpha)(2s_1\beta)\right];$$

$$\Psi_{He^*} = \frac{1}{\sqrt{2}} \begin{vmatrix} 1s_1\beta & 2s_1\beta \\ 1s_2\beta & 2s_2\beta \end{vmatrix} = \frac{1}{\sqrt{2}}\left[(1s_1\beta)(2s_2\beta) - (1s_2\beta)(2s_1\beta)\right].$$

12.27. (a) The proper antisymmetric wavefunction for He is

$\Psi(1,2) = \frac{1}{\sqrt{2}}\left[(1s_1\alpha)(1s_2\beta) - (1s_1\beta)(1s_2\alpha)\right]$, which has 2 terms. Because 2! = 2 · 1 = 2, the relationship between $n!$ and the number of terms in the wavefunction is confirmed for He.

The wavefunction given for Li in the text is $\Psi_{Li} = \frac{1}{\sqrt{6}} \begin{vmatrix} 1s_1\alpha & 1s_1\beta & 2s_1\alpha \\ 1s_2\alpha & 1s_2\beta & 2s_2\alpha \\ 1s_3\alpha & 1s_3\beta & 2s_3\alpha \end{vmatrix}$.

This 3×3 determinant can be expanded as follows:

$$\Psi_{Li} = \frac{1}{\sqrt{6}}\left[(1s_1\alpha)\{(1s_2\beta)(2s_3\alpha) - (1s_3\beta)(2s_2\alpha)\} + (1s_1\beta)\{(2s_2\alpha)(1s_3\alpha) - (2s_3\alpha)(1s_2\alpha)\}\right.$$

$\left. + (2s_1\alpha)\{(1s_2\alpha)(1s_3\beta) - (1s_3\alpha)(1s_2\beta)\}\right]$, which we can multiply out to get

$$\Psi_{Li} = \frac{1}{\sqrt{6}}\left[(1s_1\alpha)(1s_2\beta)(2s_3\alpha) - (1s_1\alpha)(1s_3\beta)(2s_2\alpha) + (1s_1\beta)(2s_2\alpha)(1s_3\alpha) - (1s_1\beta)(2s_3\alpha)(1s_2\alpha)\right.$$

$$\left. + (2s_1\alpha)(1s_2\alpha)(1s_3\beta) - (2s_1\alpha)(1s_3\alpha)(1s_2\beta)\right].$$

This expression cannot be simplified further. The wavefunction therefore has 6 terms, and since 3! = 3 · 2 · 1 = 6, the relationship is confirmed for Li as well.

Example 12.6 states that the expansion of the given Slater determinant for Be has 24 terms. Rather than list them all, we simply point out that 4! = 4 · 3 · 2 · 1 = 24, thus confirming the relationship for Be as well.

(b) Since the number of terms in the wavefunction is $n!$, where n is the number of electrons in the atom, the number of terms in the C wavefunction will be 6! = 720; the number of terms in the Na wavefunction will be 11! = 39,916,800; the number of terms in the Si wavefunction will be 14! = 87,178,291,200, and the number of terms in the P wavefunction will be 15! = 1,307,674,368,000.

12.29. Atoms that don't follow the aufbau principle exactly are marked with an asterisk in Table 12.1. They include Cr, Cu, Nb, Mo, Ru, Rh, Pd, Ag, La, Ce, Gd, Pt, Au, Ac, Th, Pa, U, Np, and Cm. None of these atoms are main group elements; they are transition metals or inner transition (i.e. lanthanide or actinide) elements.

12.31. **(a)** excited state **(b)** ground state **(c)** excited state **(d)** excited state **(e)** excited state.

(Note that the electron configuration may still be an excited state if the m_ℓ or m_s quantum numbers don't conform to the lowest-energy wavefunction, but we need more information to make that determination.)

12.33. The correction to the energy won't be an exact correction even if the integral can be solved analytically because the wavefunctions in the integral are for an ideal system, not the real system. This can be seen in equation 12.16, which forms the basis for perturbation theory: the equation is an approximation because the wavefunctions of the ideal system are used, but would be rigorously valid if the wavefunctions of the real system were used.

12.35. The integral needing to be evaluated would be $\int_{-\infty}^{+\infty}\left[\left(\frac{\alpha}{\pi}\right)^{1/4} e^{-\alpha x^2/2}\right]^* \cdot cx^3 \cdot \left(\frac{\alpha}{\pi}\right)^{1/4} e^{-\alpha x^2/2}\,dx$.

This simplifies to $\left(\frac{\alpha}{\pi}\right)^{1/2} c\int_{-\infty}^{+\infty} x^3 e^{-\alpha x^2}\,dx$. If we are trying to evaluate this integral explicitly, we are somewhat out of luck because Appendix 1 does not list any formula that would apply in this case. However, we can substitute $t = x^2$ and $dt = 2x\,dx$ so that the integral becomes $\int x^3 e^{-\alpha x^2}\,dx = \int x t e^{-\alpha t}\frac{dt}{2x} = \frac{1}{2}\int t e^{-\alpha t}\,dt$. This integral is listed in Appendix 1: $\frac{1}{2}\int t e^{-\alpha t}\,dt = \frac{1}{2}\left[e^{-\alpha t}\cdot\frac{1}{(-\alpha)^2}(-\alpha t - 1)\right] = -\frac{\alpha t + 1}{2\alpha^2 e^{+\alpha t}}$. We resubstitute $t = x^2$ and evaluate the limits: $\frac{1}{2}\left[-\frac{\alpha x^2 + 1}{\alpha^2 e^{+\alpha x^2}}\right]_{-\infty}^{+\infty} = 0 - 0 = 0$ (the exponential term in the denominator rises faster than the polynomial in the numerator for $x \rightarrow \pm\infty$; see l'Hôpital's rule). But we don't actually need to evaluate the integral explicitly. The expression in the integral is an odd function, so its value is identically zero. Therefore, defining a perturbation of cx^3 for this system does not allow us to evaluate any useful energy correction.

12.37. The wavefunction for an ideal two-dimensional rotational motion is $\Psi_m = \frac{1}{\sqrt{2\pi}}e^{im\phi}$ with energy eigenvalues of $E = \frac{m^2\hbar^2}{2I}$. For $m = 1$, the first-order correction to the energy is then $\langle E^{(1)}\rangle = \int_0^{2\pi}\left(\frac{1}{\sqrt{2\pi}}e^{i\phi}\right)^* \cdot V[1-\sin(6\phi)]\cdot\left(\frac{1}{\sqrt{2\pi}}e^{i\phi}\right)d\phi = \frac{V}{2\pi}\int_0^{2\pi} e^{-i\phi}[1-\sin(6\phi)]e^{i\phi}\,d\phi$.

The first and last term of the integrand cancel:

$$\left\langle E^{(1)}\right\rangle=\frac{V}{2\pi}\int_0^{2\pi}\left[1-\sin(6\phi)\right]d\phi=\frac{V}{2\pi}\left[\phi+\frac{1}{6}\cos(6\phi)\right]_0^{2\pi}=\frac{V}{2\pi}(2\pi+0)=V.$$ Therefore, the average energy for $m = 1$ is $\langle E\rangle=\left\langle E^{(0)}\right\rangle+\left\langle E^{(1)}\right\rangle=\frac{\hbar^2}{2I}+V$.

12.39. If we multiply the binomial out, we get a perturbation of $V=k\left(x^2-ax+\frac{a^2}{4}\right)$.

The correction to the ground-state energy is

$$\int_0^a\left(\sqrt{\frac{2}{a}}\sin\frac{\pi x}{a}\right)\cdot k\left(x^2-ax+\frac{a^2}{4}\right)\cdot\left(\sqrt{\frac{2}{a}}\sin\frac{\pi x}{a}\right)dx.$$

Let us separate this into its three parts, evaluate each part separately, then combine the answers. First term:

$$\frac{2k}{a}\int_0^a x^2\sin^2\frac{\pi x}{a}dx=\frac{2k}{a}\left[\frac{x^3}{6}-\left(\frac{ax^2}{4\pi}-\frac{a^3}{8\pi^3}\right)\sin\frac{2\pi x}{a}-\frac{a^2x}{4\pi^2}\cos\frac{2\pi x}{a}\right]_0^a=\frac{2k}{a}\left(\frac{a^3}{6}-\frac{a^3}{4\pi^2}\right)=ka^2\left(\frac{1}{3}-\frac{1}{2\pi^2}\right)$$

Second term:

$$-ka\cdot\frac{2}{a}\int_0^a x\sin^2\frac{\pi x}{a}dx=-2k\left[\frac{x^2}{4}-\frac{ax}{4\pi}\sin\frac{2\pi x}{a}-\frac{a^2}{8\pi^2}\cos\frac{2\pi x}{a}\right]_0^a=-2k\left(\frac{a^2}{4}-\frac{a^2}{8\pi^2}+\frac{a^2}{8\pi^2}\right)=-\frac{ka^2}{2}$$

Third term: $\frac{ka^2}{4}\cdot\frac{2}{a}\int_0^a\sin^2\frac{\pi x}{a}dx=\frac{ka}{2}\left[\frac{x}{2}-\frac{a}{4\pi}\sin\frac{2\pi x}{a}\right]_0^a=\frac{ka^2}{4}$

Combining the three terms (factoring out ka^2 and using a common denominator of 12 for three of the terms), we obtain $ka^2\left(\frac{4}{12}-\frac{1}{2\pi^2}-\frac{6}{12}+\frac{3}{12}\right)=ka^2\left(\frac{1}{12}-\frac{1}{2\pi^2}\right)$ as the correction to the particle-in-a-box energy. Therefore, the energy for the first level of the particle-in-a-box is $\frac{h^2}{8ma^2}+ka^2\left(\frac{1}{12}-\frac{1}{2\pi^2}\right)$. (Note that the units on k should be such that the second term has overall units of energy.)

12.41. (a) The wavefunction needs to satisfy the boundary conditions of the system: $\phi(0) = 0$ and $\phi(a) = 0$. $\phi(0) = 0$ leads to $B = 0$, so that $\phi(a) = 0$ requires that A is an odd multiple of $\pm\pi/2$. Unless A and B take on these values, the wavefunction will not be a good choice.

(b) Not a good choice because $\phi(r = 0) = 1$ (and $\phi(r = a)$ is also non-zero).

(c) Same as (b).

(d) This is an acceptable choice, since it satisfies the boundary conditions $\phi(0) = 0$ and $\phi(a) = 0$.

(e) Not a good choice because $\phi(0) = a^2$.

(f) Not a good choice because $\phi(0) = 1$ and ϕ becomes infinite for $x = a$.

(g) $\phi(0) = 0$ is satisfied, but $\phi(a) = 0$ requires that $\sin(A) \cdot \cos(A)$ is zero. This would mean that A needs to be an (even or odd) multiple of $\pm\pi/2$. Unless A takes on any of these values, the wavefunction will not be a good choice.

12.43. To evaluate the partial derivative of $\dfrac{c_{a,1}^2 H_{11} + 2c_{a,1}c_{a,2}H_{12} + c_{a,2}^2 H_{22}}{c_{a,1}^2 S_{11} + 2c_{a,1}c_{a,2}S_{12} + c_{a,2}^2 S_{22}}$ with respect to $c_{a,1}$, we need to use the quotient rule for differentiation:

$$\frac{\left(c_{a,1}^2 S_{11} + 2c_{a,1}c_{a,2}S_{12} + c_{a,2}^2 S_{22}\right)\left(2c_{a,1}H_{11} + 2c_{a,2}H_{12}\right) - \left(2c_{a,1}S_{11} + 2c_{a,2}S_{12}\right)\left(c_{a,1}^2 H_{11} + 2c_{a,1}c_{a,2}H_{12} + c_{a,2}^2 H_{22}\right)}{\left(c_{a,1}^2 S_{11} + 2c_{a,1}c_{a,2}S_{12} + c_{a,2}^2 S_{22}\right)^2}$$

This expression must equal zero, so that we only need to be concerned about the numerator. We can simplify it considerably by rewriting equation 12.29 as $c_{a,1}^2 H_{11} + 2c_{a,1}c_{a,2}H_{12} + c_{a,2}^2 H_{22} = E \cdot \left(c_{a,1}^2 S_{11} + 2c_{a,1}c_{a,2}S_{12} + c_{a,2}^2 S_{22}\right)$ and substituting:

$$\left(c_{a,1}^2 S_{11} + 2c_{a,1}c_{a,2}S_{12} + c_{a,2}^2 S_{22}\right)\left(2c_{a,1}H_{11} + 2c_{a,2}H_{12}\right) - \left(2c_{a,1}S_{11} + 2c_{a,2}S_{12}\right)E\left(c_{a,1}^2 S_{11} + 2c_{a,1}c_{a,2}S_{12} + c_{a,2}^2 S_{22}\right) = 0.$$

Now divide both sides by 2 and by the parentheses involving three S terms:

$\left(c_{a,1}H_{11} + c_{a,2}H_{12}\right) - \left(c_{a,1}S_{11} + c_{a,2}S_{12}\right)E = 0$. Multiplying out and rearranging results in $c_{a,1}H_{11} - c_{a,1}ES_{11} + c_{a,2}H_{12} - c_{a,2}ES_{12} = 0$. Factoring out $c_{a,1}$ and $c_{a,2}$, respectively, leads to the final equation: $\left(H_{11} - ES_{11}\right)c_{a,1} + \left(H_{12} - ES_{12}\right)c_{a,2} = 0$.

12.45. Let's start by normalizing the trial wavefunction:

$1 = \int_0^\infty \left(Ne^{-kr^2}\right)^* \left(Ne^{-kr^2}\right) \cdot 4\pi r^2\, dr = 4\pi N^2 \int_0^\infty r^2 e^{-2kr^2}\, dr$. This definite integral is given in Appendix 1: $\int_0^\infty r^2 e^{-2kr^2}\, dr = \dfrac{1}{2^{1+1} \cdot (2k)^1} \cdot \sqrt{\dfrac{\pi}{2k}} = \dfrac{1}{8k}\sqrt{\dfrac{\pi}{2k}}$. This means $\dfrac{4\pi N^2}{8k}\sqrt{\dfrac{\pi}{2k}} = 1$ (we will use this equation below), from which we can obtain the normalization constant: $N^2 = \left(\dfrac{2k}{\pi}\right)^{3/2}$ and $N = \left(\dfrac{2k}{\pi}\right)^{3/4}$.

To calculate the energy for this trial wavefunction, we then need to evaluate

$\langle E_{\text{trial}} \rangle = \int_0^\infty \left[Ne^{-kr^2}\right]^* \hat{H} \left[Ne^{-kr^2}\right] \cdot 4\pi r^2\, dr = 4\pi N^2 \int_0^\infty r^2 e^{-kr^2} \hat{H}\left(e^{-kr^2}\right) dr$, where

$$\hat{H} = -\frac{\hbar^2}{2\mu}\left[\frac{1}{r^2}\frac{\partial}{\partial r}\left(r^2\frac{\partial}{\partial r}\right) + \frac{1}{r^2\sin\theta}\frac{\partial}{\partial\theta}\left(\sin\theta\frac{\partial}{\partial\theta}\right) + \frac{1}{r^2\sin^2\theta}\frac{\partial^2}{\partial\phi^2}\right] - \frac{e^2}{4\pi\varepsilon_0 r}.$$

Fortunately, our trial function depends only on r, so that the angular derivatives are zero. The derivatives with respect to r are $\frac{\partial}{\partial r}\left(e^{-kr^2}\right) = -2kre^{-kr^2}$ and

$\frac{\partial}{\partial r}\left[r^2\cdot(-2kr)e^{-kr^2}\right] = -2k\frac{\partial}{\partial r}\left(r^3e^{-kr^2}\right) = -2k\left[3r^2e^{-kr^2} + r^3(-2kr)e^{-kr^2}\right]$, or

$-2ke^{-kr^2}\left(3r^2 - 2kr^4\right)$. The energy of the trial wavefunction is then

$$\langle E_{\text{trial}}\rangle = 4\pi N^2\int_0^\infty r^2e^{-kr^2}\left(-\frac{\hbar^2}{2\mu r^2}\cdot\left[-2ke^{-kr^2}\left(3r^2 - 2kr^4\right)\right] - \frac{e^2}{4\pi\varepsilon_0 r}e^{-kr^2}\right)dr,$$

which rearranges to: $\langle E_{\text{trial}}\rangle = 4\pi N^2\int_0^\infty\left(3\frac{\hbar^2kr^2}{\mu} - 2\frac{\hbar^2k^2r^4}{\mu} - \frac{e^2r}{4\pi\varepsilon_0}\right)e^{-2kr^2}dr$.

We evaluate this integral in parts:

Part 1: $4\pi N^2\cdot\frac{3\hbar^2k}{\mu}\int_0^\infty r^2e^{-2kr^2}\,dr$. We have evaluated this integral before, so that

$4\pi N^2\cdot\frac{3\hbar^2k}{\mu}\cdot\frac{1}{8k}\sqrt{\frac{\pi}{2k}}$, which is $\frac{3\hbar^2k}{\mu}$ if we make use of $\frac{4\pi N^2}{8k}\sqrt{\frac{\pi}{2k}} = 1$ (see normalization).

Part 2: $-4\pi N^2\cdot\frac{2\hbar^2k^2}{\mu}\int_0^\infty r^4e^{-2kr^2}\,dr$. Using Appendix 1:

$$-4\pi N^2\cdot\frac{2\hbar^2k^2}{\mu}\cdot\frac{1\cdot3}{2^{2+1}\cdot(2k)^2}\sqrt{\frac{\pi}{2k}} = -4\pi N^2\cdot\frac{2\hbar^2k^2}{\mu}\cdot\frac{3}{32k^2}\sqrt{\frac{\pi}{2k}}, \text{ or } -\frac{3\hbar^2k}{2\mu}.$$

Part 3: $-4\pi N^2\cdot\frac{e^2}{4\pi\varepsilon_0}\int_0^\infty re^{-2kr^2}\,dr$. Using Appendix 1:

$-4\pi N^2\cdot\frac{e^2}{4\pi\varepsilon_0}\cdot\frac{1}{2\cdot(2k)} = -\frac{e^2}{2\pi\varepsilon_0}\sqrt{\frac{2k}{\pi}}$, where we have again substituted for N^2.

Combining the three parts, we obtain the predicted energy of the ground state as

$$\langle E_{\text{trial}}\rangle = \frac{3\hbar^2k}{\mu} - \frac{3\hbar^2k}{2\mu} - \frac{e^2}{2\pi\varepsilon_0}\sqrt{\frac{2k}{\pi}} = \frac{3\hbar^2k}{2\mu} - \frac{e^2}{2\pi\varepsilon_0}\sqrt{\frac{2k}{\pi}}.$$

We now need to determine the value of k that minimizes this energy. Setting $dE_{\text{trial}}/dk =$ 0: $\frac{3\hbar^2}{2\mu} - \frac{e^2}{2\pi\varepsilon_0}\sqrt{\frac{2}{\pi}}\cdot\frac{1}{2\sqrt{k}} = 0$. We solve for k: $\sqrt{k} = \frac{\mu e^2}{6\pi\hbar^2\varepsilon_0}\sqrt{\frac{2}{\pi}}$ and $k = \frac{\mu^2e^4}{36\pi^2\hbar^4{\varepsilon_0}^2}\left(\frac{2}{\pi}\right)$

or $k = \dfrac{\mu^2 e^4}{18\pi^3\hbar^4\varepsilon_0^{\,2}}$.

The lowest value of the predicted ground state energy for the trial wavefunction is then

$$\langle E_{\text{trial}}\rangle = \frac{3\hbar^2}{2\mu}\cdot\frac{\mu^2 e^4}{18\pi^3\hbar^4\varepsilon_0^{\,2}} - \frac{e^2}{2\pi\varepsilon_0}\left(\frac{2}{\pi}\cdot\frac{\mu^2 e^4}{18\pi^3\hbar^4\varepsilon_0^{\,2}}\right)^{1/2} = \frac{\mu e^4}{12\pi^3\hbar^2\varepsilon_0^{\,2}} - \frac{\mu e^4}{6\pi^3\hbar^2\varepsilon_0^{\,2}} = -\frac{\mu e^4}{12\pi^3\hbar^2\varepsilon_0^{\,2}},$$

or $\langle E_{\text{trial}}\rangle = -\dfrac{\mu e^4}{3\pi h^2\varepsilon_0^{\,2}} = -0.106\dfrac{\mu e^4}{h^2\varepsilon_0^{\,2}}$ (where we have substituted $\hbar = \dfrac{h}{2\pi}$).

The true energy for the ground state is $E = -\dfrac{e^4\mu}{8\varepsilon_0^{\,2}h^2} = -0.125\dfrac{e^4\mu}{\varepsilon_0^{\,2}h^2}$, so the ground state of the trial wavefunction is about 15% higher in energy than the true value.

12.47. The two wavefunctions are $\phi_a = 0.996\Psi_{a,1} + 0.0896\Psi_{a,2}$ and $\phi_a' = -0.0896\Psi_{a,1} + 0.996\Psi_{a,2}$. We know already that they are properly normalized, because this was a condition for obtaining the coefficients $c_{a,1}$ and $c_{a,2}$ in Example 12.12. To show that the two wavefunctions are orthonormal, we therefore need to show that they are orthogonal. Set up the integral of $\phi_a{}^*\phi_a{}'$:

$\int\left(0.996\Psi_{a,1} + 0.0896\Psi_{a,2}\right)^*\left(-0.0896\Psi_{a,1} + 0.996\Psi_{a,2}\right)d\tau$. Distribute and evaluate:

$$\int\left(0.996\Psi_{a,1}\right)^*\left(-0.0896\Psi_{a,1}\right)d\tau + \int\left(0.0896\Psi_{a,2}\right)^*\left(-0.0896\Psi_{a,1}\right)d\tau$$
$$+\int\left(0.996\Psi_{a,1}\right)^*\left(0.996\Psi_{a,2}\right)d\tau + \int\left(0.0896\Psi_{a,2}\right)^*\left(0.996\Psi_{a,2}\right)d\tau$$
$$=(0.996)(-0.0896)\int\Psi_{a,1}{}^*\Psi_{a,1}\,d\tau + (0.0896)(-0.0896)\int\Psi_{a,2}{}^*\Psi_{a,1}\,d\tau$$
$$+(0.996)(0.996)\int\Psi_{a,1}{}^*\Psi_{a,2}\,d\tau + (0.0896)(0.996)\int\Psi_{a,2}{}^*\Psi_{a,2}\,d\tau$$

The second and third integrals are exactly zero because the original wavefunctions themselves are orthogonal. The first and fourth integrals are exactly one because the original wavefunctions are normalized. Therefore, the expression reduces to (0.996)(−0.0896) + (0.0896)(0.996) = 0. Therefore, these two new wavefunctions are orthogonal.

12.49. **(a)** According to equation 12.31, the nontrivial solution of the simultaneous equations found by minimizing the energy will be given by $\begin{vmatrix} -25 - E\cdot 1 & -0.50 - E\cdot 0 \\ -0.50 - E\cdot 0 & -20 - E\cdot 1 \end{vmatrix} = 0$.

We evaluate this secular determinant: $(-25 - E)(-20 - E) - (-0.50)(-0.50) = 0$.

Multiplying out and collecting equal terms in E gives the quadratic equation $E^2+45E+499.75=0$. Solving it for E: $E_{1/2}=\dfrac{-45\pm\sqrt{45^2-4\cdot 499.75}}{2}$, which gives $E_1\approx -19.95$ and $E_2\approx -25.05$. E_2 is therefore the ground state energy.

(b) $\begin{vmatrix} -25-E\cdot 1 & -5.0-E\cdot 0 \\ -5.0-E\cdot 0 & -20-E\cdot 1 \end{vmatrix}=0$ leads to $(-25-E)(-20-E)-(-5.0)(-5.0)=0$ and the quadratic equation $E^2+45E+475=0$. The solutions in this case are $E_1\approx -16.91$ and

$E_2\approx -28.09$. Because the energy integrals H_{12} and H_{21} are greater, the mixing is stronger and the energy levels spread further apart.

12.51. The true wavefunctions Ψ_i solve the Schrödinger equation $\hat{H}\Psi_i=E_i\Psi_i$. We need to show that $\int\phi^*\hat{H}\phi d\tau\ge E_1$ for any trial wavefunction $\phi=\sum_i c_i\Psi_i$.

We start with the left-hand side, substituting the linear combination into the integral: $\int\left(\sum_i c_i\Psi_i\right)^*\hat{H}\left(\sum_j c_j\Psi_j\right)d\tau$. We can distribute the Hamiltonian operator through the second summation: $\int\left(\sum_i c_i\Psi_i\right)^*\left(\sum_j c_j\hat{H}\Psi_j\right)d\tau$. The Ψ_j's are eigenfunctions of the Hamiltonian operator, so we can substitute $E_j\Psi_j$ in the second summation: $\int\left(\sum_i c_i\Psi_i\right)^*\left(\sum_j c_jE_j\Psi_j\right)d\tau$. Now we rearrange the summation and integral signs, bringing all constants outside of the integral: $\sum_i\sum_j c_i^*c_jE_j\int\Psi_i^*\Psi_j d\tau$. Because the ideal wavefunctions are orthogonal, the integral is exactly zero except for the terms where $i=j$. This has the effect of eliminating one of the sums and making all of the subscript labels the same. Thus, we get $\int\phi^*\hat{H}\phi d\tau=\sum_j c_j^*c_jE_j$, and we need to prove that this is always greater than or equal to E_1.

Let us focus on the sum on the right-hand side of this expression. The coefficients themselves may be positive or negative, but in either case, multiplying the coefficient by itself (or its complex conjugate) will always yield a positive value: $c_j^*c_j\ge 0$. We know that $E_j\ge E_1$ (any arbitrary energy value is either the ground-state energy or higher than the ground state energy), so we can multiply E_j and E_1 by $c_j^*c_j$, a positive number, and get $c_j^*c_jEj\ge c_j^*c_jE_1$. Furthermore, if we sum over all j's, the summation on the left will always be greater than or equal to the summation on the right because the left sum is the sum of larger individual terms: $\sum_j c_j^*c_jE_j\ge\sum_j c_j^*c_jE_1$. After factoring out E_1:

$\sum_j c_j^* c_j E_j \geq E_1 \sum_j c_j^* c_j$. Together with our expression $\int \phi^* \hat{H} \phi d\tau = \sum_j c_j^* c_j E_j$ this leads to $\int \phi^* \hat{H} \phi d\tau \geq E_1 \sum_j c_j^* c_j$.

Finally, if the trial wavefunction is normalized (which it should be), we have the following relationships: $1 = \int \phi^* \phi d\tau = \int \left(\sum_i c_i \Psi_i \right)^* \left(\sum_j c_j \Psi_j \right) d\tau$, which can again be rearranged as $1 = \sum_i \sum_j c_i^* c_j \int \Psi_i^* \Psi_j d\tau$. Because the ideal wavefunctions are orthonormal, every integral is zero except when $i = j$, in which case the integral equals 1. This again has the effect of eliminating one of the summations, and we are left with $1 = \sum_j c_j^* c_j$. Therefore, $E_1 \sum_j c_j^* c_j = E_1$, so that $\int \phi^* \hat{H} \phi d\tau \geq E_1 \sum_j c_j^* c_j$ becomes $\int \phi^* \hat{H} \phi d\tau \geq E_1$, which is the variation theorem: any trial wavefunction will have an average energy equal to or greater than then true ground-state energy of the system.

12.53. Mathematically, the Born-Oppenheimer approximation is given by equation 12.36: $\Psi_{\text{molecule}} \approx \Psi_{\text{nuc}} \times \Psi_{\text{el}}$. In words, the Born-Oppenheimer approximation is the assumption that electronic motion can be modeled as if the nuclei were not moving, because electrons move so much faster than the much heavier nuclei. If we assume that the nuclei have fixed positions, the first two nuclear kinetic energy terms of the complete Hamiltonian (equation 12.35) are zero, and the last term (repulsive potential between the two nuclei) has a fixed value. The remaining Hamiltonian now only depends on electronic coordinates, which leads directly to the mathematical assumption of separability of the molecular wavefunction.

12.55. Using the expressions in equation 12.45: $\Delta E = E_2 - E_1 = \dfrac{H_{22} - H_{12}}{1 - S_{12}} - \dfrac{H_{11} + H_{12}}{1 + S_{12}}$.

We rewrite this with a common denominator:

$$\Delta E = \frac{(H_{22} - H_{12})(1 + S_{12}) - (H_{11} + H_{12})(1 - S_{12})}{(1 + S_{12})(1 - S_{12})}.$$

Multiplying out and collecting equal terms: $\Delta E = \dfrac{H_{22} - 2H_{12} - H_{11} + H_{11}S_{12} + H_{22}S_{12}}{1 - S_{12}^2}$

For H_2^+, $H_{11} = H_{22}$, so we can simplify further: $\Delta E = \dfrac{2H_{11}S_{12} - 2H_{12}}{1 - S_{12}^2} = \dfrac{2(H_{11}S_{12} - H_{12})}{1 - S_{12}^2}$.

12.57. According to equation 12.48, the bond order for the first excited state of H_2^+ (where the only electron is in the antibonding orbitale) is $n = (0 - 1)/2 = -1/2$. Because this state is known to be a dissociative state (that is, in this electronic state the two nuclei move apart spontaneously to achieve a lower energy), we can deduce that any diatomic molecule that has a negative bond order is inherently unstable, spontaneously increasing its internuclear distance (by dissociating) to achieve a lower, more stable energy state.

12.59. Very simply, the first wavefunction in equation 12.43 is the bonding orbital because its corresponding energy is lower than that of the second wavefunction in equation 12.43.

12.61. O_2^{2+}: $(\sigma 1s)^2 (\sigma^* 1s)^2 (\sigma 2s)^2 (\sigma^* 2s)^2 (\sigma 2p_z)^2 (\pi 2p_{x,y})^4$

O_2^-: $(\sigma 1s)^2 (\sigma^* 1s)^2 (\sigma 2s)^2 (\sigma^* 2s)^2 (\sigma 2p_z)^2 (\pi 2p_{x,y})^4 (\pi^* 2p_{x,y})^3$

O_2^{2-}: $(\sigma 1s)^2 (\sigma^* 1s)^2 (\sigma 2s)^2 (\sigma^* 2s)^2 (\sigma 2p_z)^2 (\pi 2p_{x,y})^4 (\pi^* 2p_{x,y})^4$

12.63. According to Figure 12.24 and equation 12.48, NO has a bond order of

$n = (8 - 3)/2 = 5/2$.

CHAPTER 13

Introduction to Symmetry in Quantum Mecahnics

13.1. The more symmetry elements an object has, the simpler it is to mathematically describe the three–dimensional shape of an object. Thus, the phrase "higher symmetry" ultimately translates into mathematical simplicity.

13.3. **(a)** E, one fourfold axis of symmetry C_4, four vertical planes of symmetry σ_v.

(b) E, one twofold axis of symmetry C_2 through the center of the spine, two vertical planes of symmetry σ_v.

(c) E, one eightfold proper (and improper) axis of symmetry C_8 (and S_8), eight twofold proper (and improper) axes of symmetry C_2 (and S_2) perpendicular to C_8, eight vertical planes of symmetry σ_v, one horizontal plane of symmetry σ_h, one center of inversion i.

(d) E, one fourfold proper (and improper) axis of symmetry C_4 (and S_4), four twofold proper (and improper) axes of symmetry C_2 (and S_2) perpendicular to C_4, four vertical planes of symmetry σ_v, one horizontal plane of symmetry σ_h, one center of inversion i.

13.5. Consult Figure 13.15 for the flow chart on determining the symmetry of an object.

(a) D_{2h}; **(b)** C_{2v}; **(c)** D_{8h}; **(d)** D_{4h}.

13.7. **(a)** complete; **(b)** complete; **(c)** not complete (the identity operation E is missing); **(d)** not complete (the inverse operation to C_3, namely $C_3^{\,2}$, is missing).

13.9. C_2 = rotation by $\theta = 360°/2 = 180°$: $C_2 = \begin{bmatrix} \cos 180° & \sin 180° & 0 \\ -\sin 180° & \cos 180° & 0 \\ 0 & 0 & 1 \end{bmatrix} = \begin{bmatrix} -1 & 0 & 0 \\ 0 & -1 & 0 \\ 0 & 0 & 1 \end{bmatrix}$.

C_3 = rotation by $\theta = 360°/3 = 120°$: $C_3 = \begin{bmatrix} \cos 120° & \sin 120° & 0 \\ -\sin 120° & \cos 120° & 0 \\ 0 & 0 & 1 \end{bmatrix} = \begin{bmatrix} -1/2 & \sqrt{3}/2 & 0 \\ -\sqrt{3}/2 & -1/2 & 0 \\ 0 & 0 & 1 \end{bmatrix}$.

C_6 = rotation by $\theta = 360°/6 = 60°$: $C_6 = \begin{bmatrix} \cos 60° & \sin 60° & 0 \\ -\sin 60° & \cos 60° & 0 \\ 0 & 0 & 1 \end{bmatrix} = \begin{bmatrix} 1/2 & \sqrt{3}/2 & 0 \\ -\sqrt{3}/2 & 1/2 & 0 \\ 0 & 0 & 1 \end{bmatrix}$.

C_1 = rotation by θ = 360°/1 = 360°: $C_1 = \begin{bmatrix} \cos 360^\circ & \sin 360^\circ & 0 \\ -\sin 360^\circ & \cos 360^\circ & 0 \\ 0 & 0 & 1 \end{bmatrix} = \begin{bmatrix} 1 & 0 & 0 \\ 0 & 1 & 0 \\ 0 & 0 & 1 \end{bmatrix}$.

13.11. An S_n operation about the z axis change the sign of the z coordinate (but not its magnitude) due to the reflection through the xy plane. This is reflected by a value of −1 in the bottom right corner of the corresponding matrix.

13.13. Using $C_2 = \begin{bmatrix} -1 & 0 & 0 \\ 0 & -1 & 0 \\ 0 & 0 & 1 \end{bmatrix}$ from problem 13. 9, we can see that

$$\hat{C}_2\hat{C}_2 = \begin{bmatrix} -1 & 0 & 0 \\ 0 & -1 & 0 \\ 0 & 0 & 1 \end{bmatrix}\begin{bmatrix} -1 & 0 & 0 \\ 0 & -1 & 0 \\ 0 & 0 & 1 \end{bmatrix} = \begin{bmatrix} +1 & 0 & 0 \\ 0 & +1 & 0 \\ 0 & 0 & 1 \end{bmatrix} = E.$$

13.15. The number of classes can be determined by simply counting the number of rows or columns of characters from the appropriate character table in Appendix 3. The order is the total number of symmetry elements in the point group.

(a) 5 classes, order = 24; **(b)** 5 classes, order = 24; **(c)** 10 classes, order = 48.

(d) Formally: 5 classes, order = 5; technically, both are infinite; **(e)** 5 classes, order = 8.

13.17. **(a)** In T_d, an S_4^{-1} operation is equivalent to an S_4^3 operation (which is classed with the other S_4 operations);

(b) In D_{6h}, the C_2^{-1} operation is equivalent to a C_2 operation.

13.19. **(a)** C_2; **(b)** σ_v; **(c)** C_6^5; **(d)** S_4^3.

13.21. **(a)** No, symmetry operations do not necessarily satisfy a commutative law of algebra. They may for small point groups, but there is no requirement that symmetry operations commute. Take an equilateral triangle, for example, and perform a C_3 and then a σ_v operation. The outcome is different if you perform the σ_v operation first, followed by the C_3 operation (label the three corners to see that).

(b) The fact that we can write symmetry operations as matrices, and that matrix algebra isn't necessarily commutative, supports the concept that symmetry operations need not satisfy the commutative law of algebra.

13.23. If the NO_3^- ion were not undergoing resonance averaging, it would be planar with one double and two single bonds, and the symmetry would be C_{2v} (similar to the water molecule, with the two N–O single bonds taking the place of the O–H bonds, and the additional N=O double bond pointing along the C_2 axis).

13.25. As the only symmetry element that is simply a point, the center of inversion i must be at the intersection point of all symmetry elements in a point group. Otherwise, the statement that "all symmetry elements intersect in one point in space" would not hold because it wouldn't include i.

13.27. Using the scheme in Figure 13.15:

(a) *cis*-1,2-dichloroethylene has C_{2v} symmetry.

(b) *trans*-1,2-dichloroethylene has C_{2h} symmetry.

(c) Toluene has C_s symmetry (assuming that one of the methyl C−H bonds is perpendicular to the aromatic ring).

(d) 1,3-Cyclohexadiene has C_2 symmetry (the molecule is not planar; the two C atoms in the CH_2 groups are located symmetrically above and below the plane defined by the other C atoms).

13.29. **(a)** Iron pentacarbonyl has D_{3h} symmetry.

(b) The carbonate ion has D_{3h} symmetry.

(c) Staggered ethane has D_{3d} symmetry.

(d) Eclipsed ethane has D_{3h} symmetry.

13.31. 1,2-dichlorobenzene and 1,3-dichlorobenzene have C_{2v} symmetry, while 1,4-dichlorobenzene has D_{2h} symmetry.

13.33. **(a)** The wavefunctions of deuterium oxide have C_{2v} symmetry.

(b) The wavefunctions of boron trichloride have D_{3h} symmetry.

(c) The wavefunctions of phosphorus trichloride (a pyramidal molecule) have C_{3v} symmetry.

13.35. **(a)** The possible formulas would be C_4H_4 (tetrahedron), C_8H_8 (cube), and $C_{20}H_{20}$ (dodecahedron). There wouldn't be a hydrocarbon equivalent to the octahedron (all four bonds to other carbon atoms would have to bend to the same side) or icosahedron (which requires five bonds).

(b) You should be able to verify that C_4H_4 has all the symmetry elements of the T_d point group, while C_8H_8 has all of the symmetry elements of the O_h point group. Only derivatives of C_4H_4 (tetrahedrane) are known, but cubane (C_8H_8) and dodecahedrane ($C_{20}H_{20}$) have successfully been synthesized.

13.37. A permanent dipole moment requires a point group symmetry of C_s, C_n, or C_{nv} ($n > 1$).

(a) The carbonate ion has D_{3h} symmetry and will not have a permanent dipole moment.

(b) The phosphate ion has T_d symmetry and will not have a permanent dipole moment.

(c) Uranium hexafluoride has O_h symmetry will not have a permanent dipole moment.

(d) Bromine has $D_{\infty h}$ symmetry will not have a permanent dipole moment.

13.39. A permanent dipole moment requires a point group symmetry of C_s, C_n, or C_{nv} ($n > 1$).

(a) Methane has T_d symmetry and will not have a permanent dipole moment.

(b) Chloromethane has C_{3v} symmetry and will have a permanent dipole moment.

(c) Dichloromethane has C_{2v} symmetry and will have a permanent dipole moment.

(d) Trichloromethane has C_{3v} symmetry and will have a permanent dipole moment.

(e) Carbon tetrachloride has T_d symmetry and will not have a permanent dipole moment.

(f) 2,2-Dimethylpropane has T_d symmetry and will not have a permanent dipole moment.

13.41. **(a)** The only symmetry element of CHBrClF is E, so its point group is C_1.

(b) CH_2BrCl has C_s symmetry. Because it has an S_1 axis (which is equivalent to the σ plane), it is not chiral.

13.43. Assume that the N and H1 atoms are lying in the xz plane. Note that for the σ' and σ'' reflections, not only do two of the hydrogen atoms exchange their places but their x and y unit vectors are also transforming into each other.

$$\sigma(xz)(\mathrm{NH_3}) = \begin{bmatrix} 1 & 0 & 0 & 0 & 0 & 0 & 0 & 0 & 0 & 0 & 0 & 0 \\ 0 & -1 & 0 & 0 & 0 & 0 & 0 & 0 & 0 & 0 & 0 & 0 \\ 0 & 0 & 1 & 0 & 0 & 0 & 0 & 0 & 0 & 0 & 0 & 0 \\ 0 & 0 & 0 & 1 & 0 & 0 & 0 & 0 & 0 & 0 & 0 & 0 \\ 0 & 0 & 0 & 0 & -1 & 0 & 0 & 0 & 0 & 0 & 0 & 0 \\ 0 & 0 & 0 & 0 & 0 & 1 & 0 & 0 & 0 & 0 & 0 & 0 \\ 0 & 0 & 0 & 0 & 0 & 0 & 0 & 0 & 0 & 1 & 0 & 0 \\ 0 & 0 & 0 & 0 & 0 & 0 & 0 & 0 & 0 & 0 & -1 & 0 \\ 0 & 0 & 0 & 0 & 0 & 0 & 0 & 0 & 0 & 0 & 0 & 1 \\ 0 & 0 & 0 & 0 & 0 & 0 & 1 & 0 & 0 & 0 & 0 & 0 \\ 0 & 0 & 0 & 0 & 0 & 0 & 0 & -1 & 0 & 0 & 0 & 0 \\ 0 & 0 & 0 & 0 & 0 & 0 & 0 & 0 & 1 & 0 & 0 & 0 \end{bmatrix} \begin{bmatrix} x_{\mathrm{N}} \\ y_{\mathrm{N}} \\ z_{\mathrm{N}} \\ x_{\mathrm{H1}} \\ y_{\mathrm{H1}} \\ z_{\mathrm{H1}} \\ x_{\mathrm{H2}} \\ y_{\mathrm{H2}} \\ z_{\mathrm{H2}} \\ x_{\mathrm{H3}} \\ y_{\mathrm{H3}} \\ z_{\mathrm{H3}} \end{bmatrix}$$

$$\sigma'(\mathrm{NH_3}) = \begin{bmatrix} 0 & -1 & 0 & 0 & 0 & 0 & 0 & 0 & 0 & 0 & 0 & 0 \\ -1 & 0 & 0 & 0 & 0 & 0 & 0 & 0 & 0 & 0 & 0 & 0 \\ 0 & 0 & 1 & 0 & 0 & 0 & 0 & 0 & 0 & 0 & 0 & 0 \\ 0 & 0 & 0 & 0 & 0 & 0 & 0 & 0 & 0 & 0 & -1 & 0 \\ 0 & 0 & 0 & 0 & 0 & 0 & 0 & 0 & 0 & -1 & 0 & 0 \\ 0 & 0 & 0 & 0 & 0 & 0 & 0 & 0 & 0 & 0 & 0 & 1 \\ 0 & 0 & 0 & 0 & 0 & 0 & 0 & -1 & 0 & 0 & 0 & 0 \\ 0 & 0 & 0 & 0 & 0 & 0 & -1 & 0 & 0 & 0 & 0 & 0 \\ 0 & 0 & 0 & 0 & 0 & 0 & 0 & 0 & 1 & 0 & 0 & 0 \\ 0 & 0 & 0 & 0 & -1 & 0 & 0 & 0 & 0 & 0 & 0 & 0 \\ 0 & 0 & 0 & -1 & 0 & 0 & 0 & 0 & 0 & 0 & 0 & 0 \\ 0 & 0 & 0 & 0 & 0 & 1 & 0 & 0 & 0 & 0 & 0 & 0 \end{bmatrix} \begin{bmatrix} x_{\mathrm{N}} \\ y_{\mathrm{N}} \\ z_{\mathrm{N}} \\ x_{\mathrm{H1}} \\ y_{\mathrm{H1}} \\ z_{\mathrm{H1}} \\ x_{\mathrm{H2}} \\ y_{\mathrm{H2}} \\ z_{\mathrm{H2}} \\ x_{\mathrm{H3}} \\ y_{\mathrm{H3}} \\ z_{\mathrm{H3}} \end{bmatrix}$$

$$\sigma''(\mathrm{NH_3}) = \begin{bmatrix} 0 & 1 & 0 & 0 & 0 & 0 & 0 & 0 & 0 & 0 & 0 & 0 \\ 1 & 0 & 0 & 0 & 0 & 0 & 0 & 0 & 0 & 0 & 0 & 0 \\ 0 & 0 & 1 & 0 & 0 & 0 & 0 & 0 & 0 & 0 & 0 & 0 \\ 0 & 0 & 0 & 0 & 0 & 0 & 0 & 1 & 0 & 0 & 0 & 0 \\ 0 & 0 & 0 & 0 & 0 & 0 & 1 & 0 & 0 & 0 & 0 & 0 \\ 0 & 0 & 0 & 0 & 0 & 0 & 0 & 0 & 1 & 0 & 0 & 0 \\ 0 & 0 & 0 & 0 & 1 & 0 & 0 & 0 & 0 & 0 & 0 & 0 \\ 0 & 0 & 0 & 1 & 0 & 0 & 0 & 0 & 0 & 0 & 0 & 0 \\ 0 & 0 & 0 & 0 & 0 & 1 & 0 & 0 & 0 & 0 & 0 & 0 \\ 0 & 0 & 0 & 0 & 0 & 0 & 0 & 0 & 0 & 0 & 1 & 0 \\ 0 & 0 & 0 & 0 & 0 & 0 & 0 & 0 & 0 & 1 & 0 & 0 \\ 0 & 0 & 0 & 0 & 0 & 0 & 0 & 0 & 0 & 0 & 0 & 1 \end{bmatrix} \begin{bmatrix} x_{\mathrm{N}} \\ y_{\mathrm{N}} \\ z_{\mathrm{N}} \\ x_{\mathrm{H1}} \\ y_{\mathrm{H1}} \\ z_{\mathrm{H1}} \\ x_{\mathrm{H2}} \\ y_{\mathrm{H2}} \\ z_{\mathrm{H2}} \\ x_{\mathrm{H3}} \\ y_{\mathrm{H3}} \\ z_{\mathrm{H3}} \end{bmatrix}$$

13.45. We multiply the corresponding characters for the D_{2d} point group:

$A_1 \times A_1 = 1 \times 1\ 1 \times 1\ 1 \times 1\ 1 \times 1\ 1 \times 1 = 1\ 1\ 1\ 1\ 1 = A_1$

$A_1 \times A_2 = 1 \times 1\ 1 \times 1\ 1 \times 1\ 1 \times (-1)\ 1 \times (-1) = 1\ 1\ 1\ {-1}\ {-1} = A_2$

Following the same procedure, we can establish the following relationships:

$A_1 \times B_1 = B_1$ $\quad A_1 \times B_2 = B_2$ $\quad A_1 \times E = E$

$A_2 \times A_2 = A_1$ $\quad A_2 \times B_1 = B_2$ $\quad A_2 \times B_2 = B_1$ $\quad A_2 \times E = E$

$B_1 \times B_1 = A_1$ $\quad B_1 \times B_2 = A_2$ $\quad B_1 \times E = E$

$B_2 \times B_2 = A_1$ $\quad B_2 \times E = E$

The combination $E \times E$ yields the character set (4 0 4 0 0). Using the GOT to reduce it:

$$a(A_1) = \frac{1}{8}[1\cdot1\cdot4 + 2\cdot1\cdot0 + 1\cdot1\cdot4 + 2\cdot1\cdot0 + 2\cdot1\cdot0] = 1$$

$$a(A_2) = \frac{1}{8}[1\cdot1\cdot4 + 2\cdot1\cdot0 + 1\cdot1\cdot4 + 2\cdot(-1)\cdot0 + 2\cdot(-1)\cdot0] = 1$$

$$a(B_1) = \frac{1}{8}[1\cdot1\cdot4 + 2\cdot(-1)\cdot0 + 1\cdot1\cdot4 + 2\cdot1\cdot0 + 2\cdot(-1)\cdot0] = 1$$

$$a(B_2) = \frac{1}{8}[1\cdot1\cdot4 + 2\cdot(-1)\cdot0 + 1\cdot1\cdot4 + 2\cdot(-1)\cdot0 + 2\cdot1\cdot0] = 1$$

$$a(E) = \frac{1}{8}[1\cdot2\cdot4 + 2\cdot0\cdot0 + 1\cdot(-2)\cdot4 + 2\cdot0\cdot0 + 2\cdot0\cdot0] = 0$$

Therefore, $E \times E = A_1 \oplus A_2 \oplus B1 \oplus B2$, showing that this product also equals irreducible representations of the group. Therefore, the closure requirement is satisfied.

13.47. (a) $\frac{1}{h}\sum_{\text{all classes}} N\cdot\chi_A\cdot\chi_A=\frac{1}{2}\left[1\cdot1\cdot1+1\cdot1\cdot1\right]=1$

$\frac{1}{h}\sum_{\text{all classes}} N\cdot\chi_B\cdot\chi_B=\frac{1}{2}\left[1\cdot1\cdot1+1\cdot(-1)\cdot(-1)\right]=1$

(b) $\frac{1}{h}\sum_{\text{all classes}} N\cdot\chi_{A_1}\cdot\chi_{A_1}=\frac{1}{8}\left[1\cdot1\cdot1+2\cdot1\cdot1+1\cdot1\cdot1+2\cdot1\cdot1+2\cdot1\cdot1\right]=1$

$\frac{1}{h}\sum_{\text{all classes}} N\cdot\chi_{A_2}\cdot\chi_{A_2}=\frac{1}{8}\left[1\cdot1\cdot1+2\cdot1\cdot1+1\cdot1\cdot1+2\cdot(-1)\cdot(-1)+2\cdot(-1)\cdot(-1)\right]=1$

$\frac{1}{h}\sum_{\text{all classes}} N\cdot\chi_{B_1}\cdot\chi_{B_1}=\frac{1}{8}\left[1\cdot1\cdot1+2\cdot(-1)\cdot(-1)+1\cdot1\cdot1+2\cdot1\cdot1+2\cdot(-1)\cdot(-1)\right]=1$

$\frac{1}{h}\sum_{\text{all classes}} N\cdot\chi_{B_2}\cdot\chi_{B_2}=\frac{1}{8}\left[1\cdot1\cdot1+2\cdot(-1)\cdot(-1)+1\cdot1\cdot1+2\cdot(-1)\cdot(-1)+2\cdot1\cdot1\right]=1$

$\frac{1}{h}\sum_{\text{all classes}} N\cdot\chi_E\cdot\chi_E=\frac{1}{8}\left[1\cdot2\cdot2+2\cdot0\cdot0+1\cdot(-2)\cdot(-2)+2\cdot0\cdot0+2\cdot0\cdot0\right]=1$

(c) $\frac{1}{h}\sum_{\text{all classes}} N\cdot\chi_{A_{1g}}\cdot\chi_{A_{1g}}=\frac{1}{48}\left[1\cdot1\cdot1+8\cdot1\cdot1+3\cdot1\cdot1+6\cdot1\cdot1+6\cdot1\cdot1\right.$

$\left.+1\cdot1\cdot1+8\cdot1\cdot1+3\cdot1\cdot1+6\cdot1\cdot1+6\cdot1\cdot1\right]=1$

$\frac{1}{h}\sum_{\text{all classes}} N\cdot\chi_{A_{2g}}\cdot\chi_{A_{2g}}=\frac{1}{48}\left[1\cdot1\cdot1+8\cdot1\cdot1+3\cdot1\cdot1+6\cdot(-1)\cdot(-1)+6\cdot(-1)\cdot(-1)\right.$

$\left.+1\cdot1\cdot1+8\cdot1\cdot1+3\cdot1\cdot1+6\cdot(-1)\cdot(-1)+6\cdot(-1)\cdot(-1)\right]=1$

$\frac{1}{h}\sum_{\text{all classes}} N\cdot\chi_{E_g}\cdot\chi_{E_g}=\frac{1}{48}\left[1\cdot2\cdot2+8\cdot(-1)\cdot(-1)+3\cdot2\cdot2+6\cdot0\cdot0+6\cdot0\cdot0\right.$

$\left.+1\cdot2\cdot2+8\cdot(-1)\cdot(-1)+3\cdot2\cdot2+6\cdot0\cdot0+6\cdot0\cdot0\right]=1$

$\frac{1}{h}\sum_{\text{all classes}} N\cdot\chi_{T_{1g}}\cdot\chi_{T_{1g}}=\frac{1}{48}\left[1\cdot3\cdot3+8\cdot0\cdot0+3\cdot(-1)\cdot(-1)+6\cdot1\cdot1+6\cdot(-1)\cdot(-1)\right.$

$\left.+1\cdot3\cdot3+8\cdot0\cdot0+3\cdot(-1)\cdot(-1)+6\cdot1\cdot1+6\cdot(-1)\cdot(-1)\right]=1$

$\frac{1}{h}\sum_{\text{all classes}} N\cdot\chi_{T_{2g}}\cdot\chi_{T_{2g}}=\frac{1}{48}\left[1\cdot3\cdot3+8\cdot0\cdot0+3\cdot(-1)\cdot(-1)+6\cdot(-1)\cdot(-1)+6\cdot1\cdot1\right.$

$\left.+1\cdot3\cdot3+8\cdot0\cdot0+3\cdot(-1)\cdot(-1)+6\cdot(-1)\cdot(-1)+6\cdot1\cdot1\right]=1$

$\frac{1}{h}\sum_{\text{all classes}} N\cdot\chi_{A_{1u}}\cdot\chi_{A_{1u}}=\frac{1}{48}\left[1\cdot1\cdot1+8\cdot1\cdot1+3\cdot1\cdot1+6\cdot1\cdot1+6\cdot1\cdot1\right.$

$$+1\cdot(-1)\cdot(-1)+8\cdot(-1)\cdot(-1)+3\cdot(-1)\cdot(-1)+6\cdot(-1)\cdot(-1)+6\cdot(-1)\cdot(-1)\big]=1$$

$$\frac{1}{h}\sum_{\text{all classes}} N\cdot\chi_{A_{2u}}\cdot\chi_{A_{2u}}=\frac{1}{48}\big[1\cdot1\cdot1+8\cdot1\cdot1+3\cdot1\cdot1+6\cdot(-1)\cdot(-1)+6\cdot(-1)\cdot(-1)$$

$$+1\cdot(-1)\cdot(-1)+8\cdot(-1)\cdot(-1)+3\cdot(-1)\cdot(-1)+6\cdot1\cdot1+6\cdot1\cdot1\big]=1$$

$$\frac{1}{h}\sum_{\text{all classes}} N\cdot\chi_{E_u}\cdot\chi_{E_u}=\frac{1}{48}\big[1\cdot2\cdot2+8\cdot(-1)\cdot(-1)+3\cdot2\cdot2+6\cdot0\cdot0+6\cdot0\cdot0$$

$$+1\cdot(-2)\cdot(-2)+8\cdot1\cdot1+3\cdot(-2)\cdot(-2)+6\cdot0\cdot0+6\cdot0\cdot0\big]=1$$

$$\frac{1}{h}\sum_{\text{all classes}} N\cdot\chi_{T_{1u}}\cdot\chi_{T_{1u}}=\frac{1}{48}\big[1\cdot3\cdot3+8\cdot0\cdot0+3\cdot(-1)\cdot(-1)+6\cdot1\cdot1+6\cdot(-1)\cdot(-1)$$

$$+1\cdot(-3)\cdot(-3)+8\cdot0\cdot0+3\cdot1\cdot1+6\cdot(-1)\cdot(-1)+6\cdot1\cdot1\big]=1$$

$$\frac{1}{h}\sum_{\text{all classes}} N\cdot\chi_{T_{2u}}\cdot\chi_{T_{2u}}=\frac{1}{48}\big[1\cdot3\cdot3+8\cdot0\cdot0+3\cdot(-1)\cdot(-1)+6\cdot(-1)\cdot(-1)+6\cdot1\cdot1$$

$$+1\cdot(-3)\cdot(-3)+8\cdot0\cdot0+3\cdot1\cdot1+6\cdot1\cdot1+6\cdot(-1)\cdot(-1)\big]=1$$

(d) $$\frac{1}{h}\sum_{\text{all classes}} N\cdot\chi_{A_{1g}}\cdot\chi_{A_{1g}}=\frac{1}{16}\big[1\cdot1\cdot1+2\cdot1\cdot1+1\cdot1\cdot1+2\cdot1\cdot1+2\cdot1\cdot1$$

$$+1\cdot1\cdot1+2\cdot1\cdot1+1\cdot1\cdot1+2\cdot1\cdot1+2\cdot1\cdot1\big]=1$$

$$\frac{1}{h}\sum_{\text{all classes}} N\cdot\chi_{A_{2g}}\cdot\chi_{A_{2g}}=\frac{1}{16}\big[1\cdot1\cdot1+2\cdot1\cdot1+1\cdot1\cdot1+2\cdot(-1)\cdot(-1)+2\cdot(-1)\cdot(-1)$$

$$+1\cdot1\cdot1+2\cdot1\cdot1+1\cdot1\cdot1+2\cdot(-1)\cdot(-1)+2\cdot(-1)\cdot(-1)\big]=1$$

$$\frac{1}{h}\sum_{\text{all classes}} N\cdot\chi_{B_{1g}}\cdot\chi_{B_{1g}}=\frac{1}{16}\big[1\cdot1\cdot1+2\cdot(-1)\cdot(-1)+1\cdot1\cdot1+2\cdot1\cdot1+2\cdot(-1)\cdot(-1)$$

$$+1\cdot1\cdot1+2\cdot(-1)\cdot(-1)+1\cdot1\cdot1+2\cdot1\cdot1+2\cdot(-1)\cdot(-1)\big]=1$$

$$\frac{1}{h}\sum_{\text{all classes}} N\cdot\chi_{B_{2g}}\cdot\chi_{B_{2g}}=\frac{1}{16}\big[1\cdot1\cdot1+2\cdot(-1)\cdot(-1)+1\cdot1\cdot1+2\cdot(-1)\cdot(-1)+2\cdot1\cdot1$$

$$+1\cdot1\cdot1+2\cdot(-1)\cdot(-1)+1\cdot1\cdot1+2\cdot(-1)\cdot(-1)+2\cdot1\cdot1\big]=1$$

$$\frac{1}{h}\sum_{\text{all classes}} N\cdot\chi_{E_g}\cdot\chi_{E_g}=\frac{1}{16}\big[1\cdot2\cdot2+2\cdot0\cdot0+1\cdot(-2)\cdot(-2)+2\cdot0\cdot0+2\cdot0\cdot0$$

$$+1\cdot2\cdot2+2\cdot0\cdot0+1\cdot(-2)\cdot(-2)+2\cdot0\cdot0+2\cdot0\cdot0\big]=1$$

$$\frac{1}{h}\sum_{\text{all classes}} N\cdot\chi_{A_{1u}}\cdot\chi_{A_{1u}}=\frac{1}{16}\big[1\cdot1\cdot1+2\cdot1\cdot1+1\cdot1\cdot1+2\cdot1\cdot1+2\cdot1\cdot1$$

$$+1\cdot(-1)\cdot(-1)+2\cdot(-1)\cdot(-1)+1\cdot(-1)\cdot(-1)+2\cdot(-1)\cdot(-1)+2\cdot(-1)\cdot(-1)]=1$$

$$\frac{1}{h}\sum_{\text{all classes}} N\cdot\chi_{A_{2u}}\cdot\chi_{A_{2u}}=\frac{1}{16}[1\cdot1\cdot1+2\cdot1\cdot1+1\cdot1\cdot1+2\cdot(-1)\cdot(-1)+2\cdot(-1)\cdot(-1)$$

$$+1\cdot(-1)\cdot(-1)+2\cdot(-1)\cdot(-1)+1\cdot(-1)\cdot(-1)+2\cdot1\cdot1+2\cdot1\cdot1]=1$$

$$\frac{1}{h}\sum_{\text{all classes}} N\cdot\chi_{B_{1u}}\cdot\chi_{B_{1u}}=\frac{1}{16}[1\cdot1\cdot1+2\cdot(-1)\cdot(-1)+1\cdot1\cdot1+2\cdot1\cdot1+2\cdot(-1)\cdot(-1)$$

$$+1\cdot(-1)\cdot(-1)+2\cdot1\cdot1+1\cdot(-1)\cdot(-1)+2\cdot(-1)\cdot(-1)+2\cdot1\cdot1]=1$$

$$\frac{1}{h}\sum_{\text{all classes}} N\cdot\chi_{B_{2u}}\cdot\chi_{B_{2u}}=\frac{1}{16}[1\cdot1\cdot1+2\cdot(-1)\cdot(-1)+1\cdot1\cdot1+2\cdot(-1)\cdot(-1)+2\cdot1\cdot1$$

$$+1\cdot(-1)\cdot(-1)+2\cdot1\cdot1+1\cdot(-1)\cdot(-1)+2\cdot1\cdot1+2\cdot(-1)\cdot(-1)]=1$$

$$\frac{1}{h}\sum_{\text{all classes}} N\cdot\chi_{E_{u}}\cdot\chi_{E_{u}}=\frac{1}{16}[1\cdot2\cdot2+2\cdot0\cdot0+1\cdot(-2)\cdot(-2)+2\cdot0\cdot0+2\cdot0\cdot0$$

$$+1\cdot(-2)\cdot(-2)+2\cdot0\cdot0+1\cdot2\cdot2+2\cdot0\cdot0+2\cdot0\cdot0]=1$$

(e) $$\frac{1}{h}\sum_{\text{all classes}} N\cdot\chi_{A_1}\cdot\chi_{A_1}=\frac{1}{12}[1\cdot1\cdot1+2\cdot1\cdot1+2\cdot1\cdot1+1\cdot1\cdot1+3\cdot1\cdot1+3\cdot1\cdot1]=1$$

$$\frac{1}{h}\sum_{\text{all classes}} N\cdot\chi_{A_2}\cdot\chi_{A_2}=\frac{1}{12}[1\cdot1\cdot1+2\cdot1\cdot1+2\cdot1\cdot1+1\cdot1\cdot1+3\cdot(-1)\cdot(-1)+3\cdot(-1)\cdot(-1)]=1$$

$$\frac{1}{h}\sum_{\text{all classes}} N\cdot\chi_{B_1}\cdot\chi_{B_1}=\frac{1}{12}[1\cdot1\cdot1+2\cdot(-1)\cdot(-1)+2\cdot1\cdot1+1\cdot(-1)\cdot(-1)+3\cdot1\cdot1+3\cdot(-1)\cdot(-1)]=1$$

$$\frac{1}{h}\sum_{\text{all classes}} N\cdot\chi_{B_2}\cdot\chi_{B_2}=\frac{1}{12}[1\cdot1\cdot1+2\cdot(-1)\cdot(-1)+2\cdot1\cdot1+1\cdot(-1)\cdot(-1)+3\cdot(-1)\cdot(-1)+3\cdot1\cdot1]=1$$

$$\frac{1}{h}\sum_{\text{all classes}} N\cdot\chi_{E_1}\cdot\chi_{E_1}=\frac{1}{12}[1\cdot2\cdot2+2\cdot1\cdot1+2\cdot(-1)\cdot(-1)+1\cdot(-2)\cdot(-2)+3\cdot0\cdot0+3\cdot0\cdot0]=1$$

$$\frac{1}{h}\sum_{\text{all classes}} N\cdot\chi_{E_2}\cdot\chi_{E_2}=\frac{1}{12}[1\cdot2\cdot2+2\cdot(-1)\cdot(-1)+2\cdot(-1)\cdot(-1)+1\cdot2\cdot2+3\cdot0\cdot0+3\cdot0\cdot0]=1$$

13.49. At least one three–fold (or higher) proper or improper rotational axis is necessary to have an *E* (or *T*) irreducible representation.

13.51. Irreducible representations from different point groups will have different symmetry classes and orders. Therefore, the GOT can't be applied properly and the concept of orthogonality or normality is irrelevant.

13.53. In the exercise, the character for identity *E* is given as 7, which means that *j* must equal 3 in the $R_h(3)$ character table. Furthermore, we note that *f* orbitals have different phases when we reflect through the origin, so we need to have a negative character for *i*. This leaves us with only one possible representation for f orbitals (in a spherical environment), namely $D_u^{(2j+1)}$ with $j = 3$. The O_h environment has E, C_3, C_2, C_4, i, S_6, σ, and S_4 operations. We determine their characters from the $R_h(3)$ character table:

$\chi(E) = 2j + 1 = 2 \cdot 3 + 1 = 7$

$\chi(C_3) = 1 + 2\cos(120^\circ) + 2\cos(240^\circ) + 2\cos(360^\circ) = 1 + 2\cdot(-\frac{1}{2}) + 2\cdot(-\frac{1}{2}) + 2 = 1$

$\chi(C_2) = 1 + 2\cos(180^\circ) + 2\cos(360^\circ) + 2\cos(540^\circ) = 1 + 2\cdot(-1) + 2 + 2\cdot(-1) = 1 = -1$

$\chi(C_4) = 1 + 2\cos(90^\circ) + 2\cos(180^\circ) + 2\cos(270^\circ) = 1 + 0 + 2\cdot(-1) + 0 = -1$

$\chi(i) = -(2j + 1) = -7$

$\chi(S_6) = -1 + 2\cos(60^\circ) - 2\cos(120^\circ) + 2\cos(180^\circ) = -1 + 2\cdot\frac{1}{2} - 2\cdot(-\frac{1}{2}) + 2\cdot(-1) = -1$

$\chi(\sigma) = -(-1)^3 = 1$

$\chi(S_4) = -1 + 2\cos(90^\circ) - 2\cos(180^\circ) + 2\cos(270^\circ) = -1 + 0 - 2\cdot(-1) + 0 = 1$

Therefore, we have the following character set:

	E	$8\,C_3$	$3\,C_2$	$6\,C_4$	$6\,C_2'$	i	$8\,S_6$	$3\,\sigma_h$	$6\,S_4$	$6\,\sigma_d$
Γ_{combo}	7	1	−1	−1	−1	−7	−1	1	1	1

We can reduce this into its irreducible representations by using the GOT:

$$a(A_{1g}) = \frac{1}{48}\big[1\cdot1\cdot7 + 8\cdot1\cdot1 + 3\cdot1\cdot(-1) + 6\cdot1\cdot(-1) + 6\cdot1\cdot(-1)$$
$$+1\cdot1\cdot(-7) + 8\cdot1\cdot(-1) + 3\cdot1\cdot1 + 6\cdot1\cdot1 + 6\cdot1\cdot1\big] = 0$$

$$a(A_{2g}) = \frac{1}{48}\big[1\cdot1\cdot7 + 8\cdot1\cdot1 + 3\cdot1\cdot(-1) + 6\cdot(-1)\cdot(-1) + 6\cdot(-1)\cdot(-1)$$
$$+1\cdot1\cdot(-7) + 8\cdot1\cdot(-1) + 3\cdot1\cdot1 + 6\cdot(-1)\cdot1 + 6\cdot(-1)\cdot1\big] = 0$$

$$a(E_g) = \frac{1}{48}\big[1\cdot2\cdot7 + 8\cdot(-1)\cdot1 + 3\cdot2\cdot(-1) + 6\cdot0\cdot(-1) + 6\cdot0\cdot(-1)$$
$$+1\cdot2\cdot(-7) + 8\cdot(-1)\cdot(-1) + 3\cdot2\cdot1 + 6\cdot0\cdot1 + 6\cdot0\cdot1\big] = 0$$

$$a(T_{1g}) = \frac{1}{48}\big[1\cdot3\cdot7 + 8\cdot0\cdot1 + 3\cdot(-1)\cdot(-1) + 6\cdot1\cdot(-1) + 6\cdot(-1)\cdot(-1)$$
$$+1\cdot3\cdot(-7) + 8\cdot0\cdot(-1) + 3\cdot(-1)\cdot1 + 6\cdot1\cdot1 + 6\cdot(-1)\cdot1\big] = 0$$

$$a(T_{2g}) = \frac{1}{48}\big[1\cdot3\cdot7 + 8\cdot0\cdot1 + 3\cdot(-1)\cdot(-1) + 6\cdot(-1)\cdot(-1) + 6\cdot1\cdot(-1)$$
$$+1\cdot3\cdot(-7) + 8\cdot0\cdot(-1) + 3\cdot(-1)\cdot1 + 6\cdot(-1)\cdot1 + 6\cdot1\cdot1\big] = 0$$

$$a(A_{1u}) = \frac{1}{48}\big[1\cdot1\cdot7 + 8\cdot1\cdot1 + 3\cdot1\cdot(-1) + 6\cdot1\cdot(-1) + 6\cdot1\cdot(-1)$$

$$+1\cdot(-1)\cdot(-7)+8\cdot(-1)\cdot(-1)+3\cdot(-1)\cdot 1+6\cdot(-1)\cdot 1+6\cdot(-1)\cdot 1\big]=0$$

$$a(A_{2u})=\frac{1}{48}\big[1\cdot 1\cdot 7+8\cdot 1\cdot 1+3\cdot 1\cdot(-1)+6\cdot(-1)\cdot(-1)+6\cdot(-1)\cdot(-1)$$

$$+1\cdot(-1)\cdot(-7)+8\cdot(-1)\cdot(-1)+3\cdot(-1)\cdot 1+6\cdot 1\cdot 1+6\cdot 1\cdot 1\big]=1$$

$$a(E_{u})=\frac{1}{48}\big[1\cdot 2\cdot 7+8\cdot(-1)\cdot 1+3\cdot 2\cdot(-1)+6\cdot 0\cdot(-1)+6\cdot 0\cdot(-1)$$

$$+1\cdot(-2)\cdot(-7)+8\cdot 1\cdot(-1)+3\cdot(-2)\cdot 1+6\cdot 0\cdot 1+6\cdot 0\cdot 1\big]=0$$

$$a(T_{1u})=\frac{1}{48}\big[1\cdot 3\cdot 7+8\cdot 0\cdot 1+3\cdot(-1)\cdot(-1)+6\cdot 1\cdot(-1)+6\cdot(-1)\cdot(-1)$$

$$+1\cdot(-3)\cdot(-7)+8\cdot 0\cdot(-1)+3\cdot 1\cdot 1+6\cdot(-1)\cdot 1+6\cdot 1\cdot 1\big]=1$$

$$a(T_{2u})=\frac{1}{48}\big[1\cdot 3\cdot 7+8\cdot 0\cdot 1+3\cdot(-1)\cdot(-1)+6\cdot(-1)\cdot(-1)+6\cdot 1\cdot(-1)$$

$$+1\cdot(-3)\cdot(-7)+8\cdot 0\cdot(-1)+3\cdot 1\cdot 1+6\cdot 1\cdot 1+6\cdot(-1)\cdot 1\big]=1$$

Thus, the irreducible representation that describes the seven *f* orbitals in octahedral symmetry is $A_{2u} \oplus T_{1u} \oplus T_{2u}$.

13.55. In the case where the position of symmetry is the point $x = \pi/2$, the symmetry elements are E, C_2 (the $x = \pi/2$ axis), and 2 σ's (the *xy* and the *yz* planes).

13.57. According to the scheme in Figure 13.14, ethylene (C_2H_4) has D_{2h} point group symmetry. With the center of inversion being in the middle of the molecule, we can see that the inversion operation changes the sign on the wavefunction; therefore, the character of i should be −1. According to how the axes are defined, the *xz* plane is the plane of the page, and since the orbitals are being reflected onto themselves, the character of $\sigma(xz)$ should be +1. Finally, the *yz* plane cuts through the plane of the page between the two orbitals. Reflecting the orbitals through that plane reflects them onto an orbital of the same phase, so the character of $\sigma(yz)$ should also be +1. This is enough information to indicate that these orbitals have the B_{1u} irreducible representation in the D_{2h} point group.

13.59. **(a)** $a(A)=\frac{1}{2}\big[1\cdot 1\cdot 5+1\cdot 1\cdot 1\big]=3$

$$a(B)=\frac{1}{2}\big[1\cdot 1\cdot 5+1\cdot(-1)\cdot 1\big]=2$$

Therefore, in this case, $\Gamma = 3A \oplus 2B$.

(b) $a(A_1)=\frac{1}{6}\big[1\cdot 1\cdot 6+2\cdot 1\cdot 0+3\cdot 1\cdot 0\big]=1$

$$a(A_2)=\frac{1}{6}\big[1\cdot 1\cdot 6+2\cdot 1\cdot 0+3\cdot(-1)\cdot 0\big]=1$$

$$a(E)=\frac{1}{6}\left[1\cdot 2\cdot 6+2\cdot(-1)\cdot 0+3\cdot 0\cdot 0\right]=2$$

Therefore, in this case, $\Gamma = A_1 \oplus A_2 \oplus 2E$.

(c) $$a(A_1)=\frac{1}{8}\left[1\cdot 1\cdot 6+2\cdot 1\cdot(-2)+1\cdot 1\cdot 2+2\cdot 1\cdot 2+2\cdot 1\cdot(-4)\right]=0$$

$$a(A_2)=\frac{1}{8}\left[1\cdot 1\cdot 6+2\cdot 1\cdot(-2)+1\cdot 1\cdot 2+2\cdot(-1)\cdot 2+2\cdot(-1)\cdot(-4)\right]=1$$

$$a(B_1)=\frac{1}{8}\left[1\cdot 1\cdot 6+2\cdot(-1)\cdot(-2)+1\cdot 1\cdot 2+2\cdot 1\cdot 2+2\cdot(-1)\cdot(-4)\right]=3$$

$$a(B_2)=\frac{1}{8}\left[1\cdot 1\cdot 6+2\cdot(-1)\cdot(-2)+1\cdot 1\cdot 2+2\cdot(-1)\cdot 2+2\cdot 1\cdot(-4)\right]=0$$

$$a(E)=\frac{1}{8}\left[1\cdot 2\cdot 6+2\cdot 0\cdot(-2)+1\cdot(-2)\cdot 2+2\cdot 0\cdot 2+2\cdot 0\cdot(-4)\right]=1$$

Therefore, in this case, $\Gamma = A_2 \oplus 3B_1 \oplus E$.

(d) $$a(A_1)=\frac{1}{24}\left[1\cdot 1\cdot 7+8\cdot 1\cdot(-2)+3\cdot 1\cdot 3+6\cdot 1\cdot 1+6\cdot 1\cdot(-1)\right]=0$$

$$a(A_2)=\frac{1}{24}\left[1\cdot 1\cdot 7+8\cdot 1\cdot(-2)+3\cdot 1\cdot 3+6\cdot(-1)\cdot 1+6\cdot(-1)\cdot(-1)\right]=0$$

$$a(E)=\frac{1}{24}\left[1\cdot 2\cdot 7+8\cdot(-1)\cdot(-2)+3\cdot 2\cdot 3+6\cdot 0\cdot 1+6\cdot 0\cdot(-1)\right]=2$$

$$a(T_1)=\frac{1}{24}\left[1\cdot 3\cdot 7+8\cdot 0\cdot(-2)+3\cdot(-1)\cdot 3+6\cdot 1\cdot 1+6\cdot(-1)\cdot(-1)\right]=1$$

$$a(T_2)=\frac{1}{24}\left[1\cdot 3\cdot 7+8\cdot 0\cdot(-2)+3\cdot(-1)\cdot 3+6\cdot(-1)\cdot 1+6\cdot 1\cdot(-1)\right]=0$$

Therefore, in this case, $\Gamma = 2E \oplus T_1$.

13.61. (a) In D_{6h}, the product $B_{1g} \times B_{1g}$ yields the character set

1×1 (−1)×(−1) 1×1 (−1)×(−1) 1×1 (−1)×(−1) 1×1 (−1)×(−1) 1×1 (−1)×(−1) 1×1 (−1)×(−1) = 1 1 1 1 1 1 1 1 1 1 1 1, which is the A_{1g} irreducible representation.

(b) In T_d, the product $E \times T_1$ yields the character set (6 0 −2 0 0). Using the GOT to reduce this representation, we find that it equals $T_1 \oplus T_2$.

(c) In T_d, the product $T_2 \times T_2$ yields the character set (9 0 1 1 1). Using the GOT to reduce this representation, we find that it equals $A_1 \oplus E \oplus T_1 \oplus T_2$.

(d) In O_h, the product $E_g \times T_{2g}$ yields the character set (6 0 −2 0 0 6 0 −2 0 0). Using the GOT to reduce this representation, we find that it equals $T_{1g} \oplus T_{2g}$.

13.63. In order for a product of functions to be non−zero, the product of the irreducible representations must contain the all−symmetric irreducible representation, usually labeled A_1 (or A or A' or A_1' or A_{1g}). If the product of irreducible representations does not contain A_1, then the integral must be exactly zero.

(a) In C_{3v}, the product $A_1 \times A_2$ yields the character set (1 1 −1), which is the A_2 irreducible representation. Because it does not contain A_1, an integral having these irreducible representations must equal zero.

(b) In C_{6v}, the product $E_1 \times E_2$ yields the character set (4 −1 1 −4 0 0). Using the GOT to reduce this representation, we find that it equals $B_1 \oplus B_2 \oplus E_1$. Thus, it does not contain A_1, and an integral having these irreducible representations must equal zero.

(c) In D_{3h}, the product $A_2' \times A_1'' \times E''$ yields the character set (2 −1 0 2 −1 0), which is the E' irreducible representation. Thus, it does not contain A_1', and an integral having these irreducible representations must equal zero.

(d) In D_{6h}, the product $B_{2g} \times B_{2u}$ yields the character set (1 1 1 1 1 −1 −1 −1 −1 −1 −1), which is the A_{1u} irreducible representation. Thus, it does not contain A_{1g}, and an integral having these irreducible representations must equal zero.

13.65. In C_{4v}, the product $E \times E \times B_2$ yields the character set (4 0 4 0 0). Use the GOT (equation 13.6) to see if this combination contains A_1:

$$a(A_1) = \frac{1}{8}(1 \cdot 1 \cdot 4 + 2 \cdot 1 \cdot 0 + 1 \cdot 1 \cdot 4 + 2 \cdot 1 \cdot 0 + 2 \cdot 1 \cdot 0) = 1$$

Since it *does* contain A_1, the transition *can* occur (there is no symmetry reason requiring that the integral is zero). Note that we did not have to determine the existence of any other irreducible representation to answer this question. Also note that this conclusion does not guarantee that the transition will occur, or to what degree; only that it is not forbidden to occur by symmetry arguments alone.

13.67. According to the character table for the C_{2v} point group, the operator for z−polarized light has A_1 symmetry. The product $A_1 \times A_1$ yields the character set (1 1 1 1), which is A_1. Therefore, a non−zero transition integral is only obtained if the excited state wavefunction is of the same symmetry as the combination $A_1 \times A_1$, which means that it must have A_1 symmetry as well.

13.69. As explained in Example 13.10, the irreducible representation for H−like p orbitals is $D_u^{(1)}$ in $R_h(3)$. According to the $R_h(3)$ character table, the characters for O_h are then as follows:

$\chi(E) = 3$; $\chi(C_3) = 0$; $\chi(C_2) = -1$; $\chi(C_4) = 1$; $\chi(i) = -3$; $\chi(S_6) = 0$; $\chi(\sigma) = 1$; $\chi(S_4) = -1$.

Therefore, we have the following character set:

	E	$8\,C_3$	$3\,C_2$	$6\,C_4$	$6\,C_2'$	i	$8\,S_6$	$3\,\sigma_h$	$6\,S_4$	$6\,\sigma_d$
Γ	3	0	−1	1	−1	−3	0	1	−1	1

Comparison with the O_h character table shows that this equals the T_{1u} character set.

13.71. H_2S has C_{2v} symmetry. To determine the symmetry–adapted linear combination for the molecular orbitals of H_2S, we follow the scheme from Section 13.9 but using only the valence orbitals for H and S:

	$1s_{H1}$	$1s_{H2}$	$3s_S$	$3p_{x,S}$	$3p_{y,S}$	$3p_{z,S}$
E	$1s_{H1}$	$1s_{H2}$	$3s_S$	$3p_{x,S}$	$3p_{y,S}$	$3p_{z,S}$
C_2	$1s_{H2}$	$1s_{H1}$	$3s_S$	$-3p_{x,S}$	$-3p_{y,S}$	$3p_{z,S}$
σ	$1s_{H1}$	$1s_{H2}$	$3s_S$	$3p_{x,S}$	$-3p_{y,S}$	$3p_{z,S}$
σ'	$1s_{H2}$	$1s_{H1}$	$3s_S$	$-3p_{x,S}$	$3p_{y,S}$	$3p_{z,S}$

The combinations for the A_1 wavefunctions are determined by multiplying each row of functions by the character of the row's symmetry element, adding each column, and dividing by the order of the group (4, in this case). We get:

$$\Psi = \frac{1}{4}(1s_{H1} + 1s_{H2} + 1s_{H1} + 1s_{H2}) = \frac{1}{2}(1s_{H1} + 1s_{H2})$$

$$\Psi = \frac{1}{4}(1s_{H2} + 1s_{H1} + 1s_{H2} + 1s_{H1}) = \frac{1}{2}(1s_{H1} + 1s_{H2})$$

$$\Psi = \frac{1}{4}(3s_S + 3s_S + 3s_S + 3s_S) = 3s_S$$

$$\Psi = \frac{1}{4}(3p_{x,S} - 3p_{x,S} + 3p_{x,S} - 3p_{x,S}) = 0$$

$$\Psi = \frac{1}{4}(3p_{y,S} - 3p_{y,S} - 3p_{y,S} + 3p_{y,S}) = 0$$

$$\Psi = \frac{1}{4}(3p_{z,S} + 3p_{z,S} + 3p_{z,S} + 3p_{z,S}) = 3p_{z,S}$$

The only unique wavefunctions we get are $\Psi = ½(1s_{H1} + 1s_{H2})$, $\Psi = 3s_S$, and $\Psi = 3p_{z,S}$. For the A_2 set of wavefunctions:

$$\Psi = \frac{1}{4}(1s_{H1} + 1s_{H2} - 1s_{H1} - 1s_{H2}) = 0$$

$$\Psi = \frac{1}{4}(1s_{H2} + 1s_{H1} - 1s_{H2} - 1s_{H1}) = 0$$

$$\Psi = \frac{1}{4}(3s_S + 3s_S - 3s_S - 3s_S) = 0$$

$$\Psi = \frac{1}{4}(3p_{x,S} - 3p_{x,S} - 3p_{x,S} + 3p_{x,S}) = 0$$

$$\Psi = \frac{1}{4}(3p_{y,S} - 3p_{y,S} + 3p_{y,S} - 3p_{y,S}) = 0$$

$$\Psi = \frac{1}{4}\left(3p_{z,\mathrm{S}} + 3p_{z,\mathrm{S}} - 3p_{z,\mathrm{S}} - 3p_{z,\mathrm{S}}\right) = 0$$

Thus, there are no SALC wavefunctions that belong to the A_2 irreducible representation. For the B_1 set of wavefunctions:

$$\Psi = \frac{1}{4}\left(1s_{\mathrm{H1}} - 1s_{\mathrm{H2}} + 1s_{\mathrm{H1}} - 1s_{\mathrm{H2}}\right) = \frac{1}{2}\left(1s_{\mathrm{H1}} - 1s_{\mathrm{H2}}\right)$$

$$\Psi = \frac{1}{4}\left(1s_{\mathrm{H2}} - 1s_{\mathrm{H1}} + 1s_{\mathrm{H2}} - 1s_{\mathrm{H1}}\right) = \frac{1}{2}\left(1s_{\mathrm{H2}} - 1s_{\mathrm{H1}}\right)$$

$$\Psi = \frac{1}{4}\left(3s_{\mathrm{S}} - 3s_{\mathrm{S}} + 3s_{\mathrm{S}} - 3s_{\mathrm{S}}\right) = 0$$

$$\Psi = \frac{1}{4}\left(3p_{x,\mathrm{S}} + 3p_{x,\mathrm{S}} + 3p_{x,\mathrm{S}} + 3p_{x,\mathrm{S}}\right) = 3p_{x,\mathrm{S}}$$

$$\Psi = \frac{1}{4}\left(3p_{y,\mathrm{S}} + 3p_{y,\mathrm{S}} - 3p_{y,\mathrm{S}} - 3p_{y,\mathrm{S}}\right) = 0$$

$$\Psi = \frac{1}{4}\left(3p_{z,\mathrm{S}} - 3p_{z,\mathrm{S}} + 3p_{z,\mathrm{S}} - 3p_{z,\mathrm{S}}\right) = 0$$

Thus, there are two unique SALC wavefunctions that have B_1 symmetry: $\Psi = \frac{1}{2}(1s_{\mathrm{H2}} - 1s_{\mathrm{H1}})$ and $\Psi = 3p_{x,\mathrm{S}}$. For the B_2 set of wavefunctions:

$$\Psi = \frac{1}{4}\left(1s_{\mathrm{H1}} - 1s_{\mathrm{H2}} - 1s_{\mathrm{H1}} + 1s_{\mathrm{H2}}\right) = 0$$

$$\Psi = \frac{1}{4}\left(1s_{\mathrm{H2}} - 1s_{\mathrm{H1}} - 1s_{\mathrm{H2}} + 1s_{\mathrm{H1}}\right) = 0$$

$$\Psi = \frac{1}{4}\left(3s_{\mathrm{S}} - 3s_{\mathrm{S}} - 3s_{\mathrm{S}} + 3s_{\mathrm{S}}\right) = 0$$

$$\Psi = \frac{1}{4}\left(3p_{x,\mathrm{S}} + 3p_{x,\mathrm{S}} - 3p_{x,\mathrm{S}} - 3p_{x,\mathrm{S}}\right) = 0$$

$$\Psi = \frac{1}{4}\left(3p_{y,\mathrm{S}} + 3p_{y,\mathrm{S}} + 3p_{y,\mathrm{S}} + 3p_{y,\mathrm{S}}\right) = 3p_{y,\mathrm{S}}$$

$$\Psi = \frac{1}{4}\left(3p_{z,\mathrm{S}} - 3p_{z,\mathrm{S}} - 3p_{z,\mathrm{S}} + 3p_{z,\mathrm{S}}\right) = 0$$

Thus, there is one B_2 wavefunction: $\Psi = 3p_{y,\mathrm{S}}$. This gives us a full set of six molecular orbitals having the proper irreducible representations for this molecule.

13.73. For many molecules, a reasonably good approximation for a molecular orbital may in fact be an atomic orbital, especially if that particular atomic orbital is not involved in bonding. Many core orbitals—orbitals not part of the valence shell—do not participate in bonding, so by themselves they may be good representatives of molecular orbitals.

13.75. We will be using four $1s$ orbitals from the hydrogen atoms in CH_4, along with the $1s$, $2s$, $2p_x$, $2p_y$, and $2p_z$ orbitals of the carbon atom. This gives us a total of 9 atomic orbitals to construct the SALCs, and ultimately we should get 9 independent combinations for the molecular orbitals.

13.77. As indicated in Equation 13.12, the "first excited state" of H_2 is actually a set of three wavefunctions that have the same spatial part but different spin parts. Although we assume that the spin part of a wavefunction will not affect its energy, in reality the different spin functions will have a tiny effect on the energy of the overall wavefunction. A detailed enough spectrum will show that the "first excited state" will have three separate lines, not one.

13.79. $\int \left[\frac{1}{\sqrt{2}}\left(s+p_z\right)\right]^* \left[\frac{1}{\sqrt{2}}\left(s-p_z\right)\right] d\tau = \frac{1}{2}\left(\int s^* s\, d\tau - \int s^* p_z\, d\tau + \int p_z^* s\, d\tau - \int p_z^* p_z\, d\tau\right)$

The individual atomic orbitals are themselves normalized and orthogonal to each other. Therefore, the first and last integrals are 1 and the second and third integrals are 0. Subsituting:

$\frac{1}{2}\left(1-0+0-1\right)=0$. Thus, the two hybrid orbitals are orthogonal.

13.81. If the CH_3^+ ion is roughly trigonal planar, then the carbon atom should have sp^2 hybrid orbitals making bonds to the hydrogens, as this is the only hybridization scheme that yields the appropriate directions of bonds.

13.83. The character for E is 5, as all five orbitals operate onto themselves. The character for C_3 would be 2. The character for C_2 would be 1, for σ_h would be 3, for S_3 would be 0, and for σ_v would be 3. Therefore, the set of characters for the five sp^2d hybrid orbitals would be (5 2 1 3 0 3). Using the GOT, this representation reduces to $2A_1' \oplus A_2'' \oplus E'$.

13.85. The two orbitals are not orthogonal because they are on different atoms. Hence, there is no requirement that they be orthogonal. (In fact, if they were, no bond would be formed!)

13.87. The nitrogen atom not only must accommodate three bonds to hydrogen atoms, but also a lone electron pair. Thus, the nitrogen atom needs four hybrid orbitals: sp^3 orbitals.

CHAPTER 14

Rotational and Vibrational Spectroscopy

14.1. For a linear molecule, rotation about the molecular axis isn't recognized as a true rotation. Even if it were, the moment of inertia about that axis is almost identically zero, suggesting that only a negligible amount of energy is needed to promote rotation about that axis. A useful spectrum couldn't ever be measured.

14.3. Rearrange the equation $c = \lambda \cdot \nu$ to solve for ν: $\nu = c / \lambda$. With $c = 2.9979 \times 10^8$ m/s:

(a) $\nu = (2.9979 \times 10^8 \text{ m/s}) / (1.00 \text{ m}) = 3.00 \times 10^8 \text{ s}^{-1}$

(b) $\nu = (2.9979 \times 10^8 \text{ m/s}) / (4.77 \times 10^{-5} \text{ m}) = 6.28 \times 10^{12} \text{ s}^{-1}$

(c) 7894 Å = 7894×10^{-10} m, so $\nu = (2.9979 \times 10^8 \text{ m/s}) / (7894 \times 10^{-10} \text{ m}) = 3.798 \times 10^{14} \text{ s}^{-1}$

(d) $\nu = (2.9979 \times 10^8 \text{ m/s}) / (1.903 \times 10^3 \text{ m}) = 1.575 \times 10^5 \text{ s}^{-1}$

14.5. The energy of a photon of frequency ν is $E = h\nu$. Frequency and wavelength are related by $c = \lambda\nu$, so $\nu = c / \lambda$. Substituting this into the energy equation gives $E = hc / \lambda$. Finally, the wavenumber $\tilde{\nu}$ is the reciprocal wavelength, $\tilde{\nu} = 1/\lambda$, so the energy equation can also be written as $E = hc\tilde{\nu}$. Selecting the appropriate version of the energy equation:

(a) $E = h\nu = (6.626 \times 10^{-34} \text{ J·s}) (6.03 \times 10^{14} \text{ s}^{-1}) = 4.00 \times 10^{-19} \text{ J}$

(b) $E = hc / \lambda = (6.626 \times 10^{-34} \text{ J·s}) (2.9979 \times 10^8 \text{ m/s}) / (9.27 \times 10^{-9} \text{ m}) = 2.14 \times 10^{-17} \text{ J}$

(c) 4320 waves of light per cm equals 4320×10^2 waves per meter: $4320 \text{ cm}^{-1} = 4320 \times 10^2 \text{ m}^{-1}$:

$$E = hc\tilde{\nu} = (6.626 \times 10^{-34} \text{ J·s}) (2.9979 \times 10^8 \text{ m/s}) (4320 \times 10^2 \text{ m}^{-1}) = 8.581 \times 10^{-20} \text{ J}$$

(d) 1 μ equals 10^{-6} m, so

$$E = hc / \lambda = (6.626 \times 10^{-34} \text{ J·s}) (2.9979 \times 10^8 \text{ m/s}) / (5.69 \times 10^{-6} \text{ m}) = 3.49 \times 10^{-20} \text{ J}$$

14.7. Rearrange the equation $c = \lambda \cdot \nu$ to solve for λ:
$\lambda = c / \nu = (2.9979 \times 10^8 \text{ m/s}) / (8.041 \times 10^{12} \text{ s}^{-1}) = 3.728 \times 10^{-5} \text{ m}$

$E = h\nu = (6.626 \times 10^{-34} \text{ J·s}) (8.041 \times 10^{12} \text{ s}^{-1}) = 5.328 \times 10^{-21} \text{ J}$

The speed of the photon is the same as any other photon's speed (in vacuum): 2.9979×10^8 m/s.

14.9. The prefix "μ" stands for 10^{-6}, the prefix "c" for 10^{-2}. Therefore, 1 μm = 10^{-6} m and 1 cm = 10^{-2} m, which means there are 10^4 = 10,000 micrometers in 1 centimeter. A photon of light having $\tilde{\nu} = x\ \text{cm}^{-1}$ and a wavelength of $\lambda = y$ μm therefore fulfills the relationship $\frac{x}{\text{cm}} \times (y\ \mu\text{m}) = \frac{x}{\text{cm}} \times (10{,}000y\ \text{cm})$. The units cancel out, which leaves $x \times y$ = 10,000.

14.11. To evaluate the rotational energies for a diatomic molecule (equation 14.7), we first need to calculate the moment of inertia, $I = \mu r^2$. The reduced *molar* mass of $^{79}Br^{19}F$ is

(78.92 g/mol) (19.00 g/mol) / (78.92 g/mol + 19.00 g/mol) = 15.31 g/mol. We divide by Avogadro's constant to obtain the reduced mass *per molecule*, and convert g into kg:

$\mu(^{79}Br^{19}F) = (15.31\ \text{g/mol}) / (6.022\times10^{23}\ \text{mol}^{-1}) = 2.543\times10^{-23}\ \text{g} = 2.543\times10^{-26}\ \text{kg}$

$I = (2.543\times10^{-26}\ \text{kg})\ (0.1756\times10^{-9}\ \text{m})^2 = 7.841\times10^{-46}\ \text{kg·m}^2$

The rotational energies (in J) are given by $E = \frac{J(J+1)\hbar^2}{2I} = \frac{J(J+1)h^2}{4\pi^2 \cdot 2I}$. To convert them into wavenumbers, we divide by hc (remember that $E = h\nu = hc\tilde{\nu}$):

$$\tilde{\nu} = \frac{E}{hc} = \frac{J(J+1)h}{8\pi^2 Ic} = BJ(J+1).$$

To facilitate the calculations, we first evaluate the constant

$$B = \frac{h}{8\pi^2 Ic} = \frac{6.626\times10^{-34}\ \text{J}\cdot\text{s}}{8\pi^2(7.841\times10^{-46}\ \text{kg}\cdot\text{m}^2)(2.9979\times10^8\ \text{m/s})} = 35.70\ \text{m}^{-1}$$, or B = 0.3570 cm^{-1}.

$J = 0$: $E = BJ(J+1) = 0$; $J = 1$: $E = BJ(J+1) = 1\cdot2\cdot(0.3570\ \text{cm}^{-1}) = 0.7140\ \text{cm}^{-1}$;

$J = 2$: $E = 2\cdot3\cdot(0.3570\ \text{cm}^{-1}) = 2.142\ \text{cm}^{-1}$; $J = 3$: $E = 3\cdot4\cdot(0.3570\ \text{cm}^{-1}) = 4.284\ \text{cm}^{-1}$.

14.13. **(a)** prolate symmetric top; **(b)** spherical top; **(c)** spherical top; **(d)** asymmetric top; **(e)** asymmetric top; **(f)** asymmetric top.

14.15. We need to determine the reduced mass and moment of inertia for S_2. For that, we first convert the molar mass of 32.066 g/mol for S into a mass per molecule by dividing by Avogadro's constant and converting g to kg:

$(32.066\ \text{g/mol}) / (6.022\times10^{23}\ \text{mol}^{-1}) = 5.325\times10^{-23}\ \text{g} = 5.325\times10^{-26}\ \text{kg}$

$$\mu = \frac{(5.325\times10^{-26}\ \text{kg})(5.325\times10^{-26}\ \text{kg})}{(5.325\times10^{-26}\ \text{kg})+(5.325\times10^{-26}\ \text{kg})} = 2.662\times10^{-26}\ \text{kg}$$

$I = (2.662\times10^{-26}\ \text{kg})\ (1.880\times10^{-10}\ \text{m})^2 = 9.410\times10^{-46}\ \text{kg·m}^2$

As in exercise 14.11, we now evaluate B:

$$B=\frac{h}{8\pi^2 Ic}=\frac{6.626\times10^{-34}\ \text{J}\cdot\text{s}}{8\pi^2\left(9.410\times10^{-46}\ \text{kg}\cdot\text{m}^2\right)\left(2.9979\times10^{8}\ \text{m/s}\right)}=29.75\ \text{m}^{-1}\text{, or } 0.2975\ \text{cm}^{-1}.$$

Or in units of J: $B'=\frac{\hbar^2}{2I}=\frac{h^2}{8\pi^2 I}=\frac{\left(6.626\times10^{-34}\ \text{J}\cdot\text{s}\right)^2}{8\pi^2\left(9.410\times10^{-46}\ \text{kg}\cdot\text{m}^2\right)}=5.909\times10^{-24}\ \text{J}.$

14.17. $A=\frac{\left(6.626\times10^{-34}\ \text{J}\cdot\text{s}\right)^2}{8\pi^2\left(5.478\times10^{-47}\ \text{kg}\cdot\text{m}^2\right)}=1.015\times10^{-22}\ \text{J}$; divide by hc to get $A' = 5.110\ \text{cm}^{-1}$.

$$B=\frac{\left(6.626\times10^{-34}\ \text{J}\cdot\text{s}\right)^2}{8\pi^2\left(5.478\times10^{-47}\ \text{kg}\cdot\text{m}^2\right)}=1.015\times10^{-22}\ \text{J (or } 5.110\ \text{cm}^{-1}\text{)}$$

$$C=\frac{\left(6.626\times10^{-34}\ \text{J}\cdot\text{s}\right)^2}{8\pi^2\left(6.645\times10^{-47}\ \text{kg}\cdot\text{m}^2\right)}=8.368\times10^{-23}\ \text{J (or } 4.213\ \text{cm}^{-1}\text{)}$$

The two lower moments of inertia are the same, so the two higher rotational constants are the same. This is consistent with PH_3 being an oblate symmetric top.

14.19. Rearranging $B=\frac{h}{8\pi^2 Ic}$ leads to $I=\frac{h}{8\pi^2 Bc}$. Substituting:

$$I=\frac{6.626\times10^{-34}\ \text{J}\cdot\text{s}}{8\pi^2\left(1.176\times10^{2}\ \text{m}^{-1}\right)\left(2.9979\times10^{8}\ \text{m/s}\right)}=2.380\times10^{-46}\ \text{kg}\cdot\text{m}^2.$$

14.21. The first energy level has the value $E = 0$. The second energy level comes from the $J = 1$ value of the lower of the two rotational constants (see exercise 14.17):

$$E=CJ(J+1)=1\cdot2\cdot8.368\times10^{-23}\ \text{J}=1.674\times10^{-22}\ \text{J}.$$

The third energy level comes from the $J = 1$ value of the higher of the two rotational constants:

$$E=BJ(J+1)=1\cdot2\cdot1.015\times10^{-22}\ \text{J}=2.030\times10^{-22}\ \text{J}.$$

The fourth energy level comes from the $J = 2$ value of the lower of the two constants:

$$E=CJ(J+1)=2\cdot3\cdot8.368\times10^{-23}\ \text{J}=5.021\times10^{-22}\ \text{J}$$

Finally, the fifth energy level comes from the $J = 2$ value of the higher of the two constants:

$$E=BJ(J+1)=2\cdot3\cdot1.015\times10^{-22}\ \text{J}=6.090\times10^{-22}\ \text{J}.$$

14.23. **(a)** Deuterium has no permanent dipole moment; therefore, it will not have a pure rotational spectrum.

(b) Carbon monoxide has a permanent dipole moment and will have a pure rotational spectrum. **(c)** *cis*-1,2-Dichloroethylene will have a pure rotational spectrum.

(d) *trans*-1,2-Dichloroethylene will not have a pure rotational spectrum.

(e) Chloroform will have a pure rotational spectrum.

(f) Buckminsterfullerene will not have a pure rotational spectrum.

14.25. Remember that allowed transitions have $\Delta J = \pm 1$, $\Delta M_J = 0$ or ± 1 (for linear molecules), and $\Delta K = 0$ (for symmetric tops):

(a) allowed;

(b) not allowed ($J = -1$ does not exist);

(c) not allowed (J must increase or decrease by 1, but cannot remain unchanged);

(d) not allowed (J must increase or decrease by 1, but cannot decrease by 2).

14.27. Starting with $\Delta E = \frac{(J+1)(J+2)\hbar^2}{2I} - \frac{J(J+1)\hbar^2}{2I}$, we factor out $\frac{\hbar^2}{2I}$ from the right–hand side terms: $\Delta E = \frac{\hbar^2}{2I}\left[(J+1)(J+2) - J(J+1)\right]$. Multiplying out and simplifying:

$$\Delta E = \frac{\hbar^2}{2I}\left[\left(J^2 + 1J + 2J + 2\right) - \left(J^2 + J\right)\right] = \frac{\hbar^2(2J+2)}{2I}.$$ This is equation 14.21.

14.29. If the rotational spectrum consists of lines spaced by 0.114 cm^{-1}, this value represents $2B$, so $B = 0.057$ cm^{-1}. The reduced *molar* mass of $^{127}I^{35}Cl$ is (126.90 g/mol) (34.97 g/mol) / (126.90 g/mol + 34.97 g/mol) = 27.42 g/mol.

Divide by Avogadro's constant to obtain the reduced mass *per molecule*, and convert g into kg:

$\mu(^{127}I^{35}Cl) = (27.42\ \text{g/mol}) / (6.022\times10^{23}\ \text{mol}^{-1}) = 4.553\times10^{-23}\ \text{g} = 4.553\times10^{-26}\ \text{kg}$

Rearrange the wavenumber form of the equation for the rotational constant B,

$B = \frac{h}{8\pi^2\left(\mu r^2\right)c}$, to solve for r^2: $r^2 = \frac{h}{8\pi^2 \mu c B}$. Substituting the numerical values:

$$r^2 = \frac{6.626\times10^{-34}\ \text{J}\cdot\text{s}}{8\pi^2\left(4.553\times10^{-26}\ \text{kg}\right)\left(2.9979\times10^{8}\ \text{m/s}\right)\left(\tfrac{1}{2}\cdot 0.114\times10^{2}\ \text{m}^{-1}\right)} = 1.08\times10^{-19}\ \text{m}^2$$

(remember that 1 J = 1 kg · m^2/s^2). Now solve for r: $r = 3.28\times10^{-10}$ m = 3.28 A.

14.31. The easiest way to solve this question is to convert the B value into GHz (frequency) units, then use the fact that the first four absorptions will appear at $2B$, $4B$, $6B$, and $8B$. If the rotational spectrum consists of lines spaced by 15.026 cm^{-1}, this value represents $2B$, so $B = 7.513$ cm^{-1}. To convert this into frequency units, we substitute $\lambda = \frac{1}{\tilde{\nu}}$ into

$c = \lambda\nu = \frac{\nu}{\tilde{\nu}}$ and solve for ν:

$\nu = c\tilde{\nu} = (2.9979\times10^{8}\ \text{m/s})(7.513\times10^{2}\ \text{m}^{-1}) = 2.252\times10^{11}\ \text{s}^{-1}$. 1 GHz is $1\times10^{9}\ \text{s}^{-1}$, so this value equals 225.2 GHz. Therefore, the first four lines of the rotational spectrum should appear at 450.5 GHz, 900.9 GHz, 1351 GHz, and 1802 GHz.

14.33. The reduced mass of O_2 is (using the atomic mass instead of that of a particular isotope)

$$\mu = \frac{(15.9994\ \text{g/mol})(15.9994\ \text{g/mol})}{(15.9994\ \text{g/mol} + 15.9994\ \text{g/mol})} \times \left(\frac{1\ \text{mol}}{6.022\times10^{23}}\right) = 1.328\times10^{-23}\ \text{g} = 1.328\times10^{-26}\ \text{kg}.$$

Then $I = \mu r^2 = (1.328\times10^{-26}\ \text{kg})(0.1210\times10^{-9}\ \text{m})^2 = 1.945\times10^{-46}\ \text{kg}\cdot\text{m}^2$ and

$$B = \frac{\hbar^2}{2I} = \frac{h^2}{8\pi^2 I} = \frac{(6.626\times10^{-34}\ \text{J}\cdot\text{s})^2}{8\pi^2(1.945\times10^{-46}\ \text{kg}\cdot\text{m}^2)} = 2.859\times10^{-23}\ \text{J}$$. Using equation 14.23, we

find $J_{max} \approx \left(\frac{kT}{2B}\right)^{1/2} = \left[\frac{(1.381\times10^{-23}\ \text{J/K})(350\ \text{K})}{2(2.859\times10^{-23}\ \text{J})}\right]^{1/2} = 9$ (rounded to the nearest integer).

14.35. The reduced mass of HS is (using the atomic mass instead of a particular isotope)

$$\mu = \frac{(32.066\ \text{g/mol})(1.0079\ \text{g/mol})}{(32.066\ \text{g/mol} + 1.0079\ \text{g/mol})} \times \left(\frac{1\ \text{mol}}{6.022\times10^{23}}\right) = 1.623\times10^{-24}\ \text{g} = 1.623\times10^{-27}\ \text{kg}.$$

Then $I = \mu r^2 = (1.623\times10^{-27}\ \text{kg})(1.40\times10^{-10}\ \text{m})^2 = 3.18\times10^{-47}\ \text{kg}\cdot\text{m}^2$ and

$$B = \frac{\hbar^2}{2I} = \frac{h^2}{8\pi^2 I} = \frac{(6.626\times10^{-34}\ \text{J}\cdot\text{s})^2}{8\pi^2(3.18\times10^{-47}\ \text{kg}\cdot\text{m}^2)} = 1.75\times10^{-22}\ \text{J}.$$

Knowing that J_{max} is 8, we can estimate the sample temperature by rearranging equation 14.23:

$$J_{max} \approx \left(\frac{kT}{2B}\right)^{1/2},\ \text{so}\ J_{max}{}^2 \approx \frac{kT}{2B}\ \text{and}\ T \approx \frac{2BJ_{max}{}^2}{k} = \frac{2(1.75\times10^{-22}\ \text{J})(8)^2}{1.381\times10^{-23}\ \text{J}} = 1.62\times10^{3}\ \text{K} \approx 1600\ \text{K}.$$

14.37. From Table 14.2, we find that $B(\text{HCl}) = 10.59\ \text{cm}^{-1}$. Because $B = \frac{\hbar^2}{2I} = \frac{\hbar^2}{2\mu r^2}$, B is inversely proportional to μ (if we assume that everything else remains the same). Therefore, we can set up the following ratio: $\frac{B(\text{DCl})}{B(\text{HCl})} = \frac{1/\mu(\text{DCl})}{1/\mu(\text{HCl})} = \frac{\mu(\text{HCl})}{\mu(\text{DCl})}$. Because we take ratios, the units for μ don't matter and we can directly use the reduced *molar* masses: $\mu(\text{HCl}) = \frac{(1.008\ \text{g/mol})(35.4527\ \text{g/mol})}{1.008\ \text{g/mol} + 35.4527\ \text{g/mol}} = 0.9801\ \text{g/mol}$ and

$\mu(\text{DCl}) = \frac{(2.014\ \text{g/mol})(35.4527\ \text{g/mol})}{2.014\ \text{g/mol} + 35.4527\ \text{g/mol}} = 1.906\ \text{g/mol}$. (We are simply using the atomic mass of Cl, rather than that of a particular isotope.) This gives

$$\frac{B(\text{DCl})}{B(\text{HCl})} = \frac{0.9792\ \text{g/mol}}{1.906\ \text{g/mol}} = 0.5138 \text{ and}$$
$$B(\text{DCl}) = 0.5143 \cdot B(\text{HCl}) = 0.5143 \cdot 10.59\ \text{cm}^{-1} = 5.446\ \text{cm}^{-1}.$$

14.39. Equation 14.26 can be derived by using equation 14.25 and subtracting E_{rot} for J from E_{rot} for $J + 1$: $E_{J+1} - E_J = \left[B(J+1)(J+2) - D_J(J+1)^2(J+2)^2\right] - \left[BJ(J+1) - D_J J^2(J+1)^2\right]$

Collecting B and D_J terms separately and factoring out equal terms, we obtain

$$E_{J+1} - E_J = B(J+1)(J+2) - BJ(J+1) - D_J(J+1)^2(J+2)^2 + D_J J^2(J+1)^2$$

$= B(J+1)[J+2-J] - D_J(J+1)^2\left[(J+2)^2 - J^2\right]$. Multiplying out and simplifying leads to $E_{J+1} - E_J = B(J+1)[2] - D_J(J+1)^2\left[J^2 + 4J + 4 - J^2\right]$

$$= 2B(J+1) - D_J(J+1)^2[4J+4]$$

$$= 2B(J+1) - 4D_J(J+1)^2[J+1]$$

$= 2B(J+1) - 4D_J(J+1)^3$. This is equation 14.26.

14.41. First, at such high values of rotational quantum number, we expect that centrifugal distortions will make the energy levels (and hence the energy difference) deviate quite a bit from those of a rigid rotor. Second, as a nonpolar molecule, diatomic iodine wouldn't show a pure rotational spectrum.

14.43. Using equation 14.27: $D_J \approx \frac{4B^3}{\tilde{\nu}^2} = \frac{4(60.80\ \text{cm}^{-1})^3}{(4320\ \text{cm}^{-1})^2} = 0.04817\ \text{cm}^{-1} = 4.817\times10^{-2}\ \text{cm}^{-1}$,

which is only about 4% higher than the value of $4.64\times10^{-2}\ \text{cm}^{-1}$ from Table 14.2.

14.45. **(a)** Total degrees of freedom: 3 · 2 = 6. HF is linear, so vibrational degrees of freedom: 6 – 5 = 1.

(b) Total degrees of freedom: 3 · 3 = 9; vibrational degrees of freedom: 9 – 6 = 3 (nonlinear).

(c) Total degrees of freedom: 3 · 60 = 180; vibrational degrees of freedom: 180 – 6 = 174 (nonlinear).

(d) Total degrees of freedom: 3 · 23 = 69; vibrational degrees of freedom: 69 – 6 = 63 (nonlinear).

(e) Total degrees of freedom: 3 · 18 = 54; vibrational degrees of freedom: 54 – 6 = 48 (nonlinear).

(f) Total degrees of freedom: 3 · 4 = 12; vibrational degrees of freedom: 12 – 5 = 7 (linear).

(g) Total degrees of freedom: $3 \cdot 4 = 12$; vibrational degrees of freedom: $12 - 6 = 6$ (nonlinear).

14.47. The total number of normal modes equals the number of vibrational degrees of freedom. (We should differentiate the total number of normal modes from the total number of *distinct* vibrational frequencies, which may be different due to degeneracies.) Therefore, the total number of normal modes for the molecules in exercise 14.45 are **(a)** 1; **(b)** 3; **(c)** 174; **(d)** 63; **(e)** 48; **(f)** 7; and **(g)** 6.

14.49. If a hydrogen atom in CH_4 is replaced with a deuterium to make CH_3D, we would expect the number of IR–active vibrations to increase because the molecule is going from a T_d symmetry to a lower C_{3v} symmetry. In actuality, CH_3D has 6 IR–active vibrations.

14.51. We rearrange equation 14.32 to isolate the force constant k: $\nu = \frac{1}{2\pi}\sqrt{\frac{k}{\mu}}$, so $\frac{k}{\mu} = (2\pi\nu)^2$ and $k = 4\pi^2\nu^2\mu$. The reduced *molar* mass of $^1H^{35}Cl$ is 0.9722 g/mol (see Example 14.12), and the reduced mass *per molecule* is obtained by dividing by Avogadro's nuimber: $(0.9722\ \text{g/mol}) \times \left(\frac{1\ \text{mol}}{6.022 \times 10^{23}}\right) = 1.614 \times 10^{-24}\ \text{g} = 1.614 \times 10^{-27}\ \text{kg}$. With $\nu = 8.652 \times 10^{13}\ \text{s}^{-1}$, this gives

$k = 4\pi^2(8.652 \times 10^{13}\ \text{s}^{-1})^2(1.614 \times 10^{-27}\ \text{kg}) = 4.771 \times 10^2\ \text{kg/s}^2$. Remembering that $1\ \text{N} = 1\ \text{kg} \cdot \text{m/s}^2$, we can also express the units as N/m, which is more common for a force constant, so $k \approx 477$ N/m.

14.53. The force constant is given by $k = 4\pi^2\nu^2\mu$ (see 14.51). The reduced mass of BrF is

$$\frac{(79.904\ \text{g/mol})(18.9984\ \text{g/mol})}{79.904\ \text{g/mol} + 18.9984\ \text{g/mol}} \times \left(\frac{1\ \text{mol}}{6.022 \times 10^{23}}\right) = 2.549 \times 10^{-23}\ \text{g} = 2.549 \times 10^{-26}\ \text{kg}$$

(using atomic masses instead of masses for particular isotopes). This gives $k = 4\pi^2(2.007 \times 10^{13}\ \text{s}^{-1})^2(2.549 \times 10^{-26}\ \text{kg}) = 405.3\ \text{N/m}$ (remember that $1\ \text{N} = 1\ \text{kg} \cdot \text{m} / \text{s}^2$, so

$1\ \text{kg}/\text{s}^2 = 1\ \text{N}/\text{m}$). The ground–state vibrational energy is $E = h\nu\,(0 + ½) = ½\,h\nu$, which evaluates to $E = ½\,(6.626 \times 10^{-34}\ \text{J·s})(2.007 \times 10^{13}\ \text{s}^{-1}) = 6.649 \times 10^{-21}$ J.

14.55. First, we need the reduced masses of the C=O and C=S bonds. Because we are only interested in the ratio, units of g/mol for reduced masses are fine:

$$\mu(\text{C=O}) = \frac{(12.011\ \text{g/mol})(15.9994\ \text{g/mol})}{12.011\ \text{g/mol} + 15.9994\ \text{g/mol}} = 6.8606\ \text{g/mol};$$

$$\mu(\text{C=S}) = \frac{(12.011\ \text{g/mol})(32.066\ \text{g/mol})}{12.011\ \text{g/mol} + 32.066\ \text{g/mol}} = 8.7380\ \text{g/mol}.$$

The ratio of vibrational frequencies of the C=S and C=O bonds are directly given by the square root of the inverse reduced masses:

$\frac{\tilde{\nu}(\text{C=S})}{\tilde{\nu}(\text{C=O})} = \sqrt{\frac{\mu(\text{C=O})}{\mu(\text{C=S})}} = \sqrt{\frac{6.8606 \text{ g/mol}}{8.7380 \text{ g/mol}}} = 0.88609$, which means that

$\tilde{\nu}(\text{C=S}) = 0.88609 \cdot \tilde{\nu}(\text{C=O}) = 0.88609 \cdot (1338 \text{ cm}^{-1}) = 1186 \text{ cm}^{-1}$. The frequency of the C=S vibration actually appears at 859 cm^{-1}. Thus, assuming that S is an "isotope" of O is not a good assumption.

14.57. Nitrogen is used because, as a nonpolar diatomic molecule, its single vibrational motion is IR-inactive. Oxygen gas and any noble gas (He, Ne, Ar, Kr) could also be used—but they are more expensive. Thus, using nitrogen gas as a purge gas means that the infrared spectrum is measuring the vibrations of the sample, not any IR–absorbing contaminant from the atmosphere.

14.59. **(a)** CO has a permanent dipole moment that varies as the molecule vibrates. Therefore, it does have a pure vibrational spectrum.

(b) He does not have a vibrational spectrum; at least two atoms are needed for vibrations.

(c) SO_2 is a bent molecule with a permanent dipole moment that varies as the molecule vibrates. Thus, it shows a pure vibrational spectrum.

(d) SO_3 is planar and does not have a permanent dipole moment. However, some of its vibrations will produce a fleeting dipole moment, so it will have a pure vibrational spectrum.

(e) I_2, a homonuclear diatomic molecule, does not have a permanent dipole moment and no changing dipole moment as the two atoms vibrate. It will not have a pure vibrational spectrum.

14.61. Fundamental vibrations are the $\text{v} = 0 \rightarrow \text{v} = 1$ transitions.

Overtone vibrations are the $\text{v} = 0 \rightarrow \text{v} = n$ transitions, where n is any number other than 1.

Hot bands are the $\text{v} = n \rightarrow \text{v} = n + 1$ transitions, where n is any number other than 0 (that is, the vibrational transition starts in an excited vibrational state).

14.63. Starting with the expression $V = \frac{1}{2}kx^2$, we take the derivative of this expression with respect to x twice: $\frac{\partial V}{\partial x} = kx$ and $\frac{\partial^2 V}{\partial x^2} = k$, which verifies the relationship. The second derivative of energy with respect to position has units of J / m^2. Recall that 1 J = 1 N · m, so

J / m^2 is equivalent to (N · m) / m^2 or N / m, which is indeed the unit of k.

14.65. We need to go through a three–step process to determine a. First, use the data in Table 14.4 to calculate the "pure" anharmonicity constant x_e; second, use equation 14.40 to determine D_e; third, use equation 14.37 along with data given in the exercise to determine a.

For HF: $x_e = \frac{x_e\tilde{\nu}_e}{\tilde{\nu}_e} = \frac{90.07\ \text{cm}^{-1}}{4138.52\ \text{cm}^{-1}} = 0.02176$.

Rearranging equation 14.40 (recall that $c = \lambda\nu$, so $\nu = c/\lambda = c\tilde{\nu}$):

$$D_e = \frac{h\nu_e}{4x_e} = \frac{hc\tilde{\nu}_e}{4x_e} = \frac{(6.626\times10^{-34}\ \text{J}\cdot\text{s})(2.9979\times10^{10}\ \text{cm/s})(4138.52\ \text{cm}^{-1})}{4(0.02176)} = 9.443\times10^{-19}\ \text{J}.$$

$$a = \left(\frac{k}{2D_e}\right)^{1/2} = \sqrt{\frac{965.1\ \text{N/m}}{2(9.445\times10^{-19}\ \text{J})}} = 2.261\times10^{10}\ \text{m}^{-1} \simeq 2.26\ \text{Å}^{-1}$$ (1 m = 10^{10} Å, so inversely 1 $\text{m}^{-1} = 10^{-10}\ \text{Å}^{-1}$).

For HBr: $x_e = \frac{x_e\tilde{\nu}_e}{\tilde{\nu}_e} = \frac{45.21\ \text{cm}^{-1}}{2649.67\ \text{cm}^{-1}} = 0.01706$.

$$D_e = \frac{hc\tilde{\nu}_e}{4x_e} = \frac{(6.626\times10^{-34}\ \text{J}\cdot\text{s})(2.9979\times10^{10}\ \text{cm/s})(2649.67\ \text{cm}^{-1})}{4(0.01706)} = 7.712\times10^{-19}\ \text{J}.$$

$$a = \left(\frac{k}{2D_e}\right)^{1/2} = \sqrt{\frac{411.5\ \text{N/m}}{2(7.712\times10^{-19}\ \text{J})}} = 1.633\times10^{10}\ \text{m}^{-1} \simeq 1.63\ \text{Å}^{-1}.$$

Comparing these values with 1.87 Å^{-1} for HCl, we find a trend that a decreases as the size of the halogen atom increases.

14.67. Using the fact that $D_e = D_0 + \frac{1}{2}h\nu$ and $\nu = c/\lambda = c\tilde{\nu}$, we can calculate D_0 from $D_e = D_0 + \frac{1}{2}hc\tilde{\nu}_e$. The D_0 values are given in units of J per mole, so we must first convert them into J per molecule by dividing by Avogadro's number:

D_0 (HBr) = (362×10^3 J/mol) / (6.022×10^{23} mol^{-1}) = 6.01×10^{-19} J

D_0 (CO) = (1071×10^3 J/mol) / (6.022×10^{23} mol^{-1}) = 1.778×10^{-18} J

Then, for HBr:

$$D_e = 6.01\times10^{-19}\ \text{J} + \tfrac{1}{2}(6.626\times10^{-34}\ \text{J}\cdot\text{s})(2.9979\times10^{10}\ \text{cm/s})(2649.67\ \text{cm}^{-1}) = 6.27\times10^{-19}\ \text{J}$$

For CO:

$$D_e = 1.778\times10^{-18}\ \text{J} + \tfrac{1}{2}(6.626\times10^{-34}\ \text{J}\cdot\text{s})(2.9979\times10^{10}\ \text{cm/s})(2170.21\ \text{cm}^{-1}) = 1.800\times10^{-18}\ \text{J}$$

Now we determine x_e and $x_e\tilde{\nu}_e$. For HBr, using equation 14.40:

$$x_e = \frac{h\nu_e}{4D_e} = \frac{hc\tilde{\nu}_e}{4D_e} = \frac{(6.626\times10^{-34}\ \text{J}\cdot\text{s})(2.9979\times10^{10}\ \text{cm/s})(2649.67\ \text{cm}^{-1})}{4(6.27\times10^{-19}\ \text{J})} = 0.0210,$$ so

$x_e\tilde{\nu}_e = (0.0210)(2649.67\ \text{cm}^{-1}) = 55.6\ \text{cm}^{-1}$.

For CO:

$$x_e = \frac{h\nu_e}{4D_e} = \frac{hc\tilde{\nu}_e}{4D_e} = \frac{(6.626\times10^{-34}\text{ J}\cdot\text{s})(2.9979\times10^{10}\text{ cm/s})(2170.21\text{ cm}^{-1})}{4(1.800\times10^{-18}\text{ J})} = 0.005987$$

$$x_e\tilde{\nu}_e = (0.005987)(2170.21\text{ cm}^{-1}) = 12.99\text{ cm}^{-1}$$

The predicted $x_e\tilde{\nu}_e$ value for HBr is off by about 20%, while that for CO is rather close.

14.69. H_2S has $3\cdot3-6=3$ vibrational degrees of freedom, so what is given are the vibrational frequencies of these three modes. Because $D_e = D_0 + \sum_{i=1}^{3}\frac{1}{2}h\nu_i$, the difference between D_e and D_0 is given by the sum over the frequencies. Converting to wavenumbers, this gives:

$D_e - D_0 = \sum_{i=1}^{3}\frac{1}{2}hc\tilde{\nu}_i = \frac{1}{2}hc\sum_{i=1}^{3}\tilde{\nu}_i$. Substituting the numerical values:

$$D_e - D_0 = \tfrac{1}{2}(6.626\times10^{-34}\text{ J}\cdot\text{s})(2.9979\times10^{10}\text{ cm/s})(2626\text{ cm}^{-1}+2615\text{ cm}^{-1}+1183\text{ cm}^{-1})$$

$D_e - D_0 = 6.380\times10^{-20}$ J (or, by multiplying with Avogadro's number: 38.42 kJ/mol).

14.71. We need the square root of the ratio of the reduced masses of FeH and FeD. For this purpose, units of g/mol are sufficient:

For ^{56}FeH: $\mu = \dfrac{(1.008\text{ g/mol})(55.93\text{ g/mol})}{1.008\text{ g/mol}+55.93\text{ g/mol}} = 0.9902$ g/mol;

for ^{56}FeD: $\mu = \dfrac{(2.014\text{ g/mol})(55.93\text{ g/mol})}{2.014\text{ g/mol}+55.93\text{ g/mol}} = 1.944$ g/mol. Therefore, the square root of

the ratio is $\rho = \sqrt{\dfrac{0.9902\text{ g/mol}}{1.944\text{ g/mol}}} = 0.7137$. Set up the two simultaneous equations:

$$1661\text{ cm}^{-1} = \tilde{\nu}_e - 2x_e\tilde{\nu}_e \text{ and } 1203\text{ cm}^{-1} = 0.7137\tilde{\nu}_e - 2(0.7137)^2 x_e\tilde{\nu}_e$$

(note that we have converted the expressions from frequencies ν to wavenumbers $\tilde{\nu}$ by multiplying with *c*). Having these two expressions with two unknowns, we can solve for the values of the two unknowns using standard algebraic techniques. We get for FeH: $\tilde{\nu}_e = 1.75\times10^2\text{ cm}^{-1} \simeq 1750\text{ cm}^{-1}$ and $x_e\tilde{\nu}_e = 43\text{ cm}^{-1}$. Multiply each of these by the ρ factor to get the values for FeD: $\tilde{\nu}_e = 1.25\times10^2\text{ cm}^{-1} \simeq 1250\text{ cm}^{-1}$ and $x_e\tilde{\nu}_e = 31\text{ cm}^{-1}$.

14.73. It should be easy to see that ν_1 has A_1 symmetry and ν_2 has A_1 symmetry. The symmetries of ν_3 and ν_4 may be more difficult to envision, but they both have *E* symmetry. One way to justify this is to note that there is a choice regarding which hydrogens are participating in each particular motion. Each motion has a degeneracy associated with it because of this redundancy. Alternatively, by following the steps in Table 14.5 you will find that $\Gamma_{vib} = 2A_1 \oplus 2E$, where the two A_1 irreducible representations correspond to ν_1 and ν_2. It is clear that the two remaining modes will transform as *E*.

14.75. **(a)** Both HCl and Cl_2 will have the same number of vibrations (namely one), and will therefore both have just one vibrational frequency. However, the vibration in HCl will be IR-active, while that in Cl_2 will be IR-inactive.

(b) C_2H_2 will show the lesser number of different vibrational frequencies because, as a linear molecule, it has higher symmetry and some of its vibrations are degenerate. (This effect outweighs the fact that the linear C_2H_2 has $3 \cdot 4 - 5 = 7$ vibrational degrees of freedom, while the nonlinear H_2O_2 has only $3 \cdot 4 - 6 = 6$.)

(c) CH_4 will show the lesser number of different vibrational frequencies because it is tetrahedral and has therefore higher symmetry than XeF_4, which is square–planar.

(d) PF_5 should show the lesser number of vibrational frequencies because of its higher symmetry.

(e) $Ca_3(PO_4)_2$ should show the lower number of vibrational frequencies. The vibrations will arise from the covalently-bonded PO_4^{3-} ions, which are tetrahedral and of high symmetry.

14.77. To determine the number of infrared–active vibrations for benzene, the easiest thing to do is consult Figure 14.33 and the D_{6h} character table. First, the character table shows that only vibrations having the irreducible representations A_{2u} and E_{1u} are infrared-active (look for the labels x, y, and z in the last column). According to Figure 14.33, the vibration labeled as ν_4 has A_{2u} symmetry, and the vibrations labeled ν_{12}, ν_{13}, and ν_{14} have E_{1u} symmetry. Therefore, benzene should have 4 infrared–active absorptions.

14.79. **(a)** CH_4 has T_d symmetry:

	E	$8\ C_3$	$3\ C_2$	$6\ S_4$	$6\ \sigma_d$
$N_{stationary}$	5	2	1	1	3
θ	0º	120º	180º	90º	180º
$1 + 2\cos\theta$	3	0	−1	1	−1
$\chi_{tot} = \pm N_{stationary}(1 + 2\cos\theta)$	15	0	−1	−1	3
$\chi_r = (1 + 2\cos\theta)$	3	0	−1	1	−1
$\chi_t = \pm(1 + 2\cos\theta)$	3	0	−1	−1	1
$\chi_v = \chi_{tot} - \chi_r - \chi_t$	9	0	1	−1	3

Using the GOT, this reduces to $A_1 \oplus E \oplus 2T_2$. Of these, only the T_2 vibrations are infrared–active, so CH_4 should show only two IR–active vibrations.

(b) CH_3Cl has C_{3v} symmetry:

	E	$2\ C_3$	$3\ \sigma_v$
$N_{stationary}$	5	2	3
θ	0°	120°	180°
$1 + 2\cos\theta$	3	0	−1
$\chi_{tot} = \pm N_{stationary}(1 + 2\cos\theta)$	15	0	3
$\chi_r = (1 + 2\cos\theta)$	3	0	−1
$\chi_t = \pm(1 + 2\cos\theta)$	3	0	1
$\chi_v = \chi_{tot} - \chi_r - \chi_t$	9	0	3

Using the GOT, this reduces to $3A_1 \oplus 3E$, all of which are IR–active. Therefore, CH_3Cl will show 6 IR-active vibrations.

(c) CH_2Cl_2 has C_{2v} symmetry:

	E	C_2	σ_v	σ_v'
$N_{stationary}$	5	1	3	3
θ	0°	180°	180°	180°
$1 + 2\cos\theta$	3	−1	−1	−1
$\chi_{tot} = \pm N_{stationary}(1 + 2\cos\theta)$	15	−1	3	3
$\chi_r = (1 + 2\cos\theta)$	3	−1	−1	−1
$\chi_t = \pm(1 + 2\cos\theta)$	3	−1	1	1
$\chi_v = \chi_{tot} - \chi_r - \chi_t$	9	1	3	3

Using the GOT, this reduces to $4A_1 \oplus A_2 \oplus 2B_1 \oplus 2B_2$, of which the A_1, B_1, and B_2 are IR–active. Therefore, this molecule will have 8 infrared active vibrations.

(d) $CHCl_3$ will have the same number of IR-active vibrations as CH_3Cl: six.

(e) CCl_4 will have the same number of IR-active vibrations as CH_4: only two.

The answers make sense because methane and fully substituted methane are the molecules with the highest symmetry, whereas dichloromethane has the lowest symmetry.

14.81. (a) F_2O, like H_2O, has C_{2v} symmetry. Since the text indicated that the three vibrations of H_2O have A_1, A_1 and B_1 symmetry, so do those of F_2O; all are IR–active.

(b) NCl_3 has the same structure as NH_3. Since the text indicated that NH_3 has four IR-active vibrations, so does NCl_3.

(c) For $N(CH_3)_3$, assuming C_{3v} symmetry:

	E	$2\ C_3$	$3\ \sigma_v$
$N_{stationary}$	13	1	3
θ	0°	120°	180°
$1 + 2\cos\theta$	3	0	−1
$\chi_{tot} = \pm N_{stationary}(1 + 2\cos\theta)$	39	0	3
$\chi_r = (1 + 2\cos\theta)$	3	0	−1
$\chi_t = \pm(1 + 2\cos\theta)$	3	0	1
$\chi_v = \chi_{tot} - \chi_r - \chi_t$	33	0	3

Using the GOT, this reduces to $7A_1 \oplus 4A_2 \oplus 11E$. Only the A_1 and E vibrations are IR-active, so this molecule should have 18 IR–active vibrations.

14.83. Using simple trial and error, we can suggest that the 618 cm^{-1} absorption is the difference combination $\nu_2 - \nu_1$, the absorption at 2337 cm^{-1} is $\nu_3 - \nu_2 + 2\nu_1$, and the absorption at 3715 cm^{-1} is either $2\nu_1 + \nu_3$ or $\nu_2 + \nu_3$.

14.85. Overtone transitions are simply multiples of fundamental transitions, so specific regions can be expected for overtone transitions, based on where the fundamental transitions are expected to occur. Hot bands should appear in roughly the same region as the fundamental transition (since they also correspond to a change in quantum number of 1, but from an excited vibrational state). They may be slightly shifted, but should appear in a similar region. However, for combination bands, it is difficult to predict whether any particular combination will appear strongly in a spectrum, so it is hard to point to any particular region and expect that absorptions will appear there. There are certain symmetry rules according to which fundamental vibrations will combine, but because so many different combinations are possible, they can literally appear all over the spectrum, not just in a selected region.

14.87. An O–H stretch should appear between 3100 and 3800 cm^{-1}, while an O–H bending motion should appear between 1200 and 1600 cm^{-1}. C–H stretches should appear between 2800 and 3300 cm^{-1}, C–H bends between 1300 and 1500 and also between 500 and 900 cm^{-1}. C–C stretches should be present between 800 and 1150 cm^{-1}, while the C–O stretch should appear between 900 and 1300 cm^{-1}.

14.89. Yes, vibrational "branches" can exist in electronic spectra – as we will see in Chapter 15.

14.91. There is no listing for a "$P(0)$" absorption in the P branch because it doesn't exist. If the number labeling the particular P– or R–branch line represents the J quantum number of the originating energy level, then we can't have a line in the P branch originating in the $J = 0$ state because we can't go down 1 to a $J = -1$ state (which doesn't exist). Thus, the first line in the P branch corresponds to the initial J being 1.

14.93. The spectra are different in that they will appear in different regions of the electromagnetic spectrum. The Raman spectrum generated using the He–Ne laser as excitation source will appear in the red part of the visible spectrum, while the Raman spectrum generated using the green light from the Kr^+ laser will appear in the green region of the visible spectrum. However, the pattern of new absorptions and their relative intensities relative to the excitation line should be virtually the same.

14.95. If a point group contains *i*, a center of inversion, then vibrations that are infrared–active are not Raman–active, and vibrations that are Raman-active are not infrared-active. In these cases, the *x*, *y*, and *z* labels only appear for irreducible representations that have no second–order variables listed.

CHAPTER 15

Introduction to Electronic Spectroscopy and Structure

15.1. The integral is nonzero only if the product of the irreducible representations of the two wavefunctions and the operator contains the totally symmetric irreductible representation. According to Section 13.8, this is the case if the symmetry species resulting from the combin ation of two of the functions is the same as that of the third. The combination of the electronic ground state (A_{1g}) and y-polarized light (E_{1u}) is $A_{1g} \times E_{1u} = E_{1u}$, so the allowed excited states must also have E_{1u} symmetry—that's the only way that the product of the three irreducible representations will contain the totally symmetric irreductible representation A_{1g}.

15.3. When an electronic transition occurs, an electron goes from one wavefunction with its particular spatial distribution of electronic charge to another wavefunction with another spatial distribution. Because the spatial distribution of the electron undoubtedly changes, a change in the dipole moment of the species occurs as well.

15.5. The atomic mass of D is $(2.014\times10^{-3}$ kg/mol) / $(6.022\times10^{23}$ mol$^{-1}) = 3.344\times10^{-27}$ kg, so the mass of the D nucleus alone is 3.344×10^{-27} kg $- 9.109\times10^{-31}$ kg $= 3.343\times10^{-27}$ kg.

This gives a reduced mass of $\mu = \dfrac{(3.343\times10^{-27}\text{ kg})(9.109\times10^{-31}\text{ kg})}{3.343\times10^{-27}\text{ kg} + 9.109\times10^{-31}\text{ kg}} = 9.107\times10^{-31}$ kg.

In wavenumbers, the Rydberg constant for D is then (to four significant figures)

$$R_{\text{D}} = \frac{e^4\mu}{8\varepsilon_0{}^2h^3c} = \frac{(1.602\times10^{-19}\text{ C})^4(9.107\times10^{-31}\text{ kg})}{8\left(8.854\times10^{-12}\dfrac{\text{C}^2}{\text{J}\cdot\text{m}}\right)^2(6.626\times10^{-34}\text{ J}\cdot\text{s})^3(2.9979\times10^{10}\text{ cm/s})} = 109{,}700\text{ cm}^{-1}$$

We can use this to determine the positions of the four lines of the Balmer series for D (again to four significant figures):

$$\tilde{\nu} = R_{\text{D}}\left(\frac{1}{2^2}-\frac{1}{n^2}\right) = (109{,}700\text{ cm}^{-1})\left(\frac{1}{2^2}-\frac{1}{3^2}\right) = 15{,}230\text{ cm}^{-1};$$

$$\tilde{\nu} = (109{,}700\text{ cm}^{-1})\left(\frac{1}{2^2}-\frac{1}{4^2}\right) = 20{,}560\text{ cm}^{-1};$$

$$\tilde{\nu} = (109{,}700\text{ cm}^{-1})\left(\frac{1}{2^2}-\frac{1}{5^2}\right) = 23{,}030\text{ cm}^{-1};$$

$$\tilde{\nu} = (109{,}700\text{ cm}^{-1})\left(\frac{1}{2^2}-\frac{1}{6^2}\right) = 24{,}370\text{ cm}^{-1}.$$

15.7. The energy change would be zero because the two levels are degenerate. The energy of the electron in hydrogen is not dependent on the ℓ quantum number, only the n quantum number.

15.9. **(a)** The possible values of two coupled p electrons in different shells ($\ell_1 = 1$ and $\ell_2 = 1$) are $1 + 1 = 2$ through $|1 - 1| = 0$ in integral steps, so L can take on the values 2, 1, and 0.

M_L goes from $-L$ to L and so can be at most −2, −1, 0, 1, and 2 (depending on the value of L).

S can be $\frac{1}{2} + \frac{1}{2} = 1$ or $|\frac{1}{2} - \frac{1}{2}| = 0$, with M_S going from $-S$ to S (and so can be −1, 0, or 1, depending on S).

J depends on the values of L and S, but can range from 3 ($J = L + S$ with $L = 2$ and $S = 1$) to 0 (e.g. $J = |L - S|$ with $L = 0$ and $S = 0$, or with $L = 1$ and $S = 1$).

M_J will depend on the value of J, but will range from $-J$ to J, so might be −3, −2, −1, 0, 1, 2, or 3.

(b) The possible values of two coupled f electrons in different shells ($\ell_1 = 3$ and $\ell_2 = 3$) are $3 + 3 = 6$ through $|3 - 3| = 0$ in integral steps, so L can take on the values 6, 5, 4, 3, 2, 1, and 0. M_L goes from $-L$ to L and so can be at most −6 through 6 in integer steps (depending on the value of L).

S can be $\frac{1}{2} + \frac{1}{2} = 1$ or $|\frac{1}{2} - \frac{1}{2}| = 0$, with M_S going from $-S$ to S (−1, 0, or 1, depending on S).

J depends on the values of L and S, but can range from 7 ($J = L + S$ with $L = 6$ and $S = 1$) to 0 (e.g. $J = |L - S|$ with $L = 0$ and $S = 0$, or with $L = 1$ and $S = 1$).

M_J will depend on the value of J, but will range from $-J$ to J, so might be −7 through 7 in integer steps.

(c) The possible values of coupled p and d electrons ($\ell_1 = 1$ and $\ell_2 = 2$) are $1 + 2 = 3$ through

$|1 - 2| = 1$ in integral steps, so L can take on the values 3, 2, and 1.

M_L goes from $-L$ to L and so can be at most −3, −2, −1, 0, 1, 2, and 3 (depending on the value of L).

S can be $\frac{1}{2} + \frac{1}{2} = 1$ or $|\frac{1}{2} - \frac{1}{2}| = 0$, with M_S going from $-S$ to S (−1, 0, or 1, depending on S).

J depends on the values of L and S, but can range from 4 ($J = L + S$ with $L = 3$ and $S = 1$) to 0

($J = |L - S|$ with $L = 1$ and $S = 1$).

M_J will depend on the value of J, but will range from $-J$ to J, so might be −4, −3, −2, −1, 0, 1, 2, 3, or 4.

15.11. The aluminum atom has a single unpaired *p* electron in its valence shell. The spin of that single electron can show hyperfine coupling – that is, the spin can interact with the spin of the aluminum nucleus (which, by the way, has spin of 5/2) to exhibit closely–spaced but slightly different energy levels.

15.13. We need to convert the two wavelengths into energy values and evaluate the difference:

$$E = h\nu = h\cdot\frac{c}{\lambda} = \frac{(6.626\times10^{-34}\ \mathrm{J\cdot s})(2.9979\times10^{8}\ \mathrm{m/s})}{589.0\times10^{-9}\ \mathrm{m}} = 3.373\times10^{-19}\ \mathrm{J};$$

$$E = \frac{(6.626\times10^{-34}\ \mathrm{J\cdot s})(2.9979\times10^{8}\ \mathrm{m/s})}{589.6\times10^{-9}\ \mathrm{m}} = 3.369\times10^{-19}\ \mathrm{J}.$$

The difference between these two energies is 3×10^{-22} J, which corresponds to about 2×10^{2} J/mol.

15.15. Hund's rule states that electrons first occupy each degenerate orbital alone (they remain unpaired) before they pair up with another electron of opposite spin in the same orbital. Two electrons with $m_s = +½$ each or with $m_s = -½$ each give $M_S = +1$ or $M_S = -1$, respectively. They are therefore associated with the highest value for S (namely 1), meaning that the multiplicity

$2S + 1$ is maximized.

15.17. This is correct, assuming that the uppermost shell is the only partially filled shell. Completely filled subshells have no net contribution to the total angular momentum of the atom, and the principal quantum number n of the uppermost shell has no influence on the term symbol (only the quantum numbers L and S do).

15.19. The $^3P_0$ state may not be labeled with an energy because it is the ground state (it has the highest multiplicity) and serves as the energy reference for the other states.

15.21. The case for a d^3 electron configuration for the valence subshell was already treated with the Mn^{4+} ion (Example 15.9). The ground–state term symbol is $^4F_{3/2}$.

15.23. Using Table 15.1, we find that the term symbol with the highest multiplicity is ^5D. This means that $S = 2$ and $L = 2$, so J can take on values between $J = 2 + 2 = 4$ through $J = |2 - 2| = 0$, resulting in possible term symbols of $^5D_4$, $^5D_3$, $^5D_2$, $^5D_1$, and $^5D_0$. Because the valence shell is more than half–filled, the higher the J, the lower the energy, so the ground–state term symbol is $^5D_4$.

15.25. Since $\Delta S = 0$, any allowed excited state of $^3D_1$ will also have a multiplicity of 3 ($S = 1$).

$\Delta L = 0$ or ±1, so allowed excited states can be either P, D, or F terms. Finally, since $\Delta J = 0$ or ±1, allowed excited states can have a value for J of 0, 1, or 2. Combining these three items, the possible term symbols are $^3P_0$, $^3P_1$, $^3P_2$, $^3D_0$, $^3D_1$, $^3D_2$, $^3F_0$, $^3F_1$, or $^3F_2$. Some of these aren't possible because J has to be between $L + S$ and $|L - S|$, which excludes $^3D_0$, $^3F_0$, and $^3F_1$. Therefore, only $^3P_0$, $^3P_1$, $^3P_2$, $^3D_1$, $^3D_2$, and $^3F_2$ are valid term symbols.

15.27. Since $\Delta S = 0$, any allowed excited state of $^3F_2$ will also have a multiplicity of 3 ($S = 1$). $\Delta L = 0$ or ± 1, so allowed excited states can be either D, F, or G terms. Finally, since $\Delta J = 0$ or ± 1, allowed excited states can have a value for J of 1, 2, or 3. Combining these three items, the possible term symbols are $^3D_1$, $^3D_2$, $^3D_3$, $^3F_1$, $^3F_2$, $^3F_3$, $^3G_1$, $^3G_2$, or $^3G_3$. Some of these aren't possible because J has to be between $L + S$ and $| L - S |$, which excludes $^3F_1$, $^3G_1$, and $^3G_2$. Therefore, only $^3D_1$, $^3D_2$, $^3D_3$, $^3F_2$, $^3F_3$, or $^3G_3$ are valid term symbols.

15.29. A diatomic molecule is, in a first approximation, a rigid rotor. Thus, its behavior can be modeled to some degree of accuracy by the 3–dimensional rigid rotor ideal system, which ultimately predicts that the angular momentum of the system is quantized.

15.31. Because the π_u orbitals are bonding orbitals, they are antisymmetric with respect to i and are "ungerade": Π_u. The direct product $\Pi_u \times \Pi_u$ yields the exact same character set as $\Pi_g \times \Pi_g$, which is the case of O_2 and was treated in the text. Therefore, the following term symbols are possible: $^1\Sigma_g^+$, $^3\Sigma_g^-$ (ground state), and $^1\Delta_g$.

15.33. Since $\Delta S = 0$, any allowed excited state of $^2\Sigma^+$ will also have a multiplicity of 2 ($S = ½$). Also, $\Delta\Lambda = 0$ or ± 1, so allowed excited states can be either Σ or Π terms. For Σ states the allowed transitions requires $+ \rightarrow +$, so the possible term symbol for the final Σ state is $^2\Sigma^+$. The Π state

($\Lambda = 1$) is labeled with Ω, which can have values of $\Lambda + \Sigma$ through $\Lambda - \Sigma$, which evaluates to 3/2 and ½, respectivly, so the possible Π term symbols are $^2\Sigma_{3/2}$ and $^2\Sigma_{1/2}$.

15.35. A $^3\Pi$ term symbol indicates a multiplicity of 3, which means $S = 1$ and $\Sigma = 1$; the Π symbol indicates $\Lambda = 1$.

15.37. The acetylide ion, C_2^{2-}, is isoelectronic with N_2, so it should have the same molecular orbital diagram. Thus, all of its subshells are filled, which means that $S = 0$ and $\Lambda = 0$; the term symbol for the (totally symmetric) ground state is then $^1\Sigma_{\square}^+$ (or $^1\Sigma_0$).

15.39. **(a)** No unpaired electrons, so NO_3^- should be colorless.

(b) MnO_4^- has unpaired electrons in the Mn's d subshell, so it is highly likely that the ion will absorb visible light and have some color. (It is, in fact, strongly purple.)

(c) There are no unpaired electrons, so NH_4^+ should be colorless.

15.41. Using equation 15.23 as a guide, we suggest the following form for an electronic transition moment that has vibrational and rotational structure:

$M = \int \Psi^*_{\text{el,upper}} \Psi^*_{\text{vib,upper}} \Psi^*_{\text{rot,upper}} \hat{\mu} \Psi^*_{\text{el,lower}} \Psi^*_{\text{vib,lower}} \Psi^*_{\text{rot,lower}} \, d\tau$, where $\hat{\mu}$ is the transition dipole moment operator.

15.43. The ten π electrons can be placed in the first five levels of the "particle–in–a–box" molecule (each can contain two electrons). Therefore, the highest occupied energy state has $n = 5$, and thus the lowest energy transition occurs from $n_1 = 5$ to $n_2 = 6$.

The "particle–in–a–box" energies are $E_i = \frac{n_i^2 h^2}{8ma^2}$, so the energy gap between these levels is $\Delta E = \frac{h^2}{8ma^2}\left(n_2^2 - n_1^2\right) = \frac{h^2}{8ma^2}\left(6^2 - 5^2\right) = \frac{11h^2}{8ma^2}$. This energy gap equals the energy of a photon with λ = 713 nm: $\Delta E = h\nu = \frac{hc}{\lambda}$. Combining these two equations leads to $\frac{hc}{\lambda} = \frac{11h^2}{8ma^2}$ or after dividing both sides by *h*: $\frac{c}{\lambda} = \frac{11h}{8ma^2}$. We need to solve this equation for *a*, which is the length of the box. Rearranging and evaluating using the numerical values gives:

$$a^2 = \frac{11h\lambda}{8mc} = \frac{11\left(6.626 \times 10^{-34}\ \text{J}\cdot\text{s}\right)\left(713 \times 10^{-9}\ \text{m}\right)}{8\left(9.109 \times 10^{-31}\ \text{kg}\right)\left(2.9979 \times 10^{8}\ \text{m/s}\right)} = 2.38 \times 10^{-18}\ \text{m}^2.$$

This means that $a = 1.54 \times 10^{-9}$ m = 1.54 nm.

15.45. **(a)** The "particle–in–a–box" energies are $E_i = \frac{n_i^2 h^2}{8ma^2}$, so the energy gap between $n_1 = 9$ and $n_2 = 10$ is $\Delta E = \frac{h^2}{8ma^2}\left(n_2^2 - n_1^2\right) = \frac{h^2}{8ma^2}\left(10^2 - 9^2\right) = \frac{19h^2}{8ma^2}$. This energy gap equals the energy of a photon with λ = 450.2 nm: $\Delta E = h\nu = \frac{hc}{\lambda}$. Combining these two equations leads to $\frac{hc}{\lambda} = \frac{19h^2}{8ma^2}$ or after dividing both sides by *h*: $\frac{c}{\lambda} = \frac{19h}{8ma^2}$. We need to solve this equation for *a*, which is the length of the box. Rearranging and evaluating using the numerical values gives:

$$a^2 = \frac{19h\lambda}{8mc} = \frac{19\left(6.626 \times 10^{-34}\ \text{J}\cdot\text{s}\right)\left(450.2 \times 10^{-9}\ \text{m}\right)}{8\left(9.109 \times 10^{-31}\ \text{kg}\right)\left(2.9979 \times 10^{8}\ \text{m/s}\right)} = 2.594 \times 10^{-18}\ \text{m}^2.$$

This means that $a = 1.611 \times 10^{-9}$ m = 1.611 nm.

(b) Four benzene rings in a row would then have a length of 4 · 0.280 nm = 1.12 nm. The predicted length is therefore about 45% higher, which is not surprising given that the naphthacene system is better described by a particle in a two–dimensional box (with a length of about 1.12 nm and a width equal to the benzene length of 0.280 nm). The additional "room" for the electron in the 2D box has to come from extending the length of the 1D box, so we expect that length to be higher.

15.47. Nothing would change in the Hückel approximation of ethylene if deuterium atoms were substituted for hydrogen atoms in the molecule. What goes into the Hückel model (at least at this level of Hückel theory) are the wavefunctions of the 2*p* atomic orbitals of carbon, and whether they are bound to hydrogen or deuterium is therefore without consequence.

15.49. **(a)** Possible values for E are $\alpha + 1.42\ \beta$, α, and $\alpha - 1.42\ \beta$ (in order of increasing energy).

(b) Be sure to "connect" the first and last atoms in the determinant for cyclopentadiene by placing a β in the upper right and lower left corners (see benzene, equation 15.26). Possible values for E, listed in order of increasing energy, are $\alpha + 2\ \beta$, $\alpha + 0.62\ \beta$ (doubly degenerate), and $\alpha - 1.62\ \beta$ (doubly degenerate).

15.51. We will follow the IUPAC numbering sequence for napthalene to construct the determinant matrix (depending on the numbering scheme that you have used, you may obtain a slightly different determinant matrix). With the benzene rings arranged horizontally, we start with the right benzene ring and label the top carbon atom as atom 1. The remaining atoms are then numbered in a clockwise fashion following the perimeter of the naphthalene system. (According to IUPAC rules, the carbon atoms common to both benzene rings actually carry the number of the preceding atom followed by the letter "a", such as "4a" and "8a", but we will ignore this and simply continue with whole numbers.) Therefore, the two atoms connecting both rings are atoms 5 and 10; this means that atoms 5 and 10 as well as atoms 10 and 1 are connected to each other by a "β" entry at the corresponding matrix positions:

$$\begin{vmatrix}
\alpha-E & \beta & 0 & 0 & 0 & 0 & 0 & 0 & 0 & \beta \\
\beta & \alpha-E & \beta & 0 & 0 & 0 & 0 & 0 & 0 & 0 \\
0 & \beta & \alpha-E & \beta & 0 & 0 & 0 & 0 & 0 & 0 \\
0 & 0 & \beta & \alpha-E & \beta & 0 & 0 & 0 & 0 & 0 \\
0 & 0 & 0 & \beta & \alpha-E & \beta & 0 & 0 & 0 & \beta \\
0 & 0 & 0 & 0 & \beta & \alpha-E & \beta & 0 & 0 & 0 \\
0 & 0 & 0 & 0 & 0 & \beta & \alpha-E & \beta & 0 & 0 \\
0 & 0 & 0 & 0 & 0 & 0 & \beta & \alpha-E & \beta & 0 \\
0 & 0 & 0 & 0 & 0 & 0 & 0 & \beta & \alpha-E & \beta \\
\beta & 0 & 0 & 0 & \beta & 0 & 0 & 0 & \beta & \alpha-E
\end{vmatrix} = 0.$$

15.53. The seven–membered ring can lose an electron and, with now six π electrons, become aromatic; the five–membered ring can gain an electron and, with now six π electrons, become aromatic as well.

15.55. **(a)** Ideally, cyclopolyenes with a single positive charge will be aromatic if they have $4n + 3$ carbon atoms (with $4n + 2\ \pi$ electrons) in the ring; n can take on the values 0, 1, 2, etc.

(b) Ideally, cyclopolyenes with a double positive charge will be aromatic if they have $4n + 4 = 4n'$ carbon atoms (with $4n + 2\ \pi$ electrons) in the ring; n can take on the values 0, 1, 2, and so on, while $n' = 1, 2, 3$, and so on.

15.57. Heating a potentially laser–active substance excites systems (atoms and molecules) into excited states in a manner that ultimately mimics thermal equilibrium. Thermal equilibrium is described by the Boltzmann distribution (equation 15.34); the higher an energy level is, the less populated it is. Thus, it would be extremely difficult (if not impossible) to achieve population inversion by purely thermal means; some other method would be necessary.

15.59. Light emission from fireflies is a complicated phenomenon called bioluminescence. The light is produced by a chemical reaction that ultimately produces the excited, unstable intermediate oxyluciferin, which can decay to its ground state by emitting a fluorescence photon.

15.61. Collimated light propagates as an (almost) parallel beam, not spreading much in other directions. Laser light is collimated because it is formed in a cavity of mirrors that reflect the photons back and forth through the laser medium; this optical arrangement defines the direction of propagation and leads to the formation of a bundled "beam of light" exiting through the partially transmitting mirror. Monochromatic light is made up of photons/waves that all have the same frequency/wavelength (or "color"). Coherent light is made up of waves that oscillate in phase or "in sync" with each other.

15.63. $E = h\nu = h\dfrac{c}{\lambda} = \dfrac{(6.626\times10^{-34}\ \text{J}\cdot\text{s})(2.9979\times10^{8}\ \text{m/s})}{632.8\times10^{-9}\ \text{m}} = 3.139\times10^{-19}\ \text{J}$. This is the energy of a single photon. The number of photons necessary to obtain an energy of 1 J is

$(1\ \text{J}) / (3.139\times10^{-19}\ \text{J}) = 3.186\times10^{18}$.

The 1 J/s He–Ne laser emits about 3.2×10^{18} photons per second.

15.65. A wavelength of 543.5 nm corresponds to a frequency of $\nu = c/\lambda = (2.9979\times10^{8}\ \text{m/s}) / (543.5\times10^{-9}\ \text{m}) = 5.516\times10^{14}\ \text{s}^{-1}$. Therefore, according to equation 15.33:

$$\frac{A}{B} = \frac{8\pi h\nu^3}{c^3} = \frac{8\pi(6.626\times10^{-34}\ \text{J}\cdot\text{s})(5.516\times10^{14}\ \text{s}^{-1})^3}{(2.9979\times10^{8}\ \text{m/s})^3} = 1.037\times10^{-13}\ \text{kg/(m}\cdot\text{s)}.$$

15.67. Power is defined as energy per unit time, so for a 300 mJ = 300×10^{-3} J pulse in 2.50 ns = 2.50×10^{-9} s:

$$\text{power} = \frac{300\times10^{-3}\ \text{J}}{2.50\times10^{-9}\ \text{s}} = 1.20\times10^{8}\ \text{J/s} = 1.20\times10^{8}\ \text{W} = 120\ \text{MW (megawatts)}.$$

15.69. Vibrational transitions usually occur in the infrared region of light, so chemical lasers should laser in the infrared region.

CHAPTER 16

Introduction to Magnetic Spectroscopy

16.1. A magnetic field vector describes the directon and magnitude of the magnetic field formed by moving electric charges, such as the circular magnetic field produced when a current I flows through a straight wire. A magnetic dipole vector is a linear magnetic effect formed by a current I flowing in a circle. In analogy to the electrostatic dipole, its direction is defined by two poles called the north and south pole.

16.3. **(a)** If the electron had a positive charge, the value of the magnetic dipole wouldn't change because its value is based on the magnitude of the charge, not its positivity or negativity. However, because the charge is opposite, the direction of the magnetic dipole vector will be opposite its original direction.

(b) A positron ("antielectron") has the same mass as an electron but, like a proton, a positive charge of $+e$. Therefore, rather than having the electron moving around a massive, unmoving nucleus as in hydrogen, we have two equal-mass particles in orbit about a mutual center. The orbits will be, on average, equal (producing equal magnetic dipoles) but opposite – implying that the two magnetic dipole vectors will cancel each other out, and that the overall magnetic dipole moment will be zero.

16.5. According to equation 16.6, the Bohr magneton μ_B equals $\dfrac{e\hbar}{2m_e} = \dfrac{eh}{4\pi m_e}$ (recall that $\hbar = \dfrac{h}{2\pi}$). Substituting:

$$\mu_B = \frac{(1.602 \times 10^{-19}\ \text{C})(6.626 \times 10^{-34}\ \text{J}\cdot\text{s})}{4\pi(9.109 \times 10^{-31}\ \text{kg})} = 9.273 \times 10^{-24}\ \text{C}\cdot\text{J}\cdot\text{s/kg}.$$

Within truncation error, this is the same value as given in the text. However, the units are unusual. But if we recall that 1 T = 1 kg / (C · s), we note that the units of μ_B are equivalent to J / T. Thus, we verify both the value and units of the Bohr magneton.

16.7. The drawing is left to the student. In a magnetic field, the ^{1}P state ($L = 1$) splits into three sublevels (corresponding to $M_L = -1$, 0, and +1), while the ^{1}D state ($L = 2$) splits into five sublevels (corresponding to $M_L = -2$, -1, 0, +1, and +2). Both states have $S = 0$ and therefore $J = L + S = L$, which means $\Delta S = 0$ and $\Delta L = \Delta J = +1$. Thus, according to the selection rules in equation 16.8, all transitions from each ^{1}P sublevel with $\Delta M_L = 0, \pm 1$ are allowed, and we would expect to see nine different transitions. However, because the magnitude of the splitting between adjacent sublevels is the same in both states, these nine transitions collapse into only three different lines in the spectrum, with each line corresponding to three individual transitions.

16.9. The highest level of a ^1F state ($L = 3$) will have $M_L = +3$, while the lowest level will have $M_L = -3$, for a difference of 6. Thus, we want to know the value of B such that $\Delta E_{\text{mag}} = 6\mu_B \cdot B$ is equivalent to 1.0 cm^{-1}. Converting this wavenumber value into J (by multiplying with hc), we find that 1.0 cm^{-1} equals 2.0×10^{-23} J. Solving the equation above for B gives:

$$B = \frac{\Delta E_{\text{mag}}}{6\mu_B} = \frac{2.0\times10^{-23}\text{ J}}{6\left(9.274\times10^{-24}\text{ J/T}\right)} = 0.36\text{ T.}$$

16.11. **(a)** From the solution to 16.7 we can immediately see that all transitions from each ^1D sublevel with $\Delta M_L = 0, \pm1$ are allowed. This gives us one transition from $M_L = -2$ (to $M_L = -1$), two transitions from $M_L = -1$ (to $M_L = -1$ and $M_L = 0$, respectively), three transitions from $M_L = 0$ (to $M_L = -1$, 0, and +1, respectively), two transitions from $M_L = +1$ (to $M_L = 0$ and $M_L = +1$, respectively), and one transition from $M_L = +2$ (to $M_L = +1$), for a total of nine transitions.

(b) Because the magnitude of the splitting between adjacent sublevels is the same in both states, these nine transitions collapse into only three different spectral lines, with each line corresponding to three individual transitions.

16.13. The $^3G_5$ state of the Cr^{2+} ion has $S = 1$, $L = 4$, and $J = 5$. Using equation 16.13:

$$g_J = 1 + \frac{J(J+1)+S(S+1)-L(L+1)}{2J(J+1)} = 1 + \frac{5\cdot6+1\cdot2-4\cdot5}{2\cdot5\cdot6} = 1.2.$$

16.15. The $^4F_{3/2}$ ground state of V has $S = 3/2$, $L = 3$, and $J = 3/2$. Using equation 16.13:

$$g_J = 1 + \frac{J(J+1)+S(S+1)-L(L+1)}{2J(J+1)} = 1 + \frac{\frac{3}{2}\cdot\frac{5}{2}+\frac{3}{2}\cdot\frac{5}{2}-3\cdot4}{2\cdot\frac{3}{2}\cdot\frac{5}{2}} = 0.4$$

In the presence of a magnetic field having $B = 5.57\times10^3$ G = 0.557 T, the term is split into $2J + 1 = 4$ energy levels whose ΔE_{mag} values are:

$M_J = -3/2$: $\Delta E = (0.4)\ (9.274\times10^{-24}\text{ J/T})\ (-3/2)\ (0.557\text{ T}) = -3.10\times10^{-24}$ J. This corresponds to

-0.156 cm^{-1} (as you can verify by dividing the value in joules by hc).

$M_J = -1/2$: $\Delta E = (0.4)\ (9.274\times10^{-24}\text{ J/T})\ (-1/2)\ (0.557\text{ T}) = -1.03\times10^{-24}$ J (or -0.0520 cm^{-1});

$M_J = +1/2$: $\Delta E = (0.4)\ (9.274\times10^{-24}\text{ J/T})\ (+1/2)\ (0.557\text{ T}) = +1.03\times10^{-24}$ J (or $+0.0520$ cm^{-1});

$M_J = +3/2$: $\Delta E = (0.4)\ (9.274\times10^{-24}\text{ J/T})\ (+3/2)\ (0.557\text{ T}) = +3.10\times10^{-24}$ J (or $+0.156$ cm^{-1}).

16.17. Using equation 16.14 with $m_s = -½$ and $m_s = +½$ (recall that 1 G = 10^{-4} T):

$m_s = -½$:

$$\Delta E_{\text{mag}} = g_e \mu_B m_s B = (2.002)(9.274 \times 10^{-24}\text{ J/T})(-1/2)(8600 \times 10^{-4}\text{ T}) = -7.984 \times 10^{-24}\text{ J}$$

$m_s = +½$:

$$\Delta E_{\text{mag}} = g_e \mu_B m_s B = (2.002)(9.274 \times 10^{-24}\text{ J/T})(+1/2)(8600 \times 10^{-4}\text{ T}) = +7.984 \times 10^{-24}\text{ J}$$

The energy difference is therefore $\Delta E = 15.967 \times 10^{-24}$ J, which corresponds to a photon of frequency $\Delta E / h = (1.597 \times 10^{-23}\text{ J}) / (6.626 \times 10^{-34}\text{ J·s}) = 2.410 \times 10^{10}\text{ s}^{-1} = 24.10$ GHz.

16.19. The drawing is left to the student. The diagram has only two cones: one on the positive side, with a z component of $S_z = +1/2\hbar$ and a slant height of $S = \sqrt{3}/2\hbar$ (see equations 12.1 and 12.2), and one symmetrically opposite on the negative side ($S_z = -1/2\hbar$).

16.21. According to equation 16.17, an unpaired electron experiences a local magnetic field of $B_{\text{local environment}} = B_{\text{mag field}} + a \cdot m_I$ in the presence of a nuclear spin. The energy shift in the presence of a magnetic field (equation 16.14) is then $\Delta E_{\text{mag}} = g_e \mu_B m_s \left(B_{\text{mag field}} + a \cdot m_I\right)$, and the $m_s = -½ \rightarrow +½$ transition energy becomes $\Delta E = g_e \mu_B \left(B_{\text{mag field}} + a \cdot m_I\right)$. Microwaves of energy E_{ph} are absorbed when the applied magnetic field is tuned to a value such that the the microwave energy E_{ph} equals the transition energy ΔE, which is the case for $B_{\text{mag field}} = \dfrac{E_{\text{ph}}}{g_e \mu_B} - a \cdot m_I$.

In the case of a nuclear spin $I = ½$ (where $m_I = +½$ or $m_I = -½$), resonance occurs at the following two magnetic fields (see Fig. 16.11):

For $m_I = -½$ at $B_{\text{mag field}} = \dfrac{E_{\text{ph}}}{g_e \mu_B} + \dfrac{1}{2}a$.; for $m_I = +½$: $B_{\text{mag field}} = \dfrac{E_{\text{ph}}}{g_e \mu_B} - \dfrac{1}{2}a$. The difference in these magnetic fields is the hyperfine coupling:

$$\Delta B_{\text{mag field}} = \left(\frac{E_{\text{ph}}}{g_e \mu_B} + \frac{1}{2}a\right) - \left(\frac{E_{\text{ph}}}{g_e \mu_B} - \frac{1}{2}a\right) = a.$$

16.23. We will assume that the amine radical $NH_2\cdot$ only contains the ^{14}N isotope with $I = 1$. The two hydrogens have $I = 1/2$, so that the ESR line will show

$\prod_i (2N_i I_i + 1) = (2 \cdot 1 \cdot 1 + 1)\left(2 \cdot 2 \cdot \frac{1}{2} + 1\right) = 3 \cdot 3 = 9$ separated signals.

16.25. $C_7H_7\cdot$ radical, I(C) = 0, I(H) = ½: Using the $2NI + 1$ formula, we expect $2 \cdot 7 \cdot ½ + 1 = 8$ signals in the ESR spectrum of the cycloheptatrienyl radical.

16.27. We rearrange equation 16.15 to solve for B:

$$B = \frac{hc\tilde{\nu}_{\text{res}}}{g_e \mu_B} = \frac{(6.626 \times 10^{-34}\text{ J·s})(2.9979 \times 10^{10}\text{ cm/s})(8.83\text{ cm}^{-1})}{(2.0035)(9.274 \times 10^{-24}\text{ J/T})} = 9.44\text{ T}.$$

16.29. $$\tilde{\nu}_{\text{res}} = \frac{g_e \mu_B B}{hc} = \frac{(2.0058)(9.274 \times 10^{-24} \text{ J/T})(3476 \times 10^{-4} \text{ T})}{(6.626 \times 10^{-34} \text{ J}\cdot\text{s})(2.9979 \times 10^{10} \text{ cm/s})} = 0.3255 \text{ cm}^{-1}$$

16.31. To get the result in units of frequency, we use equation 16.16:

$$\nu_{\text{res}} = \frac{g_e \mu_B B}{h} = \frac{(2.0023)(9.274 \times 10^{-24} \text{ J/T})(0.035 \text{ T})}{6.626 \times 10^{-34} \text{ J}\cdot\text{s}} = 9.8 \times 10^8 \text{ s}^{-1} = 0.98 \text{ GHz}.$$

To convert this into a wavelength, we use the relationship $c = \lambda\nu$, which gives $\lambda = c / \nu = (2.9979\times10^8 \text{ m/s}) / (9.8\times10^8 \text{ s}^{-1}) = 0.31 \text{ m} = 31 \text{ cm}$.

16.33. Only NO and NO_2 would be ESR-active in their molecular form, because they have an unpaired electron. (In fact, NO and NO_2 are rare examples of stable molecules that do, in fact, have an odd number of electrons.)

16.35. Both N and D have $I = 1$, so for the ND_2 radical there will be $\prod_i (2N_i I_i + 1) = (2\cdot1\cdot1+1)(2\cdot2\cdot1+1) = 3\cdot5 = 15$ signals in the ESR spectrum. The hyperfine coupling constant for deuterium is 78 gauss, which means that the deuterium quintet lines (which are further split into triplets) will appear at 3326, 3404, 3482, 3560, and 3638 gauss.

16.37. $^{11}BH_3\cdot$ with I(H) = ½ and $I(^{11}\text{B})$ = ½ will have $\prod_i (2N_i I_i + 1) = \left(2\cdot1\cdot\frac{1}{2}+1\right)\left(2\cdot3\cdot\frac{1}{2}+1\right) = 2\cdot4 = 8$ signals in the ESR spectrum.

16.39. The NMR transitions should appear at the same wavelength. Equations 16.23 and 16.24, which give expressions for the wavenumbers and frequency of resonance, respectively, do not have a dependence on the M_I values (since $\Delta M_I = +1$ was assumed to derive them).

16.41. NMR spectra can exist for any nucleus that has a nonzero I.

Thus, NMR-active nuclei are: **(a)** ^{2}H (I = 1); **(d)** ^{19}F (I = ½); **(f)** ^{31}P (I = ½); **(g)** ^{55}Mn (I = 5/2).

NMR-inactive nuclei with I = 0 are: **(b)** ^{14}C; **(c)** ^{16}O; **(e)** ^{28}Si; **(h)** ^{238}U.

16.43. The drawings are left to the student. Recalling that the length of the nuclear spin vector is $|\mathbf{I}| = \sqrt{I(I+1)}\hbar$ and its z component $I_z = M_I\hbar$, you should get the following:

(a) Two cones with a common slant height of $\hbar\sqrt{3}/2$ and z components of $I_z = -\frac{1}{2}\hbar$ and $I_z = +\frac{1}{2}\hbar$, respectively.

(b) Four cones with a common slant height of $\hbar\sqrt{15}/2$ and z components of $I_z = -\frac{3}{2}\hbar$, $I_z = -\frac{1}{2}\hbar$, $I_z = +\frac{1}{2}\hbar$, and $I_z = +\frac{3}{2}\hbar$, respectively.

(c) Because $I = 0$, this atom has no nuclear spin vector.

(d) Three cones with a common slant height of $\sqrt{2}\hbar$ and z components of $I_z = -\hbar$, $I_z = 0$, and $I_z = +\hbar$, respectively.

16.45. Since the working frequency is given in MHz, we will use the frequency version of the resonance condition, equation 16.24. Rearranging it to solve for B leads to $B = \dfrac{h\nu_{\text{res}}}{g_N \mu_N}$.

Nuclear g_N values are found in Table 16.1 and the appendix:

For ^{2}H: $$B = \frac{(6.626 \times 10^{-34}\ \text{J}\cdot\text{s})(330 \times 10^6\ \text{s}^{-1})}{(0.857)(5.051 \times 10^{-27}\ \text{J/T})} = 50.5\ \text{T};$$

for ^{19}F: $$B = \frac{(6.626 \times 10^{-34}\ \text{J}\cdot\text{s})(330 \times 10^6\ \text{s}^{-1})}{(5.2567)(5.051 \times 10^{-27}\ \text{J/T})} = 8.24\ \text{T};$$

for ^{31}P: $$B = \frac{(6.626 \times 10^{-34}\ \text{J}\cdot\text{s})(330 \times 10^6\ \text{s}^{-1})}{(2.2634)(5.051 \times 10^{-27}\ \text{J/T})} = 19.1\ \text{T}$$

for ^{55}Mn: $$B = \frac{(6.626 \times 10^{-34}\ \text{J}\cdot\text{s})(330 \times 10^6\ \text{s}^{-1})}{(1.3875)(5.051 \times 10^{-27}\ \text{J/T})} = 31.2\ \text{T}.$$

16.47. Spin-spin coupling leads to the following effects:

(a) In butane, the NMR signal from the terminal CH_3 groups are split into a triplet, while the NMR signal from the CH_2 groups are split into a sextet, due to the five hydrogen atoms on the adjacent carbon atoms.

(b) In cyclobutane, the hydrogens on the CH_2 groups all have the same chemical environment. Therefore, there will only be a single NMR absorption line in the high-resolution spectrum.

(c) In isobutane (2-methylpropane), the signal from the three equivalent CH_3 groups will be split into a doublet, due to the single H on the adjacent, central carbon atom. The NMR signal of that single hydrogen will be split into a multiplet of 10 peaks, because there are a total of 9 hydrogens on adjacent carbons.

Therefore, it is possible to tell these compounds apart by their NMR spectra. If only a single line is present, the compound is cyclobutane; if a doublet is present, it is isobutane, otherwise it is butane.

16.49. **(a)** Chloroethane (CH_3– CH_2Cl): triplet from CH_3, quartet from CH_2Cl group.

(b) 1,1,2-Trichloroethane ($CHCl_2$– CH_2Cl): doublet from CH_2Cl, triplet from $CHCl_2$ group.

(c) 2-Chloropropane (CH_3– CHCl– CH_3): doublet from both CH_3 groups, heptet from CHCl group.

(d) 1,2-Dichloropropane (CH_2Cl– CHCl– CH_3): doublet from CH_3, doublet from CH_2Cl, sextet from CHCl group.

16.51. **(a)** An acceptable molecule for this NMR spectrum is CH_3COOCH_3, or methyl acetate.

(b) An acceptable molecule that yields this NMR spectrum is 1-bromo-2-methylpropane, or $(CH_3)_2CHCH_2Br$.

16.53. Ethyl acetate (CH_3–CO–O–CH_2–CH_3) has a single line for CH_3–C=O (δ about 2-3 ppm according to Figure 16.24), a triplet for the terminal CH_3 (δ ≈ 1 ppm) and a quartet for the CH_2–O group (δ about 3-4 ppm). This fits the spectrum shown on top.

Propyl formate (HCO–O–CH_2–CH_2–CH_3) has a single line for H–C=O, a triplet for the terminal CH_3 (δ ≈ 1 ppm), a sextet for the center CH_2 group (δ ≈ 1.5 ppm), and a triplet for the CH_2–O group (δ about 3-4 ppm). This fits the spectrum in the middle.

Methyl propionate (CH_3–CH_2–CH_2–CO–O–CH_3) has a single line for CH_3–O (δ about 3-4 ppm), a triplet for the terminal CH_3 (δ ≈ 1 ppm), a sextet for the center CH_2 group (δ ≈ 1.5 ppm), and a triplet for the CH_2–O group (δ about 3-4 ppm). This fits the bottom spectrum.

16.55. The differences between energy levels are the same, since there is a unit change in the M_I quantum number for all of the allowed transitions. Using equation 16.24:

$$\nu_{res} = \frac{g_N \mu_N B}{h} = \frac{(0.5479)(5.051 \times 10^{-27}\ \text{J/T})(3.45\ \text{T})}{6.626 \times 10^{-34}\ \text{J}\cdot\text{s}} = 1.44 \times 10^{7}\ \text{s}^{-1} = 14.4\ \text{MHz}.$$

16.57. Using the frequency formula, equation 16.24, we rearrange it to solve for B: $B = \dfrac{h\nu_{res}}{g_N \mu_N}$.

$$B = \frac{(6.626 \times 10^{-34}\ \text{J}\cdot\text{s})(2.45 \times 10^{9}\ \text{s}^{-1})}{(2.0823)(5.051 \times 10^{-27}\ \text{J/T})} = 154\ \text{T}.$$

CHAPTER 17

Statistical Thermodynamics: Introduction

17.1. The drawing is left to the student. There are six possibilities for distinguishable balls, as shown in Figure 17.2, but only three possibilities if the balls are indistinguishable.

17.3. There would be 24 different ways of putting three differently-colored balls in four boxes (allowing only one ball per box).

17.5. **(a)** $\ln(6.02\times10^{23}!) \approx (6.02\times10^{23})\ln(6.02\times10^{23}) - 6.02\times10^{23} = 3.24\times10^{25}$, which means that $6.02\times10^{23}!$ itself is $e^{3.24\times10^{25}}$.

(b) $\ln\left(10^{100}!\right) = 10^{100}\cdot\ln\left(10^{100}\right) - 10^{100} \approx 2.293\times10^{102}$. This means that $10^{100}!$ itself is $e^{2.293\times10^{102}}$.

17.7. First, we take the natural logarithm of both sides of the equation:

$\ln N! = \ln\left(\left[2\pi\right]^{1/2} N^{N+1/2} e^{-N}\right) = \ln\left(\left[2\pi\right]^{1/2}\right) + \ln\left(N^{N+1/2}\right) - N$, where in the last term the logarithm and exponential cancel each other to leave the exponent itself. Simplifying using properties of logarithms, namely that $\ln\left(a^b\right) = b\cdot\ln a$:

$\ln N! = \frac{1}{2}\ln(2\pi) + \left(N + \frac{1}{2}\right)\ln N - N$. Realizing that ½ ln(2π) ≈ 0.92 is negligibly small, and that ½ is negligible compared to N if N is large, we are left with $\ln N! = N\ln N - N$ as the final form for Stirling's approximation.

17.9. The first way to determine an average is simply add up the scores and divide by the number of scores:

$$\text{average} = \frac{78+44+74+92+85+50+74+80+80+90}{10} = \frac{747}{10} = 74.7$$

The second is to multiply all scores by their probability, then add the products. We get:

$$\text{average} = \frac{1}{10}\cdot78 + \frac{1}{10}\cdot44 + \frac{2}{10}\cdot74 + \frac{1}{10}\cdot92 + \frac{1}{10}\cdot85 + \frac{1}{10}\cdot50 + \frac{2}{10}\cdot80 + \frac{1}{10}\cdot90$$

average = 7.8 + 4.4 + 14.8 + 9.2 + 8.5 + 5.0 + 16.0 + 9.0 = 74.7 (the same value).

17.11. According to the description in the exercise, we need to solve the following expression:

$$\frac{\int_0^{12}\left[-(7-x)^2+38\right]dx}{12}.$$ We expand this:

$$\frac{1}{12}\int_0^{12}\left[-49+14x-x^2+38\right]dx=\frac{1}{12}\int_0^{12}\left(14x-x^2-11\right)dx$$ and evaluate:

$$\frac{1}{12}\left[7x^2-\frac{1}{3}x^3-11x\right]_0^{12}=\frac{1}{12}\left(7\cdot 144-\frac{1728}{3}-11\cdot 12-0\right)=25.$$ Thus, on average, there will be 25 insects per month.

17.13. For a grand canonical ensemble, equations 17.4 – 17.6 would be rewritten as $V=\sum_j V_j=j\cdot V_j;\quad T=T_j;\quad \mu=\mu_j$. The last relationship is due to the fact that chemical potential (like temperature) is an intensive property.

17.15. The drawings are left to the student. Without the energy level at 3 EU, the only possible occupation number set is $(N_0, N_1, N_2) = (0, 1, 2)$, equivalent to the third row of Figure 17.6 (without the quantum state at 3 EU, of course). There are three ways of distributing the three distinguishable particles to obtain this occupation number set. This is confirmed by equation 17.7: $W=\frac{3!}{0!\cdot 1!\cdot 2!}=\frac{6}{1\cdot 1\cdot 2}=3$.

17.17. If we refer to Figure 17.6, we see that the distribution labeled (1,0,1,1) appears six times out of twelve, making it the most probable distribution (with a probability of 0.50 = 50%). For this system, the thermodynamic properties not solely dictated by this one distribution, because the distributions (0, 1, 2, 0) and (0, 2, 0, 1) have a probability of 25% each and will therefore have some obvious effect on the overall average. But for a system that has a larger number of possible distributions, the most probable distribution has a larger and larger impact on the overall thermodynamic properties of the system as a whole.

17.19. Without the energy level at 2 EU, the only possible occupation number sets are $(N_0, N_1, N_3) = (0, 2, 1)$, equivalent to the fourth row of Figure 17.6 (without the quantum state at 2 EU, of course). There are three ways of distributing the three distinguishable particles to obtain this occupation number set. Because it is the only possible occupation number set, (0, 2, 1) is the most probable distribution (with a probability of 100%).

17.21. **(a)** $\phi = C_1\xi_1 + C_2\xi_2 + C_3\xi_3$ is the complete equation. Taking the derivative with respect to ξ_1: $\frac{\partial\phi}{\partial\xi_1}=\frac{\partial(C_1\xi_1)}{\partial\xi_1}+\frac{\partial(C_2\xi_2)}{\partial\xi_1}+\frac{\partial(C_3\xi_3)}{\partial\xi_1}=C_1+0+0=C_1$. Similar expressions exist for derivatives with respect to ξ_2 and ξ_3.

(b) The general expression is $\frac{\partial \phi}{\partial \xi_i} = C_i$. This general expression was applied three times, once for each summation in equation 17.14, to get the four remaining terms (in one, the chain rule of derivation was applied) that resulted in equation 17.15.

17.23. The partition function q is a constant because, according to equation 17.32, it is a summation of terms that involve the degeneracies g_i of energy states (which have a fixed value for a certain system), the energies ε_i themselves (which also have a fixed value for a certain system), the Boltzmann constant k, and the temperature T. Thus, the value of q for a certain system only depends on temperature – which means that it is a constant at a specified temperature.

17.25. We use equation 17.32 with the supplied numerical values (remember that in order to convert the wavenumber values into proper energies in J, you must multiply by hc because

$E = h\nu$ and $\nu = c/\lambda = c\tilde{\nu}$):

$$q = 4 \cdot \exp\left[-\frac{(6.626\times10^{-34}\ \text{J}\cdot\text{s})(2.9979\times10^{10}\ \text{cm/s})(0\ \text{cm}^{-1})}{(1.381\times10^{-23}\ \text{J/K})(2000\ \text{K})}\right]$$

$$+6 \cdot \exp\left[-\frac{(6.626\times10^{-34}\ \text{J}\cdot\text{s})(2.9979\times10^{10}\ \text{cm/s})(384.3\ \text{cm}^{-1})}{(1.381\times10^{-23}\ \text{J/K})(2000\ \text{K})}\right]$$

$$+2 \cdot \exp\left[-\frac{(6.626\times10^{-34}\ \text{J}\cdot\text{s})(2.9979\times10^{10}\ \text{cm/s})(80378.6\ \text{cm}^{-1})}{(1.381\times10^{-23}\ \text{J/K})(2000\ \text{K})}\right]$$

This evaluates to $q = (4 \cdot 1) + (6 \cdot 0.7585) + (2 \cdot 7.843\times10^{-26}) = 8.551$. The energy level with $i = 2$ is already negligible in the sum, so no additional terms are needed to calculate a precise value for q.

17.27. The energy levels of the 1D particle-in-a-box are nondegenerate and given by equation 10.12. It is customary to express energies for the calculation of q as relative energies, with the ground state being at 0 J. In our case, these relative energies are then given by

$$E'_n = (n^2-1)\frac{h^2}{8ma^2} = (n^2-1)\frac{(6.626\times10^{-34}\ \text{J}\cdot\text{s})^2}{8(9.109\times10^{-31}\ \text{kg})(1.500\times10^{-8}\ \text{m})^2} = (n^2-1)\cdot 2.678\times10^{-22}\ \text{J}.$$

The first few relative energies are: $E'_1 = 0$ J;

$E'_2 == (2^2-1)\cdot 2.678\times10^{-22}\ \text{J} = 3\cdot 2.678\times10^{-22}\ \text{J}$;

$E'_3 == (3^2-1)\cdot 2.678\times10^{-22}\ \text{J} = 8\cdot 2.678\times10^{-22}\ \text{J}$

$E'_4 == (4^2-1)\cdot 2.678\times10^{-22}\ \text{J} = 15\cdot 2.678\times10^{-22}\ \text{J}$, and so on. The question is how many of these levels we would need to calculate to obtain a good value of q. Given that the energies rise only slowly, there could be a lot! To simplify our task, we write the expression for q as follows:

$$q=\sum_n 1\cdot\exp\left[-\frac{(n^2-1)\cdot 2.678\times10^{-22}\text{ J}}{(1.381\times10^{-23}\text{ J/K})(300.0\text{ K})}\right]=\sum_n \exp\left[-(n^2-1)\cdot 0.06463\right]$$. This is now easier to calculate: $q = 1 + 0.8237 + 0.5963 + 0.3793 + 0.2120 + 0.1041 + 0.04495 + 0.01705 + 0.005682 + 0.001664 + 0.0004283 + 9.685\times10^{-5} + 1.925\times10^{-5} + 3.361\times10^{-6} + \ldots = 3.1853$

This is the result to the 4th decimal place (accuracy determined by the given temperature). You can confirm that you needed to include at least the first twelve terms in the sum for q in order to obtain a result with the same accuracy (any additional terms will only change the digits after the "3").

17.29. We first convert the wavenumber values into proper energies in J (remember that in order to do this, you must multiply by hc because $E = h\nu$ and $\nu = c/\lambda = c\tilde{\nu}$). Therefore, 16.4 cm^{-1} correspond to 3.26×10^{-22} J and 43.5 cm^{-1} correspond to 8.64×10^{-22} J. Using equation 17.21, to have twice as many atoms in the ground state as the first electronic state:

$$\frac{1}{2}=\frac{3}{1}\exp\left[-\frac{3.26\times10^{-22}\text{ J}}{(1.381\times10^{-23}\text{ J/K})T}\right]$$. Solving for T by taking the natural logarithm of both sides: $-0.693 = 1.099-(23.6\text{ K})\frac{1}{T}$, so $T = 13.2$ K.

To have equal populations in the ground state and second excited state:

$$\frac{1}{1}=\frac{5}{1}\exp\left[-\frac{8.64\times10^{-22}\text{ J}}{(1.381\times10^{-23}\text{ J/K})T}\right]$$, which leads to $0 = 1.609-(62.6\text{ K})\frac{1}{T}$ and $T = 38.9$ K.

In order to determine the temperature for equal populations in the first and second excited states, we need the energy difference between the two: 8.64×10^{-22} J – 3.26×10^{-22} J = 5.38×10^{-22} J. Solving:

$$\frac{1}{1}=\frac{5}{3}\exp\left[-\frac{5.38\times10^{-22}\text{ J}}{(1.381\times10^{-23}\text{ J/K})T}\right]$$, which leads to $0 = 0.511-(39.0\text{ K})\frac{1}{T}$ and $T = 76.3$ K.

17.31. **(a)** The state function A will always have a slightly lower value (a more negative value) than G, based on the definitions in equations 17.44 and 17.45. A will therefore have a higher absolute value than G.

(b) It is difficult to tell which is greater because the expressions are so different. One can imagine circumstances in which either is greater than the other.

17.33. Starting with $S = k\left(N\ln N + \sum_j N_j \ln\frac{g_j}{N_j}\right)$, we first perform the first suggested substitution by rewriting equation 17.20, $\frac{N_j}{N} = \frac{1}{q} g_j \exp\left(-\frac{\varepsilon_j}{kT}\right)$, to $\frac{g_j}{N_j} = \frac{q}{N}\exp\left(\frac{\varepsilon_j}{kT}\right)$.

Then, we take the logarithm of both sides: $\ln\frac{g_j}{N_j} = \ln\frac{q}{N} + \frac{\varepsilon_j}{kT}$. We can substitute this expression into the summation term of the original equation:

$$S = k\left[N\ln N + \sum N_j\left(\ln\frac{q}{N} + \frac{\varepsilon_j}{kT}\right)\right] = k\left[N\ln N + \ln\frac{q}{N}\sum N_j + \frac{1}{kT}\sum N_j \cdot \varepsilon_j\right].$$

(where we have factored out some of the constant terms in the sums). The $\Sigma N_j = N$, and the last summation, $\Sigma N_j \cdot \varepsilon_j$, is simply the total energy E of the system (it's the number of particles in an energy level times the energy of that energy level, summed up over all energy levels). We can substitute equation 17.34 for E and get

$$S = k\left[N\ln N + N\ln\frac{q}{N} + \frac{1}{kT}\cdot NkT^2\left(\frac{\partial \ln q}{\partial T}\right)_V\right] = k\left[N\ln N + N\ln q - N\ln N + NT\left(\frac{\partial \ln q}{\partial T}\right)_V\right]$$

The two terms in $N \ln N$ cancel, and we get $S = Nk\left[T\left(\frac{\partial \ln q}{\partial T}\right)_V + \ln q\right]$, which is equation 17.42.

17.35. Equation 17.44, for A, can be easily derived by recalling the original definition of A, namely $A = U - TS$ (where in this chapter, the variable E is used to represent the internal energy). Substituting the definitions of E and S from equations 17.34 and 17.43:

$$A = NkT^2\left(\frac{\partial \ln q}{\partial T}\right)_V - T\left[NkT\left(\frac{\partial \ln q}{\partial T}\right)_V + Nk\ln\frac{q}{N} + Nk\right]$$

$$= NkT^2\left(\frac{\partial \ln q}{\partial T}\right)_V - NkT^2\left(\frac{\partial \ln q}{\partial T}\right)_V - NkT\ln\frac{q}{N} - NkT$$

The first two terms cancel, and we are left with $A = -NkT\left(\ln\frac{q}{N} + 1\right)$, which is equation 17.44.

For G, we use its original definition $G = H - TS$ and use equation 17.36 along with 17.43:

$$G = NkT^2\left(\frac{\partial \ln q}{\partial T}\right)_V + NkT - T\left[NkT\left(\frac{\partial \ln q}{\partial T}\right)_V + Nk\ln\frac{q}{N} + Nk\right]$$

$$= NkT^2\left(\frac{\partial \ln q}{\partial T}\right)_V + NkT - NkT^2\left(\frac{\partial \ln q}{\partial T}\right)_V - NkT\ln\frac{q}{N} - NkT$$

Four of the five terms cancel, and we are left with $G = -NkT \ln\frac{q}{N}$, which is equation 17.45.

17.37. The "+ 1" term in A comes from a term in NkT that remains when the definitions for U (referred to as E in this chapter) and S are combined. In the definition of G, that term cancels.

17.39. Substituting only units into the expression in the exponent of $\exp\left(\frac{-h^2n^2}{8mV^{2/3}kT}\right)$:

$\frac{(\text{J}\cdot\text{s})^2}{\text{kg}\cdot(\text{m}^3)^{2/3}\cdot(\text{J/K})\cdot\text{K}} = \frac{\text{J}^2\cdot\text{s}^2}{\text{kg}\cdot\text{m}^2\cdot\text{J}} = \frac{\text{J}\cdot\text{s}^2}{\text{kg}\cdot\text{m}^2} = \frac{\text{J}}{\text{J}} = 1$. Thus, all of the units cancel and the exponent is a pure, unitless number (as it should be).

17.41. 1D box: $q = \int_{n=0}^{\infty} \exp\left(\frac{-h^2n^2}{8ma^2kT}\right) dn = \frac{1}{2}\left(\frac{8\pi ma^2kT}{h^2}\right)^{1/2} = \left(\frac{2\pi mkT}{h^2}\right)^{1/2} a$;

2D box: $q = \left[\int_{n=0}^{\infty} \exp\left(\frac{-h^2n^2}{8ma^2kT}\right) dn\right]^2 = \left[\frac{1}{2}\left(\frac{8\pi ma^2kT}{h^2}\right)^{1/2}\right]^2 = \frac{2\pi mkT}{h^2}a^2$. The latter expression can be written as $q = \left(\frac{2\pi mkT}{h^2}\right)^{2/2} A$, where A is the area of the square box. We can see that these expressions follow a pattern: for the n-dimensional box, the partition function for translation is given by the nth power of $\left(\frac{2\pi mkT}{h^2}\right)^{1/2}$ multiplied by the characteristic dimension (length a, or area A, or volume V).

17.43. If $N = N_A$ in the Sackur-Tetrode equation, the equation itself doesn't change, but the value of S calculated using that value is now the molar entropy $\overline{S}$, in units of J/(mol · K).

17.45. The derivation of equation 17.56 starts from equation 17.34: $E = NkT^2\left(\frac{1}{q}\frac{\partial q}{\partial T}\right)_V$.

Take the derivative of q with respect to T:

$$\left(\frac{\partial q}{\partial T}\right)_V = \frac{\partial}{\partial T}\left[\left(\frac{2\pi mkT}{h^2}\right)^{3/2} V\right] = \left(\frac{2\pi mk}{h^2}\right)^{3/2} V\cdot\frac{3}{2}T^{1/2}.$$

Now, dividing this by q: $\frac{\left(\frac{2\pi mk}{h^2}\right)^{3/2} V\cdot\frac{3}{2}T^{1/2}}{\left(\frac{2\pi mk}{h^2}\right)^{3/2} V\cdot T^{3/2}}$ (we have taken $T^{3/2}$ out of the expression in the denominator and written it as a separate factor to make it easier to see how the terms cancel and simplify). As complex as this looks, everything cancels except $\frac{3}{2T}$.

Substituting back into the original equation: $E = NkT^2 \cdot \frac{3}{2T} = \frac{3}{2}NkT$, which is equation 17.56.

17.47. Using equation 17.64 and the molecular mass of He in kg (see Example 17.4), we find:

$$\text{At 25 K: } \Lambda = \left[\frac{\left(6.626\times10^{-34}\ \text{J}\cdot\text{s}\right)^2}{2\pi\left(6.65\times10^{-27}\ \text{kg}\right)\left(1.381\times10^{-23}\ \text{J/K}\right)\left(25\ \text{K}\right)}\right]^{1/2} = 1.7\times10^{-10}\ \text{m};$$

$$\text{at 500 K: } \Lambda = \left[\frac{\left(6.626\times10^{-34}\ \text{J}\cdot\text{s}\right)^2}{2\pi\left(6.65\times10^{-27}\ \text{kg}\right)\left(1.381\times10^{-23}\ \text{J/K}\right)\left(500\ \text{K}\right)}\right]^{1/2} = 3.90\times10^{-11}\ \text{m}.$$

17.49. Taking the reciprocal of $\Lambda = \left(\frac{h^2}{2\pi mkT}\right)^{1/2}$, we obtain $\frac{1}{\Lambda} = \left(\frac{2\pi mkT}{h^2}\right)^{1/2}$ and $\frac{1}{\Lambda^3} = \left(\frac{2\pi mkT}{h^2}\right)^{3/2}$. Therefore, we can express $q_{\text{trans}} = \left(\frac{2\pi mkT}{h^2}\right)^{3/2} \cdot V$ as $q_{\text{trans}} = \frac{V}{\Lambda^3}$.

17.51. We start by converting the pressure of 1 atm into SI units:
1 atm = 1.01325×10^5 Pa = 1.01325×10^5 N/m^2 (assuming that "1 atm" represents *exactly* one atmosphere). The atomic masses of C, Fe, and Hg in kg are obtained by dividing the molar masses by Avogadro's number and converting the result from g to kg: m(C) = 1.9945×10^{-26} kg, m(Fe) = 9.2735×10^{-26} kg, and m(Hg) = 3.3310×10^{-25} kg. From equation 17.61, we then obtain the following numerical values for the entropy of 6.022×10^{23} atoms:

(a)

$$S = \left(6.022\times10^{23}\right)\left(1.381\times10^{-23}\ \frac{\text{J}}{\text{K}}\right)\left\{\ln\left[\left(\frac{2\pi\left(1.9945\times10^{-26}\ \text{kg}\right)\left(1.381\times10^{-23}\ \frac{\text{J}}{\text{K}}\right)\left(1000\ \text{K}\right)}{\left(6.626\times10^{-34}\ \text{J}\cdot\text{s}\right)^2}\right)^{3/2} \times \frac{\left(1.381\times10^{-23}\ \text{J/K}\right)\left(1000\ \text{K}\right)}{1.01325\times10^5\ \text{N/m}^2}\right] + \frac{5}{2}\right\} = 164.9\ \text{J/K}$$

(b)

$$S = \left(6.022\times10^{23}\right)\left(1.381\times10^{-23}\ \frac{\text{J}}{\text{K}}\right)\left\{\ln\left[\left(\frac{2\pi\left(9.2735\times10^{-26}\ \text{kg}\right)\left(1.381\times10^{-23}\ \frac{\text{J}}{\text{K}}\right)\left(3500\ \text{K}\right)}{\left(6.626\times10^{-34}\ \text{J}\cdot\text{s}\right)^2}\right)^{3/2} \times \frac{\left(1.381\times10^{-23}\ \text{J/K}\right)\left(3500\ \text{K}\right)}{1.01325\times10^5\ \text{N/m}^2}\right] + \frac{5}{2}\right\} = 210.2\ \text{J/K}$$

(c)

$$S = \left(6.022\times10^{23}\right)\left(1.381\times10^{-23}\ \frac{\text{J}}{\text{K}}\right)\left\{\ln\left[\left(\frac{2\pi\left(3.3310\times10^{-25}\ \text{kg}\right)\left(1.381\times10^{-23}\ \frac{\text{J}}{\text{K}}\right)(298\ \text{K})}{\left(6.626\times10^{-34}\ \text{J}\cdot\text{s}\right)^2}\right)^{3/2}\right.\right.$$

$$\left.\left.\times\frac{\left(1.381\times10^{-23}\ \text{J/K}\right)(298\ \text{K})}{1.01325\times10^{5}\ \text{N/m}^2}\right]+\frac{5}{2}\right\} = 175\ \text{J/K}$$

17.53. At the same volume and temperature, the ratio of partition functions of ^{12}C and ^{13}C is $\frac{q_{12}}{q_{13}} = \frac{\left(2\pi m_{^{12}\text{C}}kT/h^2\right)^{3/2}\cdot V}{\left(2\pi m_{^{13}\text{C}}kT/h^2\right)^{3/2}\cdot V} = \frac{\left(m_{^{12}\text{C}}\right)^{3/2}}{\left(m_{^{13}\text{C}}\right)^{3/2}} = \left(\frac{m_{^{12}\text{C}}}{m_{^{13}\text{C}}}\right)^{3/2}$, so they depend directly on the ratio of their respective masses. This means that ^{13}C, as the heavier atom, should have the larger partition function. Because the units cancel out, we can directly use the molar masses of these isotopes, and do not need to convert them to kg first:

$$\frac{q_{12}}{q_{13}} = \left(\frac{12.00\ \text{g/mol}}{13.00\ \text{g/mol}}\right)^{3/2} = 0.8869.$$

17.55. At 298 K, the energy is $\frac{3}{2}\left(8.314\frac{\text{J}}{\text{mol}\cdot\text{K}}\right)(298\ \text{K}) = 3.72\ \text{kJ/mol}$. At 348 K, the energy is $\frac{3}{2}\left(8.314\frac{\text{J}}{\text{mol}\cdot\text{K}}\right)(348\ \text{K}) = 4.34\ \text{kJ/mol}$. The difference is (4.34 – 3.72) kJ/mol = 0.62 kJ/mol. Using $q = nC_V\Delta T$, we calculate the energy change for 1 mol as (1 mol) (12.47 J/mol/K) (50 K) = 6.2×10^3 J = 0.62 kJ. (Recall that the constant volume heat capacity for monoatomic gases is 12.47 J/mol/K.)

17.57. The temperature remains constant during an isothermal change, so we can associate the volume V_1 with pressure p_1 and volume V_2 with pressure p_2. The change in S is then

$$\Delta S = S_2 - S_1 = Nk\left\{\ln\left[\left(\frac{2\pi mkT}{h^2}\right)^{3/2}\frac{kT}{p_2}\right]+\frac{5}{2}\right\} - Nk\left\{\ln\left[\left(\frac{2\pi mkT}{h^2}\right)^{3/2}\frac{kT}{p_1}\right]+\frac{5}{2}\right\}$$

$$= Nk\ln\left[\left(\frac{2\pi mkT}{h^2}\right)^{3/2}\frac{kT}{p_2}\right] + Nk\frac{5}{2} - Nk\ln\left[\left(\frac{2\pi mkT}{h^2}\right)^{3/2}\frac{kT}{p_1}\right] - Nk\frac{5}{2}$$

The two $Nk\cdot 5/2$ terms cancel, and using $\ln\left(\frac{a}{b}\right) = \ln a - \ln b$, we can write

$\Delta S = Nk\ln\frac{\left(\frac{2\pi mkT}{h^2}\right)^{3/2}\frac{kT}{p_2}}{\left(\frac{2\pi mkT}{h^2}\right)^{3/2}\frac{kT}{p_1}}$. All of the terms in the fraction cancel except for the pressures, leaving $\Delta S = Nk\ln\frac{p_1}{p_2}$. For one mole, $N_A \cdot k = R$, so we have $\Delta\overline{S} = R\ln\frac{p_1}{p_2}$.

This can be rewritten using equation 17.58, $pV = NkT$, as $\Delta\overline{S} = R\ln\frac{V_2}{V_1}$.

For an isochoric change, the volume is constant, so a temperature change from T_1 to T_2 is accompanied by a pressure change from p_1 to p_2:

$$\Delta S = S_2 - S_1 = Nk\left\{\ln\left[\left(\frac{2\pi mkT_2}{h^2}\right)^{3/2}\frac{kT_2}{p_2}\right] + \frac{5}{2}\right\} - Nk\left\{\ln\left[\left(\frac{2\pi mkT_1}{h^2}\right)^{3/2}\frac{kT_1}{p_1}\right] + \frac{5}{2}\right\}$$

$$= Nk\ln\left[\left(\frac{2\pi mkT_2}{h^2}\right)^{3/2}\frac{kT_2}{p_2}\right] + Nk\frac{5}{2} - Nk\ln\left[\left(\frac{2\pi mkT_1}{h^2}\right)^{3/2}\frac{kT_1}{p_2}\right] - Nk\frac{5}{2}$$

The two $Nk \cdot 5/2$ terms cancel again, and using the same property of logarithms as above leads to $\Delta S = Nk\ln\frac{T_2^{\,3/2}\cdot T_2/p_2}{T_1^{\,3/2}\cdot T_1/p_1}$ (note that the temperatures do not cancel here). Because $pV = NkT$, both T_2/p_2 and T_1/p_1 are equal to $V/(Nk)$, which is constant because volume remains constant, so they cancel as well. From $\Delta S = Nk\ln\frac{T_2^{\,3/2}}{T_1^{\,3/2}} = Nk\ln\left(\frac{T_2}{T_1}\right)^{3/2}$, using $\ln(a^b) = b \cdot \ln a$, we finally obtain $\Delta S = \frac{3}{2}Nk\ln\frac{T_2}{T_1}$. For one mole, $N_A \cdot k = R$, so

$\Delta\overline{S} = \frac{3}{2}R\ln\frac{T_2}{T_1}$. Recall that for a monoatomic gas, $\overline{C}_V = \frac{3}{2}R$, so this is indeed the desired equation.

CHAPTER 18

More Statistical Thermodynamics

18.1. Using Appendix 5 and the fact that there are $2I + 1$ nuclear spin states for the nuclear spin I (see Chapter 16):

(a) ^{12}C has a nuclear spin of zero, so there are $2 \cdot 0 + 1 = 1$ nuclear spin states. Therefore, $q_{nuc} = 1$.

(b) ^{56}Fe also has a nuclear spin of zero, so $q_{nuc} = 1$.

(c) ^{1}H has a nuclear spin of ½, so there are $2 \cdot ½ + 1 = 2$ nuclear spin states. Therefore, $q_{nuc} = 2$.

(d) ^{2}H has a nuclear spin of 1, so there are $2 \cdot 1 + 1 = 3$ nuclear spin states. Therefore, $q_{nuc} = 3$.

18.3. According to NIST's list of "Isotopic Compositions for All Elements", the following are stable isotopes of tin: ^{112}Sn, ^{114}Sn, ^{115}Sn, ^{116}Sn, ^{117}Sn, ^{118}Sn, ^{119}Sn, ^{120}Sn, ^{122}Sn, and ^{124}Sn. Most of them have nuclear spins of $I = 0$, except for ^{115}Sn, ^{117}Sn, and ^{119}Sn which have $I = ½$. We then have two different values for q_{nuc}:

$q_{nuc} = 2 \cdot 0 + 1 = 1$ for ^{112}Sn, ^{114}Sn, ^{115}Sn, ^{116}Sn, ^{117}Sn, ^{118}Sn, ^{119}Sn, ^{120}Sn, ^{122}Sn, and ^{124}Sn; and $q_{nuc} = 2 \cdot ½ + 1 = 2$ for ^{115}Sn, ^{117}Sn, and ^{119}Sn.

18.5. First, we convert the wavenumber values into J by multiplying by hc (because $E = h\nu$ and $\nu = c/\lambda = c\tilde{\nu}$). The energies for level 1, 2, and 3 are then 3.37952×10^{-21} J, 7.68488×10^{-21} J, and 1.30246×10^{-19} J, respectively. Using Equation 18.6:

$$q_{elect} = 5 + 7 \cdot \exp\left[-\frac{3.37952\times10^{-21}\text{ J}}{(1.381\times10^{-23}\text{ J/K})(3560\text{ K})}\right] + 9 \cdot \exp\left[-\frac{7.68488\times10^{-21}\text{ J}}{(1.381\times10^{-23}\text{ J/K})(3560\text{ K})}\right]$$

$$+ 3 \cdot \exp\left[-\frac{3.37952\times10^{-21}\text{ J}}{(1.381\times10^{-23}\text{ J/K})(3560\text{ K})}\right] = 19.445$$

18.7. Using all electronic states:

$$q_{elect} = 1 + 3 \cdot \exp\left[-\frac{3.26\times10^{-22}\text{ J}}{(1.381\times10^{-23}\text{ J/K})(10{,}000\text{ K})}\right] + 5 \cdot \exp\left[-\frac{8.64\times10^{-22}\text{ J}}{(1.381\times10^{-23}\text{ J/K})(10{,}000\text{ K})}\right]$$

$$+5 \cdot \exp\left[-\frac{2.02\times10^{-19}\text{ J}}{(1.381\times10^{-23}\text{ J/K})(10{,}000\text{ K})}\right] + 1 \cdot \exp\left[-\frac{4.30\times10^{-19}\text{ J}}{(1.381\times10^{-23}\text{ J/K})(10{,}000\text{ K})}\right]$$

$$= 10.16$$

18.9. Taking information from Example 18.2, we can calculate the electronic partition function for Ni at 298 K as $q_{\text{elect}} = 3 + 3\cdot\exp\left[-\dfrac{3.97\times10^{-21}\text{ J}}{(1.381\times10^{-23}\text{ J/K})(298\text{ K})}\right] = 4.143$, and at 5 K as

$$q_{\text{elect}} = 3 + 3\cdot\exp\left[-\frac{3.97\times10^{-21}\text{ J}}{(1.381\times10^{-23}\text{ J/K})(5\text{ K})}\right] = 3\text{ (exactly)}.$$

18.11. We first convert cm^{-1} into s^{-1} units by multiplying with c (recall that $\nu = c/\lambda = c\tilde{\nu}$), which gives (2.9979×10^{10} cm/s) (2358 cm^{-1}) = 7.069×10^{13} s^{-1}. Also, 945.0 kJ/mol corresponds to (945.0×10^{3} J/mol) / (6.022×10^{23} mol^{-1}) = 1.569×10^{-18} J per molecule. Rewriting Equation 18.7:

$D_e = D_0 + \frac{1}{2}h\nu$ = 1.569×10^{-18} J + ½ (6.626×10^{-34} J·s) (7.069×10^{13} s^{-1}) = 1.592×10^{-18} J (or 959.1 kJ/mol).

18.13. We first convert cm^{-1} into s^{-1} units by multiplying with c (recall that $\nu = c/\lambda = c\tilde{\nu}$), which gives (2.9979×10^{10} cm/s) (4320 cm^{-1}) = 1.295×10^{14} s^{-1}. The bond energy corresponds to the value of D_0, so D_0 = 432 kJ/mol. Per molecule of H_2, this is

(432×10^{3} J/mol) / (6.022×10^{23} mol^{-1}) = 7.17×10^{-19} J.

Rewriting Equation 18.7:

$D_e = D_0 + \frac{1}{2}h\nu$ = 7.17×10^{-19} J + ½ (6.626×10^{-34} J·s) (1.295×10^{14} s^{-1}) = 7.60×10^{-19} J.

Using this value in the definition of the partition function gives:

$$q_{\text{elect}} = 1\cdot\exp\left[\frac{(7.60\times10^{-19}\text{ J})}{(1.381\times10^{-23}\text{ J/K})(298\text{ K})}\right] = 1.70\times10^{80}.$$

This is larger than the original value of about 5.05×10^{75}, but both values are still large.

18.15. 918 kJ (per mole) is the value for D_0, which corresponds to (918×10^{3} J/mol) / (6.022×10^{23} mol^{-1}) = 1.52×10^{-18} J. Substituting into q_{elect}:

$$q_{\text{elect}} = 1\cdot\exp\left[\frac{(1.52\times10^{-18}\text{ J})}{(1.381\times10^{-23}\text{ J/K})(373\text{ K})}\right] = 3.34\times10^{128}.$$

18.17. The fluorine-atmosphered planet may well have the higher temperature because the atmosphere is composed of a molecule that can absorb infrared light due to vibrational transitions, thereby warming the planet.

18.19. 459 cm^{-1} correspond to a frequency of ν = 1.38×10^{13} s^{-1} (simply multiply by c), so according to Equation 18.16: $\theta_v = \dfrac{(6.626\times10^{-34}\text{ J}\cdot\text{s})(1.38\times10^{13}\text{ s}^{-1})}{1.381\times10^{-23}\text{ J/K}} = 660\text{ K}$. At 298 K, the vibrational partition function is $q_{\text{vib}} = \dfrac{1}{1-e^{-\frac{660\text{ K}}{298\text{ K}}}} = 1.12$.

18.21. Multiplying 1183 cm^{-1}, 2615 cm^{-1}, and 2617 cm^{-1} by c gives the frequencies $\nu = 3.547\times10^{13}\ s^{-1}$, $7.840\times10^{13}\ s^{-1}$, and $7.846\times10^{13}\ s^{-1}$, respectively. The corresponding vibrational temperatures are

$$\theta_v = \frac{(6.626\times10^{-34}\ \text{J}\cdot\text{s})(3.547\times10^{13}\ \text{s}^{-1})}{1.381\times10^{-23}\ \text{J/K}} = 1702\ \text{K},\ 3761\ \text{K, and } 3764\ \text{K}.$$

Assuming singly degenerate vibrational levels, the vibrational partition function at 310 K is $q_{vib} = \dfrac{1}{1-e^{-\frac{1702\ K}{310\ K}}}\cdot\dfrac{1}{1-e^{-\frac{3762\ K}{310\ K}}}\cdot\dfrac{1}{1-e^{-\frac{3764\ K}{310\ K}}} = (1.0041)(1.0000054)(1.0000053) = 1.0042$.

18.23. According to Table 18.2, NO_2 has vibrational temperatures of 1900 K, 1980 K, and 2330 K. The vibrational partition function at the different temperatures is then as follows:

250 K:

$$q_{vib} = \frac{1}{1-e^{-\frac{1900\ K}{250\ K}}}\cdot\frac{1}{1-e^{-\frac{1980\ K}{250\ K}}}\cdot\frac{1}{1-e^{-\frac{2330\ K}{250\ K}}} = (1.00050)(1.00036)(1.000090) = 1.00095;$$

500 K: $q_{vib} = \dfrac{1}{1-e^{-\frac{1900\ K}{500\ K}}}\cdot\dfrac{1}{1-e^{-\frac{1980\ K}{500\ K}}}\cdot\dfrac{1}{1-e^{-\frac{2330\ K}{500\ K}}} = (1.023)(1.019)(1.0096) = 1.053;$

1000 K: $q_{vib} = \dfrac{1}{1-e^{-\frac{1900\ K}{1000\ K}}}\cdot\dfrac{1}{1-e^{-\frac{1980\ K}{1000\ K}}}\cdot\dfrac{1}{1-e^{-\frac{2330\ K}{1000\ K}}} = (1.18)(1.16)(1.108) = 1.51.$

As the temperature increases and more vibrational levels are populated, the vibrational partition function increases as well.

18.25. CCl_4 has nine total vibrations ($3\cdot5-6=9$), but many are degenerate, so only four distinct vibrations (with corresponding vibrational temperatures) will be seen in a spectrum. The frequencies of the vibrations can be determined by rewriting Equation 18.16: $\nu = \dfrac{k\theta_v}{h}$.

To convert these frequencies (in s^{-1}) into wavenumbers, recall that $\nu = c/\lambda = c\tilde{\nu}$, so $\tilde{\nu} = \dfrac{k\theta_v}{hc}$:

$$\text{310 K (2 vibrations): } \tilde{\nu} = \frac{(1.381\times10^{-23}\ \text{J/K})(310\ \text{K})}{(6.626\times10^{-34}\ \text{J}\cdot\text{s})(2.9979\times10^{10}\ \text{cm/s})} = 216\ \text{cm}^{-1};$$

$$\text{450 K (3 vibrations): } \tilde{\nu} = \frac{(1.381\times10^{-23}\ \text{J/K})(450\ \text{K})}{(6.626\times10^{-34}\ \text{J}\cdot\text{s})(2.9979\times10^{10}\ \text{cm/s})} = 313\ \text{cm}^{-1};$$

660 K (1 vibration): $\tilde{\nu} = \dfrac{(1.381\times10^{-23}\text{ J/K})(660\text{ K})}{(6.626\times10^{-34}\text{ J}\cdot\text{s})(2.9979\times10^{10}\text{ cm/s})} = 459\text{ cm}^{-1}$;

1120 K (3 vibrations): $\tilde{\nu} = \dfrac{(1.381\times10^{-23}\text{ J/K})(1120\text{ K})}{(6.626\times10^{-34}\text{ J}\cdot\text{s})(2.9979\times10^{10}\text{ cm/s})} = 778.7\text{ cm}^{-1}$.

18.27. We first calculate the rotational temperature from $\theta_r = \dfrac{h^2/(2\pi)^2}{2Ik} = \dfrac{h^2}{8\pi^2 Ik}$:

$\theta_r = \dfrac{(6.626\times10^{-34}\text{ J}\cdot\text{s})^2}{8\pi^2(3.167\times10^{-46}\text{ kg}\cdot\text{m}^2)(1.381\times10^{-23}\text{ J/K})} = 1.271\text{ K}$. Because F_2 is a homonuclear diatomic molecule, we need to use Equation 18.32: $q_{rot} = T/(2\Theta_r)$, which gives

$q_{rot} = (335\text{ K}) / (2\cdot 1.271\text{ K}) = 132.$

18.29. Using equation 18.29 for the rotational partition function of a heteronuclear diatomic:

$q_{rot} = \dfrac{T}{\theta_r} = \dfrac{298\text{ K}}{12.1\text{ K}} = 24.6$ for HBr. In order for HCl to have the same q_{rot}, the temperature would have to be $T = q_{rot}\cdot\Theta_r = 24.6\cdot(15.2\text{ K}) = 374\text{ K}$.

18.31. **(a)** I_2 with a nuclear spin of $I = 5/2$ has fermion nuclei. While the number of odd rotational states is the same as that for even rotational states, odd rotational states occur only with symmetric nuclear states, which have a degeneracy of $(I+1)(2I+1)$; even rotational states occur only with antisymmetric nuclear states, which have a degeneracy of $2I^2 + I$. Therefore, the ratio of molecules in odd rotational states to even rotational states is $(I+1)(2I+1)$ to $2I^2 + I$, which gives $(5/2+1)(2\cdot 5/2+1) = 21$ to $2\cdot(5/2)^2 + 5/2 = 15$, or 7 : 5.

(b) Cl_2 has boson nuclei with $I = 2$. Therefore, the ratio is the inverse of part (a): $2I^2 + I$ to $(I+1)(2I+1)$, which gives $2\cdot 2^2 + 2 = 10$ to $(2+1)(2\cdot 2+1) = 15$, or 2 : 3.

18.33. For a molecule having an antisymmetric electronic ground state and bosonic nuclei, the $(2I^2 + I)$ antisymmetric nuclear wavefunctions will pair with the even symmetric rotational wavefunctions, while the $(I+1)(2I+1)$ symmetric nuclear wavefunctions will pair with the odd antisymmetric rotational wavefunctions. This is similar to a fermion model, but it's because of the antisymmetric electronic wavefunction.

18.35. Acetylene with one deuterium substituted for a hydrogen should not show the same intensity variations in its rovibrational spectrum. Because the molecule no longer has symmetry plane perpendicular to the molecular axis, the symmetry implications on the allowed rotational states no longer apply.

18.37. Ammonia is a symmetric top, so we use equation 18.39 and the three rotational temperatures (two of them the same) from Table 18.4:

$$q_{rot} = \frac{\pi^{1/2}}{3}\left(\frac{298\text{ K}}{13.6\text{ K}}\right)\left(\frac{298\text{ K}}{8.92\text{ K}}\right)^{1/2} = 74.8\,.$$

18.39. CCl_4 is a spherical top, so we use equation 18.36 and the rotational temperature from Table 18.4: $q_{\text{rot}} = \dfrac{\pi^{1/2}}{12}\left(\dfrac{298\text{ K}}{0.0823\text{ K}}\right)^{3/2} = 3.22 \times 10^4$.

18.41. The expression for C_p would be very similar to that for C_V in equation 18.52 except it would have one more term in Nk in it, to account for the extra NkT term in the definition of H as compared to E (see equations 18.45 and 18.43). Therefore, the expression would be $C_p = Nk\left[\dfrac{5}{2} + \dfrac{3}{2} + \displaystyle\sum_{j=1}^{3N^*-6}\left(\dfrac{\theta_j}{T}\right)^2 \cdot \dfrac{e^{-\theta_j/T}}{\left(1-e^{-\theta_j/T}\right)^2}\right]$. (The additional NkT term actually can be attributed to the translational motion, as can be easily seen by comparing the E and H terms in Table 18.5. This is why we have added it to the first term inside the square brackets.)

18.43. Technically, we have to solve the expression $p = NkT\left(\dfrac{\partial \ln Q}{\partial V}\right)_T$. According to equation 18.49, $\ln Q = \ln q_{\text{trans}} + \ln q_{\text{elect}} + \ln q_{\text{vib}} + \ln q_{\text{rot}} + \ln q_{\text{nuc}}$. When we take the partial derivative of this expression with respect to volume V, we can simplify our task by recognizing that only q_{trans} depends explicitly on V, so the derivatives for all of the other partition functions are zero. This simplifies the expression for p to

$$p = NkT\left(\frac{\partial \ln Q}{\partial V}\right)_T = NkT\left(\frac{\partial \ln q_{\text{trans}}}{\partial V}\right)_T.$$

Solving, using properties of logarithms:

$$p = NkT\left(\frac{\partial \ln q_{\text{trans}}}{\partial V}\right)_T = NkT\frac{1}{q_{\text{trans}}}\left(\frac{\partial q_{\text{trans}}}{\partial V}\right)_T = \frac{NkT}{\left(2\pi mkT/h^2\right)^{3/2}V}\cdot\left(\frac{\partial\left[\left(2\pi mkT/h^2\right)^{3/2}V\right]}{\partial V}\right)_T.$$

Most of the variables in q_{trans} are unaffected by the differentiation, so this simplifies to $p = \dfrac{NkT}{\left(2\pi mkT/h^2\right)^{3/2}V}\cdot\left(2\pi mkT/h^2\right)^{3/2} = \dfrac{NkT}{V}$, where $\left(2\pi mkT/h^2\right)^{3/2}$ has canceled.

We have left: $p = \dfrac{NkT}{V}$, which rearranges to $pV = NkT$. In molar quantities, this becomes the familiar form of the ideal gas law, $pV = nRT$.

18.45. Table 18.5 gives a useful summary of the contributions of each partition function to the thermodynamic property. Let us simply use those terms for a polyatomic molecule, looking up the values that we need from Tables 18.2 and 18.4, and solve.

For E:

$$E=\frac{3}{2}NkT-ND_0+NkT\sum_{j=1}^{3N^*-6}\left(\frac{\theta_{v,j}}{2T}+\frac{\theta_{v,j}/T}{e^{\theta_{v,j}/T}-1}\right)+\frac{3}{2}NkT$$

$$=3NkT-ND_0+NkT\sum_{j=1}^{3N^*-6}\left(\frac{\theta_{v,j}}{2T}+\frac{\theta_{v,j}/T}{e^{\theta_{v,j}/T}-1}\right)$$

To evaluate this, we break off the vibrational part (third term) from the rest. Furthermore, we assume n = 1 mol and replace Nk by nR:

$$E_{\text{non-vib}}=3(1\text{ mol})\left(8.314\frac{\text{J}}{\text{mol}\cdot\text{K}}\right)(298\text{ K})\text{-}(1\text{ mol})(1163\text{ kJ/mol})=-1156\text{ kJ}.$$

$$E_{\text{vib}}=(1\text{ mol})\left(8.314\frac{\text{J}}{\text{mol}\cdot\text{K}}\right)(298\text{ K})$$

$$\times\left[3\cdot\left(\frac{1870\text{ K}}{2(298\text{ K})}+\frac{(1870\text{ K})/(298\text{ K})}{e^{(1870\text{ K})/(298\text{ K})}-1}\right)+2\cdot\left(\frac{2180\text{ K}}{2(298\text{ K})}+\frac{(2180\text{ K})/(298\text{ K})}{e^{(2180\text{ K})/(298\text{ K})}-1}\right)\right.$$

$$\left.+1\cdot\left(\frac{4170\text{ K}}{2(298\text{ K})}+\frac{(4170\text{ K})/(298\text{ K})}{e^{(4170\text{ K})/(298\text{ K})}-1}\right)+3\cdot\left(\frac{4320\text{ K}}{2(298\text{ K})}+\frac{(4320\text{ K})/(298\text{ K})}{e^{(4320\text{ K})/(298\text{ K})}-1}\right)\right]$$

E_{vib} = (2.48 kJ) (9.44 + 7.32 + 6.99 + 21.73) = 113 kJ

E = −1156 kJ + 113 kJ = −1043 kJ

For H:

$$H=\frac{5}{2}NkT-ND_0+NkT\sum_{j=1}^{3N^*-6}\left(\frac{\theta_{v,j}}{2T}+\frac{\theta_{v,j}/T}{e^{\theta_{v,j}/T}-1}\right)+\frac{3}{2}NkT$$

Comparison shows that this expression is identical to $E + NkT$, so that we obtain:

$$H=-1043\text{ kJ}+(1\text{ mol})\left(8.314\frac{\text{J}}{\text{mol}\cdot\text{K}}\right)(298\text{ K})=-1040\text{ kJ}$$

For S:

The ground electronic state is singly degenerate, so the electronic participation is zero. Again, let us separate the vibrational component from the non-vibrational components:

$$S_{\text{non-vib}}=Nk\ln\left[\left(\frac{2\pi mkT}{h^2}\right)^{3/2}\frac{kT}{p}\right]+\frac{5}{2}Nk+Nk\ln\left[\frac{\pi^{1/2}}{\sigma}\left(\frac{T^3}{\theta_A\theta_B\theta_C}\right)^{1/2}\right]+\frac{3}{2}Nk$$

$$=Nk\ln\left[\left(\frac{2\pi mkT}{h^2}\right)^{3/2}\frac{kT}{p}\right]+4Nk+Nk\ln\left[\frac{\pi^{1/2}}{\sigma}\left(\frac{T^3}{\theta_A\theta_B\theta_C}\right)^{1/2}\right]$$

$$S_{\text{vib}}=Nk\sum_{j=1}^{3N^*-6}\left[\frac{\theta_{v,j}/T}{e^{\theta_{v,j}/T}-1}-\ln\left(1-e^{-\theta_{v,j}/T}\right)\right]$$

Using p = 1 atm = 1.01325×10^5 Pa, m = (16.043 g/mol) / (6.022×10^{23} mol^{-1}) = 2.6640×10^{-23} g or 2.6640×10^{-26} kg, and the same numerical values as above:

$$S_{\text{non-vib}} = (1\text{ mol})\left(8.314\frac{\text{J}}{\text{mol}\cdot\text{K}}\right)$$

$$\times\ln\left(\left[\frac{2\pi(2.6640\times10^{-26}\text{ kg})(1.381\times10^{-23}\text{ J/K})(298\text{ K})}{(6.626\times10^{-34}\text{ J}\cdot\text{s})^2}\right]^{3/2}\frac{(1.381\times10^{-23}\text{ J/K})(298\text{ K})}{(1.0\times10^5\text{ Pa})}\right)$$

$$+4(1\text{ mol})\left(8.314\frac{\text{J}}{\text{mol}\cdot\text{K}}\right)+(1\text{ mol})\left(8.314\frac{\text{J}}{\text{mol}\cdot\text{K}}\right)\ln\left(\frac{\pi^{1/2}}{12}\left[\frac{(298\text{ K})^3}{(7.54\text{ K})^3}\right]^{1/2}\right)$$

$S_{\text{non-vib}}$ = 123 J/K + 33.256 J/K + 30.0 J/K = 186 J/K

$$S_{\text{vib}} = (1\text{ mol})\left(8.314\frac{\text{J}}{\text{mol}\cdot\text{K}}\right)$$

$$\times\left\{3\cdot\left[\frac{(1870\text{ K})/(298\text{ K})}{e^{(1870\text{ K})/(298\text{ K})}-1}-\ln\left(1-e^{-(1870\text{ K})/(298\text{ K})}\right)\right]+2\cdot\left[\frac{(2180\text{ K})/(298\text{ K})}{e^{(2180\text{ K})/(298\text{ K})}-1}-\ln\left(1-e^{-(2180\text{ K})/(298\text{ K})}\right)\right]\right.$$

$$\left.+1\cdot\left[\frac{(4170\text{ K})/(298\text{ K})}{e^{(4170\text{ K})/(298\text{ K})}-1}-\ln\left(1-e^{-(4170\text{ K})/(298\text{ K})}\right)\right]+3\cdot\left[\frac{(4320\text{ K})/(298\text{ K})}{e^{(4320\text{ K})/(298\text{ K})}-1}-\ln\left(1-e^{-(4320\text{ K})/(298\text{ K})}\right)\right]\right\}$$

S_{vib} = (8.314 J/K) (0.0413 + 0.01111 + 1.26×10^{-5} + 2.369×10^{-5}) = 0.436 J/K

S = 186 J/K + 0.436 J/K = 186 J/K

Experimentally, $S°$ is 188.66 J/(mol·K) at 298 K according to Appendix 2.

For G, we simply make use of $G = H - TS$, so G = -1041 kJ – (298 K)(186.3 J/K) = −1096 kJ.

18.47. Let us calculate the H and S values for each species, then do the "products-minus-reactants" calculation. Using the formulas from Table 18.5, we obtain for H_2 and O_2 (remember that for homonuclear diatomics, $\sigma = 2$):

$$H = \frac{5}{2}NkT - ND_0 + NkT\left(\frac{\theta_v}{2T}+\frac{\theta_v/T}{e^{\theta_v/T}-1}\right)+NkT$$

$$= \frac{7}{2}NkT - ND_0 + NkT\left(\frac{\theta_v}{2T}+\frac{\theta_v/T}{e^{\theta_v/T}-1}\right)$$

$$S = Nk\ln\left[\left(\frac{2\pi mkT}{h^2}\right)^{3/2}\frac{kT}{p}\right]+\frac{5}{2}Nk+Nk\ln g_1+Nk\left[\frac{\theta_v/T}{e^{\theta_v/T}-1}-\ln\left(1-e^{-\theta_v/T}\right)\right]+Nk\ln\frac{T}{\sigma\theta_r}+Nk$$

$$= Nk\ln\left[\left(\frac{2\pi mkT}{h^2}\right)^{3/2}\frac{kT}{p}\right]+\frac{7}{2}Nk+Nk\ln g_1+Nk\left[\frac{\theta_v/T}{e^{\theta_v/T}-1}-\ln\left(1-e^{-\theta_v/T}\right)\right]+Nk\ln\frac{T}{\sigma\theta_r}$$

For H_2O:

$$H = \frac{5}{2}NkT - ND_0 + NkT\sum_{j=1}^{3N^*-6}\left(\frac{\theta_{v,j}}{2T} + \frac{\theta_{v,j}/T}{e^{\theta_{v,j}/T}-1}\right) + \frac{3}{2}NkT$$

$$= 4NkT - ND_0 + NkT\sum_{j=1}^{3N^*-6}\left(\frac{\theta_{v,j}}{2T} + \frac{\theta_{v,j}/T}{e^{\theta_{v,j}/T}-1}\right)$$

To make the equation for $S(H_2O)$ more managable, we break it down into non-vibrational (translational, electronic, rotational) and vibrational components:

$$S_{\text{non-vib}} = Nk\ln\left[\left(\frac{2\pi mkT}{h^2}\right)^{3/2}\frac{kT}{p}\right] + \frac{5}{2}Nk + Nk\ln g_1 + Nk\ln\left[\frac{\pi^{1/2}}{\sigma}\left(\frac{T^3}{\theta_A\theta_B\theta_C}\right)^{1/2}\right] + \frac{3}{2}Nk$$

$$= Nk\ln\left[\left(\frac{2\pi mkT}{h^2}\right)^{3/2}\frac{kT}{p}\right] + 4Nk + Nk\ln g_1 + Nk\ln\left[\frac{\pi^{1/2}}{\sigma}\left(\frac{T^3}{\theta_A\theta_B\theta_C}\right)^{1/2}\right]$$

$$S_{\text{vib}} = Nk\sum_{j=1}^{3N^*-6}\left[\frac{\theta_{v,j}/T}{e^{\theta_{v,j}/T}-1} - \ln\left(1-e^{-\theta_{v,j}/T}\right)\right]$$

Now we are ready to substitute the numerical values from Tables 18.1 and 18.2. Replacing Nk by nR, and using $D_0(H_2) = 431.6$ kJ/mol, we obtain the following:

For 1 mol H_2:

$$H = \frac{7}{2}(1\text{ mol})\left(8.314\frac{\text{J}}{\text{mol}\cdot\text{K}}\right)(298\text{ K}) - (1\text{ mol})(431.6\text{ kJ/mol})$$
$$+(1\text{ mol})\left(8.314\frac{\text{J}}{\text{mol}\cdot\text{K}}\right)(298\text{ K})\left[\frac{6215\text{ K}}{2(298\text{ K})} + \frac{(6215\text{ K})/(298\text{ K})}{e^{(6215\text{ K})/(298\text{ K})}-1}\right]$$

$H = 8.67\times10^3\text{ J} - 431.6\times10^3\text{ J} + 25.8\times10^3\text{ J} = -397.1\text{ kJ}$

For 1 mol O_2 (using $D_0(O_2) = 497.0$ kJ/mol):

$$H = \frac{7}{2}(1\text{ mol})\left(8.314\frac{\text{J}}{\text{mol}\cdot\text{K}}\right)(298\text{ K}) - (1\text{ mol})(497.0\text{ kJ/mol})$$
$$+(1\text{ mol})\left(8.314\frac{\text{J}}{\text{mol}\cdot\text{K}}\right)(298\text{ K})\left[\frac{2230\text{ K}}{2(298\text{ K})} + \frac{(2230\text{ K})/(298\text{ K})}{e^{(2230\text{ K})/(298\text{ K})}-1}\right]$$

$H = 8.67\times10^3\text{ J} - 497.0\times10^3\text{ J} + 9.28\times10^3\text{ J} = -479.0\text{ kJ}$

For 1 mol H_2O (using $D_0(H_2O) = 864.4$ kJ/mol):

$$H = 4(1\text{ mol})\left(8.314\frac{\text{J}}{\text{mol}\cdot\text{K}}\right)(298\text{ K}) - (1\text{ mol})(864.4\text{ kJ/mol})$$

$$+(1\text{ mol})\left(8.314\frac{\text{J}}{\text{mol}\cdot\text{K}}\right)(298\text{ K}) \times \left\{\left[\frac{2287\text{ K}}{2(298\text{ K})} + \frac{(2287\text{ K})/(298\text{ K})}{e^{(2287\text{ K})/(298\text{ K})} - 1}\right]\right.$$

$$\left. + \left[\frac{5163\text{ K}}{2(298\text{ K})} + \frac{(5163\text{ K})/(298\text{ K})}{e^{(5163\text{ K})/(298\text{ K})} - 1}\right] + \left[\frac{5350\text{ K}}{2(298\text{ K})} + \frac{(5350\text{ K})/(298\text{ K})}{e^{(5350\text{ K})/(298\text{ K})} - 1}\right]\right\}$$

$$H = 9.91\times10^3\text{ J} - 864.4\times10^3\text{ J} + 53.22\times10^3\text{ J} = -801.3\text{ kJ}$$

The reaction enthalpy is $\Delta H^\circ = 2\ (-801.3\text{ kJ}) - [2\ (-397.1\text{ kJ}) + 1\ (-479.0\text{ kJ})] = -329.3\text{ kJ}$.

As for entropy, recall that the ground electronic state for H_2 and H_2O is a singlet state, while it is a triplet state for O_2. We obtain (using Tables 18.1 through 18.4) for 1 mol H_2:

$$S = (1\text{ mol})\left(8.314\frac{\text{J}}{\text{mol}\cdot\text{K}}\right)$$

$$\times \ln\left(\left[\frac{2\pi(3.3474\times10^{-27}\text{ kg})(1.381\times10^{-23}\text{ J/K})(298\text{ K})}{(6.626\times10^{-34}\text{ J}\cdot\text{s})^2}\right]^{3/2}\frac{(1.381\times10^{-23}\text{ J/K})(298\text{ K})}{(1.01325\times10^5\text{ Pa})}\right)$$

$$+\frac{7}{2}(1\text{ mol})\left(8.314\frac{\text{J}}{\text{mol}\cdot\text{K}}\right) + (1\text{ mol})\left(8.314\frac{\text{J}}{\text{mol}\cdot\text{K}}\right)\ln(1)$$

$$+(1\text{ mol})\left(8.314\frac{\text{J}}{\text{mol}\cdot\text{K}}\right)\left[\frac{(6215\text{ K})/(298\text{ K})}{e^{(6215\text{ K})/(298\text{ K})} - 1} - \ln\left(1 - e^{-(6215\text{ K})/(298\text{ K})}\right)\right]$$

$$+(1\text{ mol})\left(8.314\frac{\text{J}}{\text{mol}\cdot\text{K}}\right)\ln\frac{298\text{ K}}{2(85.4\text{ K})}$$

$$S = 96.7\text{ J/K} + 29.10\text{ J/K} + 0 + 1.59\times10^{-7}\text{ J/K} + 4.63\text{ J/K} = 130.4\text{ J/K}$$

For 1 mol O_2:

$$S = (1\text{ mol})\left(8.314\frac{\text{J}}{\text{mol}\cdot\text{K}}\right)$$

$$\times \ln\left(\left[\frac{2\pi(5.31365\times10^{-26}\text{ kg})(1.381\times10^{-23}\text{ J/K})(298\text{ K})}{(6.626\times10^{-34}\text{ J}\cdot\text{s})^2}\right]^{3/2}\frac{(1.381\times10^{-23}\text{ J/K})(298\text{ K})}{1.01325\times10^5\text{ Pa}}\right)$$

$$+\frac{7}{2}(1\text{ mol})\left(8.314\frac{\text{J}}{\text{mol}\cdot\text{K}}\right) + (1\text{ mol})\left(8.314\frac{\text{J}}{\text{mol}\cdot\text{K}}\right)\ln(3)$$

$$+(1\text{ mol})\left(8.314\frac{\text{J}}{\text{mol}\cdot\text{K}}\right)\left[\frac{(2230\text{ K})/(298\text{ K})}{e^{(2230\text{ K})/(298\text{ K})} - 1} - \ln\left(1 - e^{-(2230\text{ K})/(298\text{ K})}\right)\right]$$

$$+(1\text{ mol})\left(8.314\frac{\text{J}}{\text{mol}\cdot\text{K}}\right)\ln\frac{298\text{ K}}{2(2.07\text{ K})}$$

S = 131 J/K + 29.10 J/K + 9.134 J/K + 0.0397 J/K + 35.6 J/K = 205 J/K

For 1 mol H_2O:

$$S_{\text{non-vib}} = (1\text{ mol})\left(8.314\frac{\text{J}}{\text{mol}\cdot\text{K}}\right)$$

$$\times\ln\left[\left(\frac{2\pi(2.99156\times10^{-26}\text{ kg})(1.381\times10^{-23}\text{ J/K})(298\text{ K})}{(6.626\times10^{-34}\text{ J}\cdot\text{s})^2}\right)^{3/2}\frac{(1.381\times10^{-23}\text{ J/K})(298\text{ K})}{1.01325\times10^5\text{ Pa}}\right]$$

$$+4(1\text{ mol})\left(8.314\frac{\text{J}}{\text{mol}\cdot\text{K}}\right)+(1\text{ mol})\left(8.314\frac{\text{J}}{\text{mol}\cdot\text{K}}\right)\ln(1)$$

$$+(1\text{ mol})\left(8.314\frac{\text{J}}{\text{mol}\cdot\text{K}}\right)\ln\left(\frac{\pi^{1/2}}{2}\left[\frac{(298\text{ K})^3}{(13.4\text{ K})(20.9\text{ K})(40.1\text{ K})}\right]^{1/2}\right)$$

$S_{\text{non-vib}}$ = 124 J/K + 33.26 J/K + 0 + 31.3 J/K = 189 J/K

$$S_{\text{vib}} = (1\text{ mol})\left(8.314\frac{\text{J}}{\text{mol}\cdot\text{K}}\right)\left\{1\cdot\left[\frac{(2287\text{ K})/(298\text{ K})}{e^{(2287\text{ K})/(298\text{ K})}-1}-\ln\left(1-e^{-(2287\text{ K})/(298\text{ K})}\right)\right]\right.$$

$$\left.+1\cdot\left[\frac{(5163\text{ K})/(298\text{ K})}{e^{(5163\text{ K})/(298\text{ K})}-1}-\ln\left(1-e^{-(5163\text{ K})/(298\text{ K})}\right)\right]+1\cdot\left[\frac{(5350\text{ K})/(298\text{ K})}{e^{(5350\text{ K})/(298\text{ K})}-1}-\ln\left(1-e^{-(5350\text{ K})/(298\text{ K})}\right)\right]\right\}$$

S_{vib} = (8.314 J/K) (0.00403 + 5.48×10^{-7} + 3.03×10^{-7}) = 0.0335 J/K

S = 189 J/K + 0.0335 J/K = 189 J/K

Now we can evaluate the ΔS of the reaction:

$\Delta S°$ = 2 (189 J/K) – 2 (130.4 J/K) – 205 J/K = -89 J/K

Using experimental values from Appendix 2:

$\Delta S°$ = 2 (188.83 J/K) – 2 (130.68 J/K) – 205.14 J/K = −88.84 J/K.

18.49. H_2S is a nonlinear polyatomic molecule, so we use equation 18.52, but we first calculate the vibrational temperatures from the given frequencies using equation 18.16:

$\theta_v = \frac{h\nu}{k} = \frac{hc\tilde{\nu}}{k}$. For 1183 cm^{-1}:

$$\theta_v = \frac{(6.626\times10^{-34}\text{ J}\cdot\text{s})(2.9979\times10^{10}\text{ cm/s})(1183\text{ cm}^{-1})}{1.381\times10^{-23}\text{ J/K}} = 1702\text{ K; similarly, 2615 and}$$

2617 cm^{-1} correspond to 3761 and 3764 K, respectively.

$$C_V = \left(6.022\times10^{23}\right)\left(1.381\times10^{-23}\text{ J/K}\right)\left\{\frac{3}{2}+\frac{3}{2}+\left(\frac{1701\text{ K}}{298\text{ K}}\right)^2\frac{e^{-(1701\text{ K})/(298\text{ K})}}{\left(1-e^{-(1701\text{ K})/(298\text{ K})}\right)^2}\right.$$

$$\left.+\left(\frac{3761\text{ K}}{298\text{ K}}\right)^2\frac{e^{-(3761\text{ K})/(298\text{ K})}}{\left(1-e^{-(3761\text{ K})/(298\text{ K})}\right)^2}+\left(\frac{3764\text{ K}}{298\text{ K}}\right)^2\frac{e^{-(3764\text{ K})/(298\text{ K})}}{\left(1-e^{-(3764\text{ K})/(298\text{ K})}\right)^2}\right\}$$

$$C_V = 8.316\text{ J/K }(1.5 + 1.5 + 0.109 + 5.25\times10^{-4} + 5.21\times10^{-4}) = 25.85\text{ J/K}$$

18.51. At equilibrium, the net amount of each species in the reaction remains constant. Therefore, N_i (the number of atoms or molecules of each species) is constant in all expressions on the right side of equation 18.59. The coefficients ν_i are characteristic of the balanced chemical equation and also constant. Thus, at equilibrium each partition function is constant, and the expression of partition functions in equation 18.59 is also constant.

18.53. We first calculate the q contributions for each species, then substitute the numerical values into the expression for K. By doing so, we will neglect the nuclear contribution, and consider the electronic part separately, because individually some very large numbers would be generated. At 1000 K, we can also use the high-temperature form of the rotational partition function. At 1000 K and 1 atm, the molar volume of an ideal gas is 0.08205 m^3.

For 1 mol H_2, the non-electronic contributions are:

$$q_{\text{non-elect}} = q_{\text{trans}}\cdot q_{\text{vib}}\cdot q_{\text{rot}} = \left[\frac{2\pi\left(3.3474\times10^{-27}\text{ kg}\right)\left(1.381\times10^{-23}\text{ J/K}\right)(1000\text{ K})}{\left(6.626\times10^{-34}\text{ J}\cdot\text{s}\right)^2}\right]^{3/2}\left(0.08205\text{ m}^3\right)$$

$$\times\left[\frac{1}{1-e^{-(6215\text{ K})/(1000\text{ K})}}\right]\times\left[\frac{1000\text{ K}}{2(85.4\text{ K})}\right]$$

$$q_{\text{non-elect}} = 8.19\times10^{30}$$

For O_2:

$$q_{\text{non-elect}} = \left[\frac{2\pi\left(5.31364\times10^{-26}\text{ kg}\right)\left(1.381\times10^{-23}\text{ J/K}\right)(1000\text{ K})}{\left(6.626\times10^{-34}\text{ J}\cdot\text{s}\right)^2}\right]^{3/2}\left(0.08205\text{ m}^3\right)$$

$$\times\left[\frac{1}{1-e^{-(2230\text{ K})/(1000\text{ K})}}\right]\times\left[\frac{1000\text{ K}}{2(2.07\text{ K})}\right]$$

$$q_{\text{non-elect}} = 2.39\times10^{34}$$

For H_2O:

$$q_{\text{non-elect}} = \left[\frac{2\pi(2.99156\times10^{-26}\text{ kg})(1.381\times10^{-23}\text{ J/K})(1000\text{ K})}{(6.626\times10^{-34}\text{ J}\cdot\text{s})^2}\right]^{3/2}(0.08205\text{ m}^3)$$

$$\times\left[\frac{1}{1-e^{-(2287\text{ K})/(1000\text{ K})}}\right]\left[\frac{1}{1-e^{-(5163\text{ K})/(1000\text{ K})}}\right]\left[\frac{1}{1-e^{-(5350\text{ K})/(1000\text{ K})}}\right]$$

$$\times\frac{\pi^{1/2}}{2}\left[\frac{(1000\text{ K})^3}{(13.4\text{ K})(20.9\text{ K})(40.1\text{ K})}\right]^{1/2}$$

$$q_{\text{non-elect}} = 1.11\times10^{34}$$

Each of these q_i's needs to be multiplied by the respective $g_1e^{D_0/(kT)}$, which is the electronic contribution, in order to obtain Q_i. The expression for the thermodynamic K is then $K = \frac{(Q_{H_2O})^2}{(Q_{H_2})^2(Q_{O_2})} = \frac{(q_{\text{non-elect},H_2O})^2}{(q_{\text{non-elect},H_2})^2\cdot q_{\text{non-elect},O_2}}\times\frac{1^2\cdot e^{2D_0(H_2O)/(kT)}}{1^2\cdot e^{2D_0(H_2)/(kT)}\cdot 3\cdot e^{D_0(O_2)/(kT)}}$ (recall that the ground state of oxygen is a triplet state). The fraction of exponentials in terms of the dissociation energies can be rewritten as $\frac{1}{3}\exp\left(\frac{2D_0(H_2O)-2D_0(H_2)-D_0(O_2)}{kT}\right)$; the numerator of the exponent is 2 (864.4 kJ/mol) – 2 (431.6 kJ/mol) – (497.0 kJ/mol) = 368.6 kJ/mol. This is a molar quantity, so we need to divide by N_A to convert it into a molecular quantity:

$$\frac{1}{3}\exp\left(\frac{(368.6\times10^3\text{ J/mol})/(6.022\times10^{23}\text{ mol}^{-1})}{(1.381\times10^{-23}\text{ J/K})(1000\text{ K})}\right) = \frac{1.774\times10^{19}}{3} = 5.912\times10^{18}$$

We can now substitute:

$$K = \frac{(1.11\times10^{34})^2}{(8.19\times10^{30})^2(2.39\times10^{34})}\cdot5.912\times10^{18} = 4.54\times10^{-10}$$

This compares with $\Delta H° = 2\ (-241.8\text{ kJ/mol}) = -483.6\text{ kJ/mol}$ and

$\Delta S° = 2\ (188.83) - 2\ (130.68) - 205.14 = -88.84\text{ J/(K·mol)}$, so that

$\Delta G° = \Delta H° - T\Delta S° = -394.8\text{ kJ/mol}$, or $\ln K = -\Delta G°/(RT) = 47.48$ and $K \approx 4\times10^{20}$.

The variation is huge, but the $\Delta H°$ and $S°$ values in Appendix 2 are for 298 K, not 1000 K. We would expect $\Delta H°$ and $\Delta S°$ values for H_2, O_2, and H_2O to be substantially different at 1000 K than at 298 K.

18.55. **(a)** Using $\nu = \dfrac{1}{2\pi}\sqrt{\dfrac{k}{m}}$ and the mass of a single Al atom, we rearrange to solve for k:

$$k = (2\pi\nu)^2 m = \left[2\pi\left(4.5\times10^{12}\ \text{s}^{-1}\right)\right]^2\left(4.48049\times10^{-26}\ \text{kg}\right) = 36\ \text{kg/s}^2 = 36\ \text{N/m}$$

(b) The photon will have frequency of $4.5\times10^{12}\ \text{s}^{-1}$ as well. We convert this to a wavenumber unit: $\tilde{\nu} = \dfrac{\nu}{c} = \dfrac{4.5\times10^{12}\ \text{s}^{-1}}{2.9979\times10^{10}\ \text{cm/s}} = 1.5\times10^{2}\ \text{cm}^{-1} \approx 150\ \text{cm}^{-1}$

(c) Yes, because this wavenumber is near one end of the infrared region of the spectrum.

CHAPTER 19

The Kinetic Theory of Gases

19.1. A postulate is a statement that is assumed but not proven. Other statements can be based on them and supported or not supported by them, but the postulate itself is only supported by the results (and their agreement with experiments) that are derived from it.

19.3. According to equation 19.11, the molar kinetic energy of $SO_2(g)$ (or any other ideally behaving gas, for that matter) at 298 K is 3/2 RT = 3/2 (8.314 J/mol/K) (298 K) = 3.72 kJ/mol.

19.5. **(a)** The mass of 1.00 mol He is (1.00 mol) (4.0026 g/mol) = 4.00 g or 4.00×10^{-3} kg. Rearranging equation 19.8: $v_{avg}^2 = \frac{3pV}{Nm} = \frac{3(1\ \text{bar})(22.8\ \text{L})}{4.00 \times 10^{-3}\ \text{kg}} = 17.1 \times 10^3\ \text{L} \cdot \text{bar/kg}$. We convert the units into SI units by multiplying with 0.001 m^3/L and 1×10^5 Pa/bar: $v_{avg}^2 = 1.71 \times 10^6\ \text{m}^3 \cdot \text{Pa/kg}$. Recalling that 1 Pa = 1 N/m^2 and 1 N = 1 kg m/s^2, these units simplify to m^2/s^2. After taking the square root: $v_{avg} = 1.31 \times 10^3\ \text{m/s} = 1.31\ \text{km/s}$.

(b) The mass of 1.00 mol NO_2 is (1.00 mol) (46.0055 g/mol) = 46.0 g or 46.0×10^{-3} kg. $v_{avg}^2 = \frac{3pV}{Nm} = \frac{3(1\ \text{bar})(22.8\ \text{L})}{46.0 \times 10^{-3}\ \text{kg}} = 1.49 \times 10^3\ \text{L} \cdot \text{bar/kg} = 149 \times 10^3\ \text{m}^2/\text{s}^2$. After taking the square root: $v_{avg} = 386\ \text{m/s}$.

19.7. $p = \frac{(0.500 \times 10^{-3}\ \text{kg})(985\ \text{m/s})^2}{3(15.6\ \text{L})(0.001\ \text{m}^3/\text{L})} = 10.4 \times 10^3 \frac{\text{kg}}{\text{m} \cdot \text{s}^2}$. Multiplying the units in numerator and denominator by m/s^2 gives units of (kg·m/s^2) / (m^2), or N / m^2 = Pa. Thus, p = 10.4 kPa.

19.9. The drawing is left to the student.

19.11. Let us rewrite equation 19.8 in such a way as to get the expression for kinetic energy all by itself on one side: $\frac{1}{2} m v_{avg}^2 = \frac{3pV}{2N}$. Note that the right side of this expression has pressure, volume, and number of gas molecules – three of the four variables that the conditions of a gas depend on. The fourth variable is temperature, and by comparing the right side to the ideal gas law, we can argue that $\frac{3pV}{2N}$ is a function of temperature only, and that this is true even for real gases. Therefore $\frac{1}{2} m v^2 = f(T)$, leading to the conclusion that the (average) kinetic energy of a gas is dictated solely by the temperature of the gas.

19.13. We rearrange equation 19.13 to solve for T: $v_{\text{rms}} = \sqrt{\dfrac{3RT}{M}} \Rightarrow v_{\text{rms}}^2 = \dfrac{3RT}{M}$. With a molar mass of 83.80 g/mol = 83.80×10^{-3} kg/mol for krypton we obtain:

$$T = \frac{Mv_{\text{rms}}^2}{3R} = \frac{\left(83.80\times10^{-3}\text{ kg/mol}\right)\left(855\text{ m/s}\right)^2}{3\left(8.314\dfrac{\text{J}}{\text{mol}\cdot\text{K}}\right)} = 2.46\times10^3\text{ K} \approx 2460\text{ K}$$

(recall that $1\text{ J} = 1\text{ N}\cdot\text{m} = 1\text{ kg}\cdot\text{m}^2/\text{s}^2$).

19.15. As in the previous exercise, we use $T = \dfrac{Mv_{\text{rms}}^2}{3R}$ with the given speeds. The molar mass of cesium is 132.9055 g/mol = 132.9055×10^{-3} kg/mol:

For 200 m/s: $$T = \frac{\left(132.9055\times10^{-3}\text{ kg/mol}\right)\left(200\text{ m/s}\right)^2}{3\left(8.314\dfrac{\text{J}}{\text{mol}\cdot\text{K}}\right)} = 213\text{ K};$$

for 400 m/s: $$T = \frac{\left(132.9055\times10^{-3}\text{ kg/mol}\right)\left(400\text{ m/s}\right)^2}{3\left(8.314\dfrac{\text{J}}{\text{mol}\cdot\text{K}}\right)} = 853\text{ K};$$

for 600 m/s: $$T = \frac{\left(132.9055\times10^{-3}\text{ kg/mol}\right)\left(600\text{ m/s}\right)^2}{3\left(8.314\dfrac{\text{J}}{\text{mol}\cdot\text{K}}\right)} = 1.92\text{ K} \approx 1920\text{ K};$$

for 800 m/s: $$T = \frac{\left(132.9055\times10^{-3}\text{ kg/mol}\right)\left(800\text{ m/s}\right)^2}{3\left(8.314\dfrac{\text{J}}{\text{mol}\cdot\text{K}}\right)} = 3.41\times10^3\text{ K} \approx 3410\text{ K};$$

for 1000 m/s: $$T = \frac{\left(132.9055\times10^{-3}\text{ kg/mol}\right)\left(1000\text{ m/s}\right)^2}{3\left(8.314\dfrac{\text{J}}{\text{mol}\cdot\text{K}}\right)} = 5329\text{ K}.$$

The temperature is increasing quadratically with the rms-average speed, as can be seen from the general expression given above ($T \propto v_{\text{rms}}^2$).

19.17. As in exercise 19.13, we use $T = \frac{Mv_{\text{rms}}^2}{3R}$. With a molar mass of 1.0079 g/mol = 1.0079×10^{-3} kg/mol for H gas, we then have

$$T = \frac{(1.0079 \times 10^{-3} \text{ kg/mol})(3.00 \times 10^8 \text{ m/s})^2}{3\left(8.314 \frac{\text{J}}{\text{mol}\cdot\text{K}}\right)} = 3.64 \times 10^{12} \text{ K}$$, a temperature that is impossible to achieve (the temperature at the core of the sun is "only" on the order of 10^7 K).

19.19. We need to solve the integral $\int_{-\infty}^{+\infty} Ae^{(1/2)Kv_x^2}\, dv_x = 1$ for the proper value of A (this is equivalent to normalizing a wavefunction). This is an even function centered about zero, so we can rewrite this integral as $2\int_0^{+\infty} Ae^{(1/2)Kv_x^2}\, dv_x = 1$ which, after factoring out the constant A, gives $2A\int_0^{+\infty} e^{(1/2)Kv_x^2}\, dv_x = 1$. The integral is of the form $\int_0^{+\infty} e^{-bx^2}\, dx = \frac{1}{2}\left(\frac{\pi}{b}\right)^{1/2}$ (see Appendix 1). In our case, the constant is b = -½K, so we have: $2A\left[\frac{1}{2}\left(\frac{\pi}{-\frac{1}{2}K}\right)^{1/2}\right] = 1$.

Simplifying leads to $A\left(-\frac{2\pi}{K}\right)^{1/2} = 1$ or $A = \left(-\frac{K}{2\pi}\right)^{1/2}$. This final expression is equation 19.25.

19.21. We start with equation 19.30: $v_{\text{avg},x}^2 = 2\left(\frac{-K}{2\pi}\right)^{1/2} \int_0^{\infty} v_x^2 \cdot e^{(1/2)Kv_x^2}\, dv_x$.

If we set $v_x = x$ and b = -½K, we see that this integral is similar to the integral $\int_{-\infty}^{\infty} x^2 e^{-bx^2}\, dx$ given in Appendix 1, except that the lower limits are different. However, because the argument is an even function centered about zero, we can rewrite this as $2\int_0^{\infty} x^2 e^{-bx^2}\, dx$, so that we obtain: $\int_0^{\infty} v_x^2 \cdot e^{(1/2)Kv_x^2}\, dv_x = \frac{1}{2}\cdot\frac{1}{2}\left[\frac{\pi}{(-\frac{1}{2}K)^3}\right]^{1/2}$. The right-hand side can be simplified to $\frac{1}{4}\left[\frac{\pi}{(-K/2)^3}\right]^{1/2} = \frac{1}{4}\left[\frac{2^3\pi}{(-K)^3}\right]^{1/2} = \left[\frac{2^3\pi}{4^2(-K)^3}\right]^{1/2} = \left[\frac{\pi}{2(-K)^3}\right]^{1/2}$.

Substituting this for the integral in the original expression gives

$$v_{\text{avg},x}^2 = 2\left(\frac{-K}{2\pi}\right)^{1/2}\left[\frac{\pi}{2(-K)^3}\right]^{1/2}.$$

This simplifies to:

$$2\left[\frac{(-K)\pi}{2(2\pi)(-K)^3}\right]^{1/2} = 2\left[\frac{1}{4(-K)^2}\right]^{1/2} = \left[\frac{2^2}{4(-K)^2}\right]^{1/2} = \left[\frac{1}{(-K)^2}\right]^{1/2} = \frac{1}{-K}$$. Setting this equal to kT/m yields equation 19.31 (see text).

19.23. We need to take the derivative of equation 19.33 with respect to v, set it equal to zero, then solve for v from that expression. Using the product rule for differentiation:

$$\frac{\partial}{\partial v}\left[4\pi\left(\frac{m}{2\pi kT}\right)^{3/2} v^2 e^{-mv^2/(2kT)}\right] = 8\pi\left(\frac{m}{2\pi kT}\right)^{3/2} ve^{-mv^2/(2kT)} + 4\pi\left(\frac{m}{2\pi kT}\right)^{3/2} v^2 e^{-mv^2/(2kT)} \cdot \frac{-2mv}{2kT} = 0$$

Much can be divided out: 4π, the $[m/(2\pi kT)]^{3/2}$, the exponential, and one v. What is left is then

$2 + v \cdot \frac{-mv}{kT} = 0$. This equals $2 - \frac{mv^2}{kT} = 0$, and solving for v^2 gives $v^2 = \frac{2kT}{m}$, so $v = \sqrt{\frac{2kT}{m}}$.

The last expression is equation 19.34.

19.25. From equations 19.13 and 19.34: $\frac{v_{rms}}{v_{most\ prob}} = \frac{\sqrt{3RT/M}}{\sqrt{2RT/M}} = \sqrt{\frac{3}{2}} \approx 1.2247$.

19.27. To solve these problems, estimate $dv \approx \Delta v = 10$ m/s, and use the center value of the specified interval as the average value for v itself. Substitute into equation 19.33 and evaluate.

The molar mass for O_2 is 31.9988 g/mol, so the mass of one O_2 molecule is $(31.9988\times10^{-3}\text{ kg}) / (6.022\times10^{23}\text{ mol}^{-1}) = 5.314\times10^{-26}$ kg.

(a) For the interval 10 – 20 m/s:

$$G = 4\pi\left[\frac{5.314\times10^{-26}\text{ kg}}{2\pi(1.381\times10^{-23}\text{ J/K})(300\text{ K})}\right]^{3/2}\left(15\ \frac{\text{m}}{\text{s}}\right)^2 \exp\left[\frac{-(5.314\times10^{-26}\text{ kg})(15\text{ m/s})^2}{2(1.381\times10^{-23}\text{ J/K})(300\text{ K})}\right]\cdot\left(10\ \frac{\text{m}}{\text{s}}\right)$$

We numerically evaluate all constant factors in this expression (except for the speeds) and obtain

$$G = (3.66\times10^{-8}\text{ s}^3/\text{m}^3)(15\text{ m/s})^2 e^{-(6.41\times10^{-6}\text{ s}^2/\text{m}^2)(15\text{ m/s})^2}(10\text{ m/s}) = 8.23\times10^{-5}$$

(b) For the interval 100 – 110 m/s:

$$G = (3.66\times10^{-8}\text{ s}^3/\text{m}^3)(105\text{ m/s})^2 e^{-(6.41\times10^{-6}\text{ s}^2/\text{m}^2)(105\text{ m/s})^2}(10\text{ m/s}) = 3.76\times10^{-3}$$

(c) For the interval 1000 – 1010 m/s:

$$G=\left(3.66\times10^{-8}\ \text{s}^3/\text{m}^3\right)\left(1005\ \text{m/s}\right)^2 e^{-\left(6.41\times10^{-6}\ \text{s}^2/\text{m}^2\right)\left(1005\ \text{m/s}\right)^2}\left(10\ \text{m/s}\right)=5.69\times10^{-4}$$

(d) For the interval 5000 – 5010 m/s:

$$G=\left(3.66\times10^{-8}\ \text{s}^3/\text{m}^3\right)\left(5005\ \text{m/s}\right)^2 e^{-\left(6.41\times10^{-6}\ \text{s}^2/\text{m}^2\right)\left(5005\ \text{m/s}\right)^2}\left(10\ \text{m/s}\right)=1.58\times10^{-69}$$

(e) For the interval 10,000 – 10,010 m/s:

$$G=\left(3.66\times10^{-8}\ \text{s}^3/\text{m}^3\right)\left(10005\ \text{m/s}\right)^2 e^{-\left(6.41\times10^{-6}\ \text{s}^2/\text{m}^2\right)\left(10005\ \text{m/s}\right)^2}\left(10\ \text{m/s}\right)=6.04\times10^{-278}\approx0$$

(many calculators won't be able to even register such a magnitude of number!). Note the trend: G starts small, increases, then decreases to virtually zero as the velocity increases.

19.29. The integral we need to evaluate is $\int_0^\infty v^2 G(v)\,dv$. Substituting, we get:

$\int_0^\infty v^2\cdot4\pi\left(\frac{m}{2\pi kT}\right)^{3/2} v^2\cdot e^{-mv^2/(2kT)}dv=4\pi\left(\frac{m}{2\pi kT}\right)^{3/2}\int_0^\infty v^4 e^{-mv^2/(2kT)}\,dv$. This integral has a known form, as seen from Appendix 1: $\int_0^\infty x^{2n}e^{-bx^2}\,dx=\frac{1\cdot3\cdot5\cdot\cdots\cdot(2n-1)}{2^{n+1}\cdot b^n}\sqrt{\frac{\pi}{b}}$. In our case, $n=2$ and $b=m/(2kT)$, so we get: $\frac{1\cdot3}{2^3\cdot\left(\frac{m}{2kT}\right)^2}\sqrt{\frac{\pi\cdot2kT}{m}}=\frac{3\cdot4k^2T^2}{8m^2}\sqrt{\frac{2\pi kT}{m}}=\frac{3k^2T^2}{2m^2}\sqrt{\frac{2\pi kT}{m}}$.

Substituting this for the integral, we finally have $\overline{v^2}=4\pi\left(\frac{m}{2\pi kT}\right)^{3/2}\frac{3k^2T^2}{2m^2}\sqrt{\frac{2\pi kT}{m}}$, which simplifies to $\overline{v^2}=\frac{4\pi m^{3/2}}{2^{3/2}\pi^{3/2}k^{3/2}T^{3/2}}\cdot\frac{3k^2T^2}{2m^2}\cdot\frac{2^{1/2}\pi^{1/2}k^{1/2}T^{1/2}}{m^{1/2}}=\frac{3kT}{m}$. Taking the square root gives us $v_{\text{rms}}=\sqrt{\overline{v^2}}=\sqrt{\frac{3kT}{m}}=\sqrt{\frac{3RT}{M}}$, which is the same equation we derived previously for the root-mean-square velocity.

19.31. **(a)** A monoatomic ideal gas has $\overline{C}_V=\frac{3}{2}R$ and $\overline{C}_p=\frac{5}{2}R$ (equations 2.39 and 2.40), so $\gamma=5/3$. Using this value in the given equation results in $v_{\text{sound}}=\left(\frac{3RT}{2M}\right)^{1/2}$. Or, using the numerical value for R and $T=298$ K:

$$v_{\text{sound}}=\left[\frac{3}{2}\left(8.314\frac{\text{J}}{\text{mol}\cdot\text{K}}\right)(298\ \text{K})\right]^{1/2}\left(\frac{1}{M}\right)^{1/2}=61.0\left(\frac{\text{J}}{\text{mol}}\right)^{1/2}\cdot\left(\frac{1}{M}\right)^{1/2}.$$

(b) $v_{sound} = \left(\frac{3RT}{2M}\right)^{1/2} = \left[\frac{3(8.314\ \text{J/mol/K})(273\ \text{K})}{2(4.0026\times10^{-3}\ \text{kg/mol})}\right]^{1/2} = 922$ m/s, only about 5% lower than the experimental value. (At 298 K we obtain 964 m/s, so we could speculate that the experimental value was actually measured at 298 K, not 273 K.)

19.33. We rearrange equation 19.13 to solve for T: $v_{rms} = \sqrt{\frac{3RT}{M}} \Rightarrow v_{rms}^2 = \frac{3RT}{M}$, so $T = \frac{Mv_{rms}^2}{3R}$.

For N_2 ($M = 28.0134\times10^{-3}$ kg/mol):

$$T = \frac{(28.0134\times10^{-3}\ \text{kg/mol})(11.2\times10^3\ \text{m/s})^2}{3\left(8.314\frac{\text{J}}{\text{mol}\cdot\text{K}}\right)} = 141\times10^3\ \text{K};$$

for H_2 ($M = 2.0158\times10^{-3}$ kg/mol):

$$T = \frac{(2.0158\times10^{-3}\ \text{kg/mol})(11.2\times10^3\ \text{m/s})^2}{3\left(8.314\frac{\text{J}}{\text{mol}\cdot\text{K}}\right)} = 10.1\times10^3\ \text{K}.$$

While very few molecules will have the required kinetic energy, it is nevertheless more likely that H_2 molecules escape the Earth's atmosphere, while the heavier N_2 molecules will mostly remain confined to it.

19.35. The hard sphere radius of Ne is 140 pm, so its diameter is 280 pm = 280×10^{-12} m. Using the conversion of 100 J = 1 L · bar = 0.001 m^3 · bar with equation 19.40, we obtain

$$\lambda = \frac{kT}{\sqrt{2}\pi d^2 p} = \frac{(1.381\times10^{-23}\ \text{J/K})(300.0\ \text{K})}{\sqrt{2}\pi(280\times10^{-12}\ \text{m})^2(0.987\ \text{bar})}\times\frac{0.001\ \text{m}^3\cdot\text{bar}}{100\ \text{J}} = 1.21\times10^{-7}\ \text{m}$$

(or 121 nm).

19.37. We rearrange equation 19.40 to solve for T/p: $\frac{T}{p} = \lambda\frac{\sqrt{2}\pi d^2}{k}$. Using the numerical values: $\frac{T}{p} = \lambda\frac{\sqrt{2}\pi d^2}{k} = (130\times10^{-12}\ \text{m})\frac{\sqrt{2}\pi(65\times10^{-12}\ \text{m})^2}{1.381\times10^{-23}\ \text{J/K}} = 1.8\times10^{-7}\ \text{m}^3\cdot\text{K/J}$.

Recall that 1 J = 1 N · m and that N/m^2 is the SI unit for pressure (Pa), the units of the ratio T/p are indeed those of temperature over pressure (K/Pa). This ratio is not easy to obtain, because very low temperatures and very high pressures are necessary (for example at a pressure of 100 bar = 10^7 Pa we would need a temperature of 1.8 K, which is that of superfluid liquid helium).

19.39. We first determine the "pressure" of hydrogen in interstellar space from the ideal gas law $pV = NkT$: $p = \frac{N}{V}kT = \frac{10}{\left(1\times10^{-2}\text{ m}\right)^3}\left(1.381\times10^{-23}\text{ J/K}\right)\left(2.7\text{ K}\right) = 3.7\times10^{-16}$ Pa. Using this value in equation 19.40: $\lambda = \frac{\left(1.381\times10^{-23}\text{ J/K}\right)\left(2.7\text{ K}\right)}{\sqrt{2}\pi\left(1.10\times10^{-10}\text{ m}\right)^2\left(3.7\times10^{-16}\text{ N/m}^2\right)} = 1.9\times10^{12}$ m.

(That's almost two *billion* kilometers, more than twelve times the distance between the earth and the sun!)

19.41. Let's first convert psi into Pa: $\left(2400\text{ psi}\right)\times\frac{1\text{ atm}}{14.70\text{ psi}}\times\frac{101325\text{ Pa}}{1\text{ atm}} = 1.654\times10^{7}$ Pa.

$$\lambda = \frac{\left(1.381\times10^{-23}\text{ J/K}\right)\left(298\text{ K}\right)}{\sqrt{2}\pi\left(3.20\times10^{-10}\text{ m}\right)^2\left(1.654\times10^{7}\text{ N/m}^2\right)} = 5.47\times10^{-10}\text{ m}$$

(or 5.47 Å, which is only about 70% more than the nominal diameter of a nitrogen molecule).

19.43. We use the ideal gas law in the form $pV = NkT$ to calculate the particle density of Hg atoms under the given conditions:

$$\rho = \frac{N}{V} = \frac{p}{kT} = \frac{\left(0.001426\text{ mmHg}\right)\times\left(\frac{133.3\text{ Pa}}{\text{mmHg}}\right)}{\left(1.381\times10^{-23}\text{ J/K}\right)\left(295.2\text{ K}\right)} = 4.664\times10^{19}\text{ m}^{-3}.$$

Now, to determine the average collision frequency, we use equation 19.41:

$$z = \frac{4\pi\left(4.664\times10^{19}\text{ m}^{-3}\right)\left(2.4\times10^{-10}\text{ m}\right)^2\sqrt{\left(1.381\times10^{-23}\text{ J/K}\right)\left(295.2\text{ K}\right)}}{\sqrt{\pi\cdot3.331\times10^{-25}\text{ kg}}} = 2.1\times10^{3}\text{ s}^{-1}.$$

Even at that low pressure, there are over two thousand collisions per second for any one mercury atom.

19.45. Xe gas (hard sphere diameter: 4.00 Å) at a temperature of 298 K and having an average collision frequency of 1 per second has a particle density of

$$\rho = \frac{z\sqrt{\pi m}}{4\pi d^2\sqrt{kT}} = \frac{\left(1\text{ s}^{-1}\right)\sqrt{\pi\cdot2.180\times10^{-25}\text{ kg}}}{4\pi\left(4.00\times10^{-10}\text{ m}\right)^2\sqrt{\left(1.381\times10^{-23}\text{ J/K}\right)\left(298\text{ K}\right)}} = 6.42\times10^{15}\text{ m}^{-3}.$$

According to equation 19.42, this corresponds to a total number of collisions per second and cubic meter of $Z = \frac{1}{2}\left(1\text{ s}^{-1}\right)\left(6.42\times10^{15}\text{ m}^{-3}\right) = 3.21\times10^{15}\text{ s}^{-1}\cdot\text{m}^{-3}$.

19.47. Referring back to exercise 19.45, we calculate the volume of 1.00 mol Xe (6.02×10^{23} atoms) with a particle density of 6.42×10^{15} m^{-3}: $V = (6.02\times10^{23}) / (6.42\times10^{15}\text{ m}^{-3}) = 9.39\times10^{7}\text{ m}^3$. Therefore, the total number of collisions per second is

$Z\cdot V = (3.21\times10^{15}\text{ s}^{-1}\text{ m}^{-3})\,(9.39\times10^{7}\text{ m}^3) = 3.01\times10^{23}\text{ s}^{-1}$.

19.49. In a 50:50 mixture, the number densities of Ar (0.5 atm) and Xe (0.5 atm) are the same:

$$\rho = \frac{N}{V} = \frac{p}{kT} = \frac{0.5\text{ atm}}{\left(1.381\times10^{-23}\text{ J/K}\right)\left(273\text{ K}\right)} = 1.33\times10^{20}\text{ atm/J}.$$ With 1 atm = 101325 Pa:

$\rho = 1.34\times10^{25}$ m^{-3}. The atomic masses of Ar and Xe are 6.634×10^{-26} kg and 2.180×10^{-25} kg, respectively.

(a) The mean free path for an argon atom hitting a xenon atom is

$$\lambda_{\text{Ar}\to\text{Xe}} = \sqrt{\frac{2.180\times10^{-25}\text{ kg}}{6.634\times10^{-26}\text{ kg}+2.180\times10^{-25}\text{ kg}}}\frac{1}{\sqrt{2}\pi\left[\frac{(2.60+4.00)\times10^{-10}\text{ m}}{2}\right]^2\left(1.34\times10^{25}\text{ m}^{-3}\right)}$$

$\lambda_{\text{Ar}\to\text{Xe}} = 1.35\times10^{-7}$ m.

The mean free path for a xenon atom hitting an argon atom is

$$\lambda_{\text{Xe}\to\text{Ar}} = \sqrt{\frac{6.634\times10^{-26}\text{ kg}}{6.634\times10^{-26}\text{ kg}+2.180\times10^{-25}\text{ kg}}}\frac{1}{\sqrt{2}\pi\left[\frac{(2.60+4.00)\times10^{-10}\text{ m}}{2}\right]^2\left(1.34\times10^{25}\text{ m}^{-3}\right)}$$

$\lambda_{\text{Xe}\to\text{Ar}} = 7.43\times10^{-8}$ m.

(b) The average collision frequencies of argon with xenon and xenon with argon atoms are the same because $\rho_{\text{Ar}} = \rho_{\text{Xe}}$ (see equation 19.45).

With a reduced mass of $\mu = \frac{\left(6.634\times10^{-26}\text{ kg}\right)\left(2.180\times10^{-25}\text{ kg}\right)}{6.634\times10^{-26}\text{ kg}+2.180\times10^{-25}\text{ kg}} = 5.086\times10^{-26}$ kg, we obtain:

$$z = \frac{4\pi\left(1.34\times10^{25}\text{ m}^{-3}\right)\left[\frac{(2.60+4.00)\times10^{-10}\text{ m}}{2}\right]^2\sqrt{\left(1.381\times10^{-23}\text{ J/K}\right)\left(273\text{ K}\right)}}{\sqrt{\pi\cdot\left(5.086\times10^{-26}\text{ kg}\right)}} = 2.82\times10^{9}\text{ s}^{-1}$$

(c) The total number of collisions per second and per cubic meter is (equation 19.47):

$$Z = \frac{2\pi\left(1.34\times10^{25}\text{ m}^{-3}\right)^2\left[\frac{(2.60+4.00)\times10^{-10}\text{ m}}{2}\right]^2\sqrt{\left(1.381\times10^{-23}\text{ J/K}\right)\left(273\text{ K}\right)}}{\sqrt{\pi\cdot\left(5.086\times10^{-26}\text{ kg}\right)}} = 1.90\times10^{34}\text{ s}^{-1}\cdot\text{m}^{-3}.$$

19.51. The particle density of Xe atoms under standard conditions (1 atm = 101,325 Pa, 273.15 K) is $\rho = \frac{N}{V} = \frac{p}{kT} = \frac{101325 \text{ Pa}}{(1.381\times10^{-23} \text{ J/K})(273.15 \text{ K})} = 2.686\times10^{25} \text{ m}^{-3}$. With an atomic mass of 2.180×10^{-25} kg and d = 4.00 Å (see exercise 19.49), the collision rate is then

$$z = \frac{4\pi(2.686\times10^{25} \text{ m}^{-3})(4.00\times10^{-10} \text{ m})^2 \sqrt{(1.381\times10^{-23} \text{ J/K})(273.15 \text{ K})}}{\sqrt{\pi\cdot(2.180\times10^{-25} \text{ kg})}} = 4.01\times10^{9} \text{ s}^{-1}.$$

This means that the average time between collisions is $1 / (4.01\times10^{9} \text{ s}^{-1}) = 2.50\times10^{-10}$ s.

19.53. Briefly, effusion and diffusion are the same in that they relate to the movement of gas particles in space. They are different in that effusion describes gas particles escaping through a hole or holes into vacuum, while diffusion describes gas particles moving through another gas.

19.55. This is indeed counter-intuitive – as the temperature increases, we expect the average speed and therefore rate of effusion to increase as well. But equation 19.51 actually predicts this, because as temperature increases for a gas of fixed volume, so does its pressure. The increase in pressure is (for an ideal gas) directly proportional to temperature, so overall the rate of effusion increases with $T / T^{1/2} = T^{1/2}$.

19.57. Because 1 mm = 10^{-3} m, an area of 0.001 mm^2 equals 0.001×10^{-6} m^2 or 1×10^{-9} m^2.

A molar mass of 28.8 g/mol is $(28.8\times10^{-3} \text{ kg/mol}) / (6.022\times10^{-23} \text{ mol}^{-1}) = 4.78\times10^{-26}$ kg per "air" molecule. At a pressure of 2.00 atm = 2.0 · 101,325 Pa = 2.03×10^{5} Pa, the rate of effusion is

$$\frac{dN}{dt} = (1\times10^{-9} \text{ m}^2)(2.03\times10^{5} \text{ Pa})\left[\frac{1}{2\pi(4.78\times10^{-26} \text{ kg})(1.381\times10^{-23} \text{ J/K})(295 \text{ K})}\right]^{1/2}, \text{ so}$$

$\frac{dN}{dt} = 5.80\times10^{18} \text{ s}^{-1}$. Assuming that the balloon deflates while keeping its shape and volume, the loss of pressure corresponds to a loss of

$$n = \frac{pV}{RT} = \frac{(1.00 \text{ atm})(1.00 \text{ L})}{\left(0.08205 \frac{\text{L}\cdot\text{atm}}{\text{mol}\cdot\text{K}}\right)(295 \text{ K})} = 0.0413 \text{ mol of air}$$

(or 2.49×10^{22} molecules). At the rate calculated above, it would take

$(2.49\times10^{22}) / (5.80\times10^{18} \text{ s}^{-1})\ 4.30\times10^{3} \text{ s} \approx 4300$ s or 71.6 min for the balloon to deflate.

19.59. Generally, we can use equation 19.51 for this, but again we need to watch our units. Pressure should be expressed in Pa units: $0.100 \text{ torr} \times \frac{1 \text{ atm}}{760 \text{ torr}} \times \frac{101325 \text{ Pa}}{1 \text{ atm}} = 13.3 \text{ Pa}$.

If the diameter of the tube is 0.01625 inches, its radius is 0.00813 inches, or 0.2064 mm (1 inch is defined as exactly 25.4 mm). Thus, its area is

$\pi r^2 = \pi\ (0.2064 \text{ mm})^2 = 0.1338 \text{ mm}^2 = 1.338\times10^{-7} \text{ m}^2$. Now substituting into equation 19.51:

$$\frac{dN}{dt} = \left(1.338\times10^{-7} \text{ m}^2\right)\left(13.3 \text{ Pa}\right)\left[\frac{1}{2\pi\left(6.634\times10^{-26} \text{ kg}\right)\left(1.381\times10^{-23} \text{ J/K}\right)\left(300 \text{ K}\right)}\right]^{1/2}$$

$\frac{dN}{dt} = 4.29\times10^{16} \text{ s}^{-1}$. 4.29×10^{16} Ar atoms are $(4.29\times10^{16}) / (6.022\times10^{23} \text{ mol}^{-1}) = 7.13\times10^{-8}$ mol, so $(7.13\times10^{-8} \text{ mol})\ (39.948 \text{ g/mol}) = 2.85\times10^{-6} \text{ g} = 2.85\ \mu\text{g}$ of argon per second are entering the vacuum chamber.

19.61. Substituting only the units into the expression for D_{12} in equation 19.54:

$$\sqrt{\frac{\left(\frac{\text{J}}{\text{mol}\cdot\text{K}}\right)(\text{K})}{\text{kg/mol}}}\cdot\frac{1}{(\text{m})^2\cdot\left(\text{m}^{-3}\right)} = \sqrt{\frac{\text{kg}\cdot\text{m}^2/\text{s}^2}{\text{kg}}}\cdot\text{m} = \text{m}^2/\text{s}$$

(Recall that ρ is the particle density of the gases, meaning that it is the number of particles per cubic meter (units of m^{-3}). The kelvin and mol units in the square root term cancel, and we can decompose the joule unit to cancel the kilogram unit: $1 \text{ J} = 1 \text{ N}\cdot\text{m} = 1 \text{ kg m/s}^2\cdot\text{m} = \text{kg m}^2/\text{s}^2$.)

As mentioned in the text, D_{12} values are more often expressed in cm^2/s units.

19.63. The particle density of air under these conditions ($1.00 \text{ atm} = 1.01\times10^5 \text{ Pa}$) is

$\rho = \frac{N}{V} = \frac{p}{kT} = \frac{1.01\times10^5 \text{ Pa}}{\left(1.381\times10^{-23} \text{ J/K}\right)\left(298 \text{ K}\right)} = 2.45\times10^{25} \text{ m}^{-3}$. Using equation 19.53:

$$D = \frac{3}{8\left(190\times10^{-12} \text{ m}\right)^2\left(2.45\times10^{25} \text{ m}^{-3}\right)}\sqrt{\frac{\left(8.314\frac{\text{J}}{\text{mol}\cdot\text{K}}\right)(298 \text{ K})}{\pi\cdot\left(28.8\times10^{-3} \text{ kg/mol}\right)}} = 7.00\times10^{-5} \text{ m}^2/\text{s}.$$

In more common units, this is $D = 0.700 \text{ cm}^2/\text{s}$.

19.65. Let us assume that "normal atmospheric pressure and room temperature" are 1 atm = 101,325 Pa and 25 °C (298 K), respectively. Under those conditions, the particle density is $\rho = \frac{N}{V} = \frac{p}{kT} = \frac{101325 \text{ Pa}}{(1.381 \times 10^{-23} \text{ J/K})(298 \text{ K})} = 2.46 \times 10^{25} \text{ m}^{-3}$. The reduced mass of air+ammonia is $\mu = \frac{(28.8 \text{ g/mol})(17.0 \text{ g/mol})}{(28.8 + 17.0) \text{ g/mol}} = 10.7 \text{ g/mol} = 0.0107 \text{ kg/mol}$. Now, using equation 19.54:

$$D_{12} = \frac{3}{8}\sqrt{\frac{\left(8.314 \frac{\text{J}}{\text{mol}\cdot\text{K}}\right)(298 \text{ K})}{2\pi(0.0107 \text{ kg/mol})}} \cdot \frac{1}{\left[(1.60 + 1.90) \times 10^{-10} \text{ m}\right]^2 \left(2.46 \times 10^{25} \text{ m}^{-3}\right)} = 2.39 \times 10^{-5} \frac{\text{m}^2}{\text{s}}$$

or $D_{12} = 0.239 \text{ cm}^2\text{/s}$ (which is not far off from the value given in example 19.9).

19.67. The average speed of NH_3 is $\bar{v} = \sqrt{\frac{8\left(8.314 \frac{\text{J}}{\text{mol}\cdot\text{K}}\right)(295 \text{ K})}{\pi\cdot\left(17.0304 \times 10^{-3} \text{ kg/mol}\right)}} = 606 \text{ m/s}$ according to equation 19.36. The total distance traveled in 1 min is therefore (606 m/s) (60 s) = 3.63×10^4 m = 36.3 km!

19.69. We first calculate the D values for exercise 19.62. At standard pressure (1 atm = 101325 Pa) and 0.0 °C (273.2 K), the particle density is

$\rho = \frac{N}{V} = \frac{p}{kT} = \frac{101325 \text{ Pa}}{(1.381 \times 10^{-23} \text{ J/K})(273.2 \text{ K})} = 2.686 \times 10^{25} \text{ m}^{-3}$. Using equation 19.53:

$$D_{\text{He}} = \frac{3}{8\left(2.65 \times 10^{-10} \text{ m}\right)^2 \left(2.686 \times 10^{25} \text{ m}^{-3}\right)} \sqrt{\frac{\left(8.314 \frac{\text{J}}{\text{mol}\cdot\text{K}}\right)(273.2 \text{ K})}{\pi\cdot\left(4.0026 \times 10^{-3} \text{ kg/mol}\right)}} = 8.45 \times 10^{-5} \text{ m}^2\text{/s}$$

$$D_{\text{Xe}} = \frac{3}{8\left(4.00 \times 10^{-10} \text{ m}\right)^2 \left(2.686 \times 10^{25} \text{ m}^{-3}\right)} \sqrt{\frac{\left(8.314 \frac{\text{J}}{\text{mol}\cdot\text{K}}\right)(273.2 \text{ K})}{\pi\cdot\left(131.29 \times 10^{-3} \text{ kg/mol}\right)}} = 6.47 \times 10^{-6} \text{ m}^2\text{/s}.$$

The 3-D displacement can now be calculated from equation 19.56:

For He: $(\text{3-D displacement})^2_{\text{avg}} = 6\left(8.45 \times 10^{-5} \text{ m}^2\text{/s}\right)(1 \text{ s}) = 5.07 \times 10^{-4} \text{ m}^2$, so the average displacement is 0.0225 m or 2.25 cm.

For Ar: $(\text{3-D displacement})^2_{\text{avg}} = 6\left(6.47 \times 10^{-6} \text{ m}^2\text{/s}\right)(1 \text{ s}) = 3.88 \times 10^{-5} \text{ m}^2$, so the average displacement is 0.00623 m or 6.23 mm.

19.71. Graham's law states that the rate of effusion is proportional to $1/\sqrt{\text{mass}}$, so we can set up the following expression: $\dfrac{\text{rate (HCl)}}{\text{rate (NH}_3)} = \dfrac{1/\sqrt{36.4606 \text{ g/mol}}}{1/\sqrt{17.0304 \text{ g/mol}}} = \dfrac{0.165611}{0.242319} = 0.683440$.

If, on the other hand, we use the reduced masses $\mu(\text{HCl+air})$ and $\mu(\text{NH}_3\text{+air})$, with air having an average molar mass of 28.8 g/mol:

$$\mu(\text{HCl} + \text{air}) = \frac{(36.4606 \text{ g/mol})(28.8 \text{ g/mol})}{(36.4606 \text{ g/mol}) + (28.8 \text{ g/mol})} = 16.1 \text{ g/mol};$$

$$\mu(\text{NH}_3 + \text{air}) = \frac{(17.0304 \text{ g/mol})(28.8 \text{ g/mol})}{(17.0304 \text{ g/mol}) + (28.8 \text{ g/mol})} = 10.7 \text{ g/mol}.$$

Using the reduced molar masses: $\dfrac{\text{rate (HCl)}}{\text{rate (NH}_3)} = \dfrac{1/\sqrt{16.1 \text{ g/mol}}}{1/\sqrt{10.7 \text{ g/mol}}} = \dfrac{0.249}{0.306} = 0.816$.

Using the reduced masses of the gases gets us closer to the experimentally determined ratio of about 0.77, so this seems to be the better model.

CHAPTER 20

Kinetics

20.1. For the reaction $aA + bB \rightarrow cC + dD$, we can write several other forms of the rate:

$$-\frac{d[A]}{dt} = +\frac{a}{c}\frac{d[C]}{dt} \text{ and } -\frac{d[A]}{dt} = +\frac{a}{d}\frac{d[D]}{dt}$$, or, in terms of the concentration of B:

$$-\frac{d[B]}{dt} = +\frac{b}{c}\frac{d[C]}{dt} \text{ and } -\frac{d[B]}{dt} = +\frac{b}{d}\frac{d[D]}{dt}.$$

There are several other possibilities in terms of $d[C]/dt$ and $d[D]/dt$ as well.

20.3. The rate with respect to H^+ is $-\frac{1.00 \times 10^{-3}\text{ mol}}{153.8\text{ s}} = -6.50 \times 10^{-6}\text{ mol/s}$.

If 1.00 mmol of H^+ is consumed, then $\frac{5}{16}(1.00\text{ mmol}) = 0.313$ mmol of Fe (s) and $\frac{2}{16}(1.00\text{ mmol}) = 0.125$ mmol of MnO_4^- (aq) are consumed. Therefore, the rate with respect to Fe (s) is $-\frac{0.313 \times 10^{-3}\text{ mol}}{153.8\text{ s}} = -2.03 \times 10^{-6}\text{ mol/s}$ and the rate with respect to MnO_4^- (aq) is $-\frac{0.125 \times 10^{-3}\text{ mol Fe}}{153.8\text{ s}} = -8.13 \times 10^{-7}\text{ mol/s}$.

At the same time, $\frac{2}{16}(1.00\text{ mmol}) = 0.125$ mmol of Mn^{2+} (aq), $\frac{5}{16}(1.00\text{ mmol}) = 0.313$ mmol of Fe^{2+} (aq), and $\frac{8}{16}(1.00\text{ mmol}) = 0.500$ mmol of H_2O (ℓ) are produced.

Therefore, the rate with respect to Mn^{2+} (aq) is $+\frac{0.125 \times 10^{-3}\text{ mol}}{153.8\text{ s}} = 8.13 \times 10^{-7}\text{ mol/s}$, the rate with respect to Fe^{2+} (aq) is $\frac{0.313 \times 10^{-3}\text{ mol Fe}}{153.8\text{ s}} = 2.03 \times 10^{-6}\text{ mol/s}$, and the rate with respect to H_2O (ℓ) is $\frac{0.500 \times 10^{-3}\text{ mol}}{153.8\text{ s}} = 3.25 \times 10^{-6}\text{ mol/s}$. All of these rates differ from the invariant reaction rate.

20.5. The order with respect to NO is 2, the order with respect to O_2 is 1, and the overall order is 2 + 1 = 3.

20.7. We assume a rate law of the form $\text{rate} = k[\text{A}]^m[\text{B}]^n[\text{C}]^o$.

To determine the order with respect to A, use the first and second trials:

$$\frac{6.76\times10^{-6}\ \text{M/s}}{9.82\times10^{-7}\ \text{M/s}}=\frac{k(0.550\ \text{M})^m(0.200\ \text{M})^n(1.15\ \text{M})^o}{k(0.210\ \text{M})^m(0.200\ \text{M})^n(1.15\ \text{M})^o}$$. We cancel the units and simplify:

$6.88 = 2.62^m$. Using logarithms: $\ln 6.88 = m \ln 2.62$, so $m=\frac{\ln 6.88}{\ln 2.62}=2.00=2$.

To determine the order with respect to B, use the second and third trials:

$$\frac{9.82\times10^{-7}\ \text{M/s}}{1.68\times10^{-6}\ \text{M/s}}=\frac{k(0.210\ \text{M})^m(0.200\ \text{M})^n(1.15\ \text{M})^o}{k(0.210\ \text{M})^m(0.333\ \text{M})^n(1.15\ \text{M})^o}$$, or $0.585 = 0.601^n$. This gives

$\ln 0.585 = n \ln 0.601$ and $n=\frac{\ln 0.585}{\ln 0.601}=1.05$, so within the experimental error: $n = 1$.

To determine the order with respect to C, use the second and fourth trials:

$$\frac{9.82\times10^{-7}\ \text{M/s}}{9.84\times10^{-7}\ \text{M/s}}=\frac{k(0.210\ \text{M})^m(0.200\ \text{M})^n(1.15\ \text{M})^o}{k(0.210\ \text{M})^m(0.200\ \text{M})^n(1.77\ \text{M})^o}$$, or $0.998 = 0.650^o$. This gives

$\ln 0.998 = o \ln 0.650$ and $o=\frac{\ln 0.998}{\ln 0.650}=0.00472$, so within the experimental error: $o = 0$.

To determine the value of k, use one set of data with the orders as determined. Using trial one:

$6.76\times10^{-6}\ \text{M/s} = k\,(0.550\ \text{M})^2\,(0.200\ \text{M})^1\,(1.15\ \text{M})^0 = k\cdot 0.0605\ \text{M}^3$, so $k = 1.12\times10^{-4}\ \text{M}^{-2}\cdot\text{s}^{-1}$.

20.9. The species could be a catalyst, or perhaps an inert gas or solvent that provides collisional energy to the reactants or products.

20.11. We assume a rate law of the form $\text{rate} = k[\text{A}]^m[\text{B}]^n$. First, let us convert the times into rates by dividing the number of moles of A reacted by the seconds elapsed. We get $\frac{0.10\ \text{M}}{36.8\ \text{s}}=2.7\times10^{-3}\ \text{M/s}$, $\frac{0.10\ \text{M}}{25.0\ \text{s}}=4.0\times10^{-3}\ \text{M/s}$, and $\frac{0.10\ \text{M}}{10.0\ \text{s}}=1.0\times10^{-2}\ \text{M/s}$, respectively, for the three trials. To determine the order with respect to A, use the second and third trials: $\frac{4.0\times10^{-3}\ \text{M/s}}{1.0\times10^{-2}\ \text{M/s}}=\frac{k(0.20\ \text{M})^m(0.60\ \text{M})^n}{k(0.50\ \text{M})^m(0.60\ \text{M})^n}$. Cancel the units and simplify:

$0.40 = 0.40^m$, so $m = 1$.

To determine the order with respect to B, use the first and second trials:

$$\frac{2.7\times10^{-3}\ \text{M/s}}{4.0\times10^{-2}\ \text{M/s}}=\frac{k(0.20\ \text{M})^m(0.40\ \text{M})^n}{k(0.20\ \text{M})^m(0.60\ \text{M})^n}$$. After simplifying: $0.68 = 0.67^n$, so $n = 1$.

To determine the value of k, use one set of data with the orders as determined. Using trial one:

2.7×10^{-3} M/s = k (0.20 M)^1 (0.40 M)^1 = $k \cdot 0.080$ M^2, so $k = 0.034$ M$^{-1} \cdot$ s^{-1}.

20.13. The units for the rate (left–hand side of the equation) are M/s. Therefore, in terms of units: M/s = $k \cdot$ M$^2 \cdot$ M^2. The rate constant k has therefore units of $\dfrac{\text{M/s}}{\text{M}^2 \cdot \text{M}^2} = \text{M}^{-3} \cdot \text{s}^{-1}$.

20.15. We use equation 20.15:

After 100 s: $[\text{A}]_t = (3.4\times10^{-3}\ \text{M})\exp\left[-(7.66\times10^{-3}\ \text{s}^{-1})(100\ \text{s})\right] = 1.6\times10^{-3}\ \text{M}$;

after 1000 s: $[\text{A}]_t = (3.4\times10^{-3}\ \text{M})\exp\left[-(7.66\times10^{-3}\ \text{s}^{-1})(1000\ \text{s})\right] = 1.6\times10^{-6}\ \text{M}$.

20.17. For this exercise, it is most convenient to use equation 20.14:

$\ln\dfrac{5.5\ \mu\text{M}}{1.0\ \mu\text{M}} = (3.09\times10^{-2}\ \text{s}^{-1})\cdot t$, so $t = 1.7 / (3.09\times10^{-2}\ \text{s}^{-1}) = 55$ s.

20.19. Let us start with equation 20.14: $\ln\dfrac{[\text{A}]_0}{[\text{A}]_t} = kt$. If we take the inverse natural logarithm (exponential) of both sides, we get $\dfrac{[\text{A}]_0}{[\text{A}]_t} = e^{kt}$. We now multiply the $[\text{A}]_t$ term to the right side and divide through by the exponential: $\dfrac{[\text{A}]_0}{e^{kt}} = [\text{A}]_t$. Finally, we can bring the exponential into the numerator by changing the sign on the exponent, yielding equation 20.15: $[\text{A}]_t = [\text{A}]_0 \cdot e^{-kt}$.

20.21. If Newton's law of cooling follows first–order kinetics, then equation 20.14 is followed, but with temperature as the variable, not concentration. Thus, equation 20.14 can be rewritten as $\ln\dfrac{T_0}{T_t} = k \cdot t$, and solving for t:

$$t = \frac{\ln(T_0/T_t)}{k} = \frac{\ln\left(^{1000\ \text{K}}/_{298\ \text{K}}\right)}{0.0344\ \text{s}^{-1}} = \frac{1.21}{0.0344\ \text{s}^{-1}} = 35.2\ \text{s}.$$

20.23. First, we need to determine how many grams of HgO needs to decompose in order to make 1.00 mL and 10.0 mL of O_2 gas at STP. Using the fact that at STP there are 22.4 L of gas volume per mole of gas:

$$1.00\ \text{mL} \times \frac{1\ \text{L}}{1000\ \text{mL}} \times \frac{1\ \text{mol}}{22.4\ \text{L}} \times \frac{2\ \text{mol HgO}}{1\ \text{mol O}_2} \times \frac{216.59\ \text{g HgO}}{\text{mol HgO}} = 0.0193\ \text{g HgO}$$

must decompose in order to form 1.00 mL of O_2. Of course, for 10.0 mL, ten times that amount, or 0.1934 g HgO, need to be decomposed. If we start with exactly 1 g, then there are

1 g – 0.0193 g = 0.9807 grams left over (= $[\text{A}]_t$) after 1.00 mL of oxygen are made, and

1 g – 0.1934 g = 0.8066 g of HgO left over (= $[A]_t$) after 10.0 mL of oxygen are made. Using equation 20.20 and assuming that the second–order k has units of 1/(g · s):

(a) $\frac{1}{0.9807\text{ g}} - \frac{1}{1\text{ g}} = (6.02 \times 10^{-4}\text{ g}^{-1}\text{s}^{-1}) \cdot t$, which gives t = 33 s.

(b) $\frac{1}{0.8066\text{ g}} - \frac{1}{1.00\text{ g}} = (6.02 \times 10^{-4}\text{ g}^{-1}\text{s}^{-1}) \cdot t$, which gives t = 398 s.

20.25. A half–life is that amount of time when $[A]_t$ has become ½ $[A]_0$. Substitute this in equation 20.20: $\frac{1}{\frac{1}{2}\cdot[A]_0} - \frac{1}{[A]_0} = kt_{1/2}$. This equals $\frac{2}{[A]_0} - \frac{1}{[A]_0} = kt_{1/2}$ or $\frac{1}{[A]_0} = kt_{1/2}$, which can be solved for $t_{1/2}$: $t_{1/2} = \frac{1}{k[A]_0}$. This last expression is equation 20.22.

20.27. (a) A third order reaction is of the form $-\frac{d[A]}{dt} = k[A]^3$. We first need to find the integrated version of the rate law. To do that, we separate the variables, $-\frac{d[A]}{[A]^3} = k\,dt$, and integrate: $\int_{[A]_0}^{[A]_t} -\frac{d[A]}{[A]^3} = \int_0^t k\,dt$. The integrated rate law is then $\frac{1}{2[A]_t^2} - \frac{1}{2[A]_0^2} = kt$.

The time $t_{1/2}$ is reached when $[A]_t$ = ½ $[A]_0$: $\frac{1}{2\cdot\left(\frac{1}{2}[A]_0\right)^2} - \frac{1}{2[A]_0^2} = kt_{1/2}$, which simplifies to $\frac{4}{2[A]_0^2} - \frac{1}{2[A]_0^2} = kt_{1/2}$ and $\frac{3}{2[A]_0^2} = kt_{1/2}$, so $t_{1/2} = \frac{3}{2\text{k}[A]_0^2}$.

(b) For a reaction of order –1: $-\frac{d[A]}{dt} = k[A]^{-1} = \frac{k}{[A]}$. Separating variables and integrating: $\int_{[A]_0}^{[A]_t} -[A]\,d[A] = \int_0^t k\,dt$, so the integrated rate law is $-\frac{1}{2}[A]_t^2 + \frac{1}{2}[A]_0^2 =$ kt. The time $t_{1/2}$ is reached for $[A]_t$ = ½ $[A]_0$: $-\frac{1}{2}\left(\frac{1}{2}[A]_0\right)^2 + \frac{1}{2}[A]_0^2 = kt_{1/2}$, so $-\frac{1}{8}[A]_0^2 + \frac{1}{2}[A]_0^2 = \frac{3}{8}[A]_0^2 = kt_{1/2}$. This yields $t_{1/2} = \frac{3[A]_0^2}{8k}$.

(c) For a reaction of order ½: $-\frac{d[\mathrm{A}]}{dt}=k[\mathrm{A}]^{1/2}$. We obtain the integrated rate law from

$\int_{[\mathrm{A}]_0}^{[\mathrm{A}]_t}-\frac{d[\mathrm{A}]}{[\mathrm{A}]^{1/2}}=\int_0^t k\,dt$, which gives $-2[\mathrm{A}]_t^{1/2}+2[\mathrm{A}]_0^{1/2}=kt$. The time $t_{1/2}$ is reached for

$[\mathrm{A}]_t = ½\,[\mathrm{A}]_0$: $-2\left(\frac{1}{2}[\mathrm{A}]_0\right)^{1/2}+2[\mathrm{A}]_0^{1/2}=kt_{1/2}$, so

$-\frac{2}{\sqrt{2}}[\mathrm{A}]_0^{1/2}+2[\mathrm{A}]_0^{1/2}=\left(2-\frac{2}{\sqrt{2}}\right)[\mathrm{A}]_0^{1/2}=kt_{1/2}$. This yields $t_{1/2}=\left(2-\frac{2}{\sqrt{2}}\right)\frac{[\mathrm{A}]_0^{1/2}}{k}$.

We can rewrite this by multiplying the second term in parenthesis by $\sqrt{2}/\sqrt{2}$:

$$t_{1/2}=\left(2-\frac{2\sqrt{2}}{\sqrt{2}\cdot\sqrt{2}}\right)\frac{[\mathrm{A}]_0^{1/2}}{k}=\left(2-\frac{2\sqrt{2}}{2}\right)\frac{[\mathrm{A}]_0^{1/2}}{k}=\left(2-\sqrt{2}\right)\frac{[\mathrm{A}]_0^{1/2}}{k}.$$

20.29. If we define $\frac{1}{b[\mathrm{A}]_0-a[\mathrm{B}]_0}$ in equation 20.27 as some constant K, then we can rewrite the equation as $K\ln\frac{[\mathrm{B}]_0}{[\mathrm{A}]_0}-K\ln\frac{[\mathrm{B}]_t}{[\mathrm{A}]_t}=kt$. Then, we rearrange by isolating the $[\mathrm{A}]_t$ and $[\mathrm{B}]_t$ terms on one side, getting $\ln\frac{[\mathrm{B}]_t}{[\mathrm{A}]_t}=\ln\frac{[\mathrm{B}]_0}{[\mathrm{A}]_0}-\frac{k}{K}t$. According to this expression, if we plot the natural logarithm of the ratio $[\mathrm{B}]_t/[\mathrm{A}]_t$ versus time, the slope will be $-k/K$ and the y–intercept will be $\ln([\mathrm{B}]_0/[\mathrm{A}]_0)$.

20.31. We solve equation 20.24 for t, $t=\frac{[\mathrm{A}]_0-[\mathrm{A}]_t}{k}$, and substitute the numerical values:

(a) $t=\frac{(1.00\text{ M})-(0.75\text{ M})}{1.04\times10^{-1}\text{ M/s}}=2.4\text{ s}$;

(b) $t=\frac{(1.00\text{ M})-(0.50\text{ M})}{1.04\times10^{-1}\text{ M/s}}=4.8\text{ s}$;

(c) $t=\frac{(1.00\text{ M})-(0.00\text{ M})}{1.04\times10^{-1}\text{ M/s}}=9.62\text{ s}$.

20.33. For a zeroth–order reaction, $[\mathrm{A}]_0-[\mathrm{A}]_t=kt$ (equation 20.24) and we want to know the value of t when $[\mathrm{A}]_t=0$: $[\mathrm{A}]_0-0=kt$, so $t=[\mathrm{A}]_0/k$. This equals two times the half-life given in equation 20.25, as expected for a zeroth-order reaction.

20.35. When a reaction uses H_2O as a solvent, the concentration of H_2O is typically so high that it can be considered as constant throughout the course of the reaction, even if H_2O itself takes part in the reaction. This 'constant' concentration of H_2O can be incorporated into the value of the rate law constant, so that we obtain a pseudo first order reaction: rate = $(k \cdot [H_2O]) \cdot [A] = k' \cdot [A]$, where k' has units of s^{-1}.

20.37. If the amount of reactant is to decrease by 5%, then $[A]_t = 0.95[A]_0$. Let us make this substitution into the integrated rate laws.

For zeroth-order kinetics (equation 20.24): $[A]_0 - 0.95[A]_0 = kt$, so $t = \dfrac{0.05[A]_0}{k}$.

For first-order kinetics (equation 20.14): $\ln\dfrac{[A]_0}{0.95[A]_0} = kt$, which equals $-\ln 0.95 = kt$

[recall that $\ln(a/b) = \ln a - \ln b$ and $\ln(1) = 0$], so $t = \dfrac{-\ln 0.95}{k} \approx \dfrac{0.05129}{k}$.

For second–order kinetics (equation 20.20): $\dfrac{1}{0.95[A]_0} - \dfrac{1}{[A]_0} = kt$, which equals

$\dfrac{1/0.95 - 1}{[A]_0} = kt$, so $t = \dfrac{1/0.95 - 1}{k[A]_0} \approx \dfrac{0.05263}{k[A]_0}$.

Which type of kinetics loses the first 5% of reactant fastest? Unfortunately, by comparing the forms of the three expressions, we see that the expressions for zeroth- and second-order kinetics depend on $[A]_0$, so a definite answer cannot be given.

20.39. Yes, a plot of the base-10 logarithm of the concentration versus time will give a straight line graph for a reaction having first–order kinetics, but now the slope will be $k / 2.303$. This is easy to see; recall that because $y = 10^{\log y}$, we can write $\ln y = \ln(10^{\log y})$, which is the same as

$\ln y = \log y \cdot \ln(10) \simeq \log y \cdot 2.303$. That is, the base–10 logarithm and natural logarithm are proportional to each other, and we can substitute the ln's in equation 20.16: $2.303\log[A]_t = 2.303\log[A]_0 - kt$, or after dividing by 2.303:

$$\log[A]_t = \log[A]_0 - \frac{k}{2.303}t.$$

20.41. This is incorrect, as should be clear by looking at Figures 20.6 and 20.7, where five data points are not even sufficient to make a decision about the order of the reaction! In addition, a straight line is defined by exactly two points, so the student will always obtain a perfect straight–line fit with only two data points – no matter which plot (first order, second order, zeroth order, etc.) he uses to test for the reaction order. It is clear that more information is needed to determine the reaction order.

20.43. Using equation 20.36: $K = \dfrac{2.8 \times 10^{-2}\ M^{-1}s^{-1}}{3.6 \times 10^{-4}\ M^{-1}s^{-1}} = 78$.

20.45. We first need to determine the order of the forward and reverse reactions and the values of their rate constants. We assume rate laws of the form rate = $k_f \cdot [EtAc]^m \cdot [H_2O]^n$ and rate = $k_r \cdot [EtOH]^p \cdot [HOAc]^p$ for the forward and reverse reaction, respectively.

For the forward reaction, we determine the order with respect to EtAc using trials one and two:

$$\frac{3.68 \times 10^{-3}\ \text{M/s}}{5.37 \times 10^{-3}\ \text{M/s}} = \frac{k_f(0.678\ \text{M})^m(0.500\ \text{M})^n}{k_f(0.987\ \text{M})^m(0.500\ \text{M})^n},$$ which simplifies to $0.685 = 0.687^m$,

so $m = 1$.

The order with respect to H_2O is determined using trials two and three:

$$\frac{5.37 \times 10^{-3}\ \text{M/s}}{3.28 \times 10^{-3}\ \text{M/s}} = \frac{k_f(0.987\ \text{M})^m(0.500\ \text{M})^n}{k_f(0.987\ \text{M})^m(0.309\ \text{M})^n},$$ which simplifies to $1.64 = 1.62^n$, so $n = 1$.

Using trial one and rate = $k_f \cdot [EtAc]^1 \cdot [H_2O]^1$, we can solve for k_f:

$$k_f = \frac{3.68 \times 10^{-3}\ \text{M/s}}{(0.678\ \text{M})(0.500\ \text{M})} = 1.09 \times 10^{-2}\ \text{M}^{-1}\text{s}^{-1}.$$

For the reverse reaction, we determine the order with respect to EtOH using trials one and three: $\frac{1.77 \times 10^{-4}\ \text{M/s}}{2.29 \times 10^{-4}\ \text{M/s}} = \frac{k_r(0.241\ \text{M})^p(0.115\ \text{M})^q}{k_r(0.300\ \text{M})^p(0.115\ \text{M})^q}$, which simplifies to $0.773 = 0.803^p$,

so $p \approx 1$.

The order with respect to HOAc is determined using trials one and two:

$$\frac{1.77 \times 10^{-4}\ \text{M/s}}{6.10 \times 10^{-4}\ \text{M/s}} = \frac{k_r(0.241\ \text{M})^p(0.115\ \text{M})^q}{k_r(0.241\ \text{M})^p(0.395\ \text{M})^q},$$ which simplifies to $0.290 = 0.291^q$,

so $q = 1$.

Using trial one and rate = $k_r \cdot [EtOH]^1 \cdot [HOAc]^1$, we can solve for k_r:

$$k_r = \frac{1.77 \times 10^{-4}\ \text{M/s}}{(0.241\ \text{M})(0.115\ \text{M})} = 6.39 \times 10^{-3}\ \text{M}^{-1}\text{s}^{-1}.$$

The equilibrium constant is then $K = \frac{1.09 \times 10^{-2}\ \text{M}^{-1}\text{s}^{-1}}{6.39 \times 10^{-3}\ \text{M}^{-1}\text{s}^{-1}} = 1.70$.

20.47. If there were three parallel reactions, the rate of loss of A would be written as

$-\frac{d[A]_t}{dt} = (k_1 + k_2 + k_3)[A]_t$, where k_1, k_2, and k_3 refer to the rate law constants for each individual reaction.

20.49. For the reaction scheme presented, we are dealing with two reactions where a single reactant molecule is converted to a single product molecule. This means that at any point in the process (even at equilibrium) the total number of molecules always equals the number of molecules present initially, represented as $[A]_0$. Consider the sum of the three equilibrium amounts in equation 20.44:

$$[A]_{eq}+[B]_{eq}+[C]_{eq}=\frac{[A]_0}{K_1+K_2+1}+\frac{K_1[A]_0}{K_1+K_2+1}+\frac{K_2[A]_0}{K_1+K_2+1}=\frac{[A]_0+K_1[A]_0+K_2[A]_0}{K_1+K_2+1}$$

Factoring out $[A]_0$ in the numerator: $[A]_{eq}+[B]_{eq}+[C]_{eq}=\dfrac{[A]_0(1+K_1+K_2)}{K_1+K_2+1}=[A]_0$.

This shows that the total number of molecules at equilibrium equals the number of molecules initially, as required by the law of conservation of matter.

20.51. Starting with equation 20.38, $[A]_t=[A]_0e^{-(k_1+k_2)t}$, we take the logarithm of both sides:

$\ln [A]_t = \ln [A]_0 - (k_1 + k_2)\,t$. This equation is in the form $y = mx + b$, the form for a straight line. In this case, the slope would be given by the expression $-(k_1 + k_2)$. Unfortunately, the slope is the sum of the two rate constants, and it is therefore not possible to determine the individual values of the two rate constants from this type of plot; other experiments (and/or plots) are necessary.

20.53. Although straight–line plots of equations 20.41 and 20.42 would be difficult to define and/or interpret, the best way to determine k_1 and k_2 may be by least–squares fitting of the experimental data to equations 20.41 and 20.42. Analytic solutions for the two rate constants may not be forthcoming.

20.55., It should be clear that the amount of ^{210}Bi will be at maximum at $t = 0$ (the time before any of it decays). The amount of ^{206}Pb will be at maximum when all of ^{210}Bi and the intermediate ^{210}Po have decayed, meaning at $t = \infty$.

20.57. **(a)** A is ethyl alcohol that has been ingested, B is ethyl alcohol that has been absorbed by the body, and C is acetaldehyde.

(b) The integrated form for the standard, first–order rate equation is $[A]_t=[A]_0e^{-k_1t}$. We can substitute this for the $[A]_t$ term in the second rate equation. The integrated form of the second equation is then obtained by moving the dt over to the other side and integrating:

$d[B]=\left(k_1[A]_0e^{-k_1t}-k_2\right)dt$, so $\displaystyle\int_0^{[B]_t}d[B]=\int_0^t\left(k_1[A]_0e^{-k_1t}-k_2\right)dt$. The primitive function on the left is [B], and is evaluated at 0 and $[B]_t$, which is simply $[B]_t$. The right–hand integral looks more complex, but is just a simple exponential and a zeroth–order power series. We can evaluate it term by term:

$\displaystyle\int_0^t\left(k_1[A]_0e^{-k_1t}-k_2\right)dt=-\frac{1}{k_1}k_1[A]_0e^{-k_1t}-k_2t\Big|_0^t=-[A]_0\left(e^{-k_1t}-e^0\right)-k_2(t-0)$. This can be written as $\int_0^t \left(k_1[A]_0 e^{-k_1 t} - k_2\right)dt = [A]_0\left(1-e^{-k_1 t}\right) - k_2 t$. Thus we have, for the integrated form of the rate law for the intermediate product: $[B]_t = [A]_0\left(1-e^{-k_1 t}\right) - k_2 t$.

(c) We move the dt over to the other side, $d[C] = k_2 dt$, and integrate, $\int_0^{[C]_t} d[C] = \int_0^t k_2\,dt$, to obtain: $[C]_t = k_2 t$.

(d) The graphing is left to the student.

20.59. Equation 20.48 implies that if ln K is plotted on a y axis and $1/T$ is plotted on the x axis, a straight line is produced with a slope given by $-\Delta_{rxn}H/R$.

20.61. First, we rewrite the derivative to bring d(1/T) to the other side equation 20.53:

$$d\ln k = \left(-\frac{E_A}{R} - mT\right)d(1/T).$$

Now we integrate both sides: $\int d\ln k = \int\left(-\frac{E_A}{R} - mT\right)d(1/T)$.

The integral on the left reduces to ln k plus an integration constant C_1, while the integral on the right is best solved by rewriting it in terms of $x = 1/T$:

$\int\left(-\frac{E_A}{R} - \frac{m}{x}\right)dx = -\frac{E_A}{R}x - m\ln x + C_2$. Resubstituting $x = 1/T$:

$$\int\left(-\frac{E_A}{R} - mT\right)d(1/T) = -\frac{E_A}{RT} - m\ln\frac{1}{T} + C_2 = -\frac{E_A}{RT} + m\ln T + C_2,$$

where in the final expression we have taken advantage of the properties of logarithms to remove the fraction: ln(1/T) = ln(1) – ln T = 0 – ln T. After combining both constants into a single one:

$\ln k = -\frac{E_A}{RT} + m\ln T + C$. Now take the exponential of both sides, recalling that a ln b = ln(b^a):

$k = e^{-\frac{E_A}{RT} + m\ln T + C} = e^{-\frac{E_A}{RT}} e^{\ln(T^m)} e^C$. Defining e^C as some lead constant A and simplifying the term in T^m, we get as our final expression (after reordering the factors): $k = AT^m e^{-\frac{E_A}{RT}}$.

20.63. Half of an O–H bond energy is ½ · 498 kJ/mol = 249 kJ/mol = 249×10^3 J/mol. Rearrange the Arrhenius equation (equation 20.50) to solve for A (remembering that dividing both sides by e^{-x} is the same as multiplying them by e^{+x}):

$$A = k\exp\left(+\frac{E_A}{RT}\right) = \left(1\times10^{11}\ \text{s}^{-1}\right)\exp\left[+\frac{249\times10^3\ \text{J/mol}}{\left(8.314\,\frac{\text{J}}{\text{mol·K}}\right)(295\ \text{K})}\right] = 1\times10^{55}\ \text{s}^{-1}.$$

20.65. **(a)** The order can be determined from the units on k. In order for the rate to have the units M/s, the rate constant must be multiplied by concentration twice. Therefore, it must be a second-order reaction.

(b) We rearrange equation 20.52 to solve for one of the rate constants (for example, k_1):

$$\ln k_1 = \ln k_2 - \frac{E_A}{R}\left(\frac{1}{T_1} - \frac{1}{T_2}\right) = \ln\left(1.77 \times 10^{-6}\right) - \frac{2.00 \times 10^3 \text{ J/mol}}{8.314 \frac{\text{J}}{\text{mol·K}}}\left(\frac{1}{373 \text{ K}} - \frac{1}{298.2 \text{ K}}\right) = -13.1$$

Here, we made use of the fact that $\ln(a/b) = \ln a - \ln b$. The rate constant k_1 will have the same units as k_2, namely 1/(M·s) or $M^{-1}·s^{-1}$, so:
$k_1 = e^{-13.1} \text{ M}^{-1}\text{s}^{-1} = 2.08 \times 10^{-6} \text{ M}^{-1}\text{s}^{-1}$ at 100 °C.

20.67. The nature of the products and reactants suggests that if the Cl atom in methyl chloride were oriented near the incoming Na atom, the reaction would probably proceed more easily. Na itself has spherical symmetry, so its "orientation" is without influence.

20.69. **(a)** $k = \left(7.9 \times 10^{11} \frac{\text{cm}^3}{\text{mol}·\text{s}}\right) · \exp\left[-\frac{10.5 \times 10^3 \text{ J/mol}}{\left(8.314 \frac{\text{J}}{\text{mol·K}}\right)(298 \text{ K})}\right] = 1.1 \times 10^{10} \frac{\text{cm}^3}{\text{mol}·\text{s}}$

(b) The units on k suggest that the reaction is second–order, and the fact that two concentrations are given suggest the rate law rate = k [NO] [O_3]. Substituting:

$$\text{rate} = \left(1.1 \times 10^{10} \frac{\text{cm}^3}{\text{mol}·\text{s}}\right)\left(2.0 \times 10^{-12} \frac{\text{mol}}{\text{cm}^3}\right)\left(5.4 \times 10^{-12} \frac{\text{mol}}{\text{cm}^3}\right) = 1.2 \times 10^{-13} \frac{\text{mol}}{\text{cm}^3·\text{s}}.$$

20.71. We start with the steady–state assumption made in equation 20.60; rearranging it leads to $0 = k_1[\text{A}] - (k_{-1} + k_2)[\text{B}]$, so $[\text{B}] = \frac{k_1}{k_{-1} + k_2}[\text{A}]$. Substituting this in the rate law, equation 20.61, yields directly equation 20.62.

20.73. If the mechanism of the chlorination of methane presented in the text is correct and the first step is the RDS, then the rate law is simply rate = k [Cl_2].

20.75. If the first step is the RDS, then the overall rate would be simply rate = k [Cl_2] [CH_4]. To determine if this mechanism is potentially correct, we need to determine experimentally how the rate depends on the concentrations of Cl_2 and CH_4. If both concentrations are directly proportional to the rate, this would be consistent with the proposed mechanism. If they are not, the mechanism is incorrect.

20.77. If the first step is the RDS, the the rate law is simply rate = k [CH_3OH] [H^+].

20.79. If the second step is the RDS, then the rate is rate = k [O_3*]. This means that enough O_3^* should accumulate to make the back reaction of the first step probable, and we can assume that O_3^* reaches a steady-state concentration:

$$\frac{d[\text{O}_3^*]}{dt} = 0 = k_1[\text{O}_3][\text{M}] - k_{-1}[\text{O}_3^*][\text{M}] - k_2[\text{O}_3^*].$$

We solve this equation for [O_3^*]: $\left[O_3^*\right]=\dfrac{k_1[O_3][M]}{k_{-1}[M]+k_2}$. Substitute this into the original rate law to obtain as the final proposed rate law: $\text{rate}=k_2\left[O_3^*\right]=\dfrac{k_1k_2[O_3][M]}{k_{-1}[M]+k_2}$.

This rate law no longer has a simple form but is not impossible. We can see that it depends on [M], and that the rate should increase if [M] increases (the numerator increases at the same "rate" that [M] increases, but the denominator increases more slowly due to the presence of a constant term). Thus, the addition of an inert gas (like argon) should increase the rate of reaction.

20.81. Recall that a plot of 1/rate versus 1/[S] should yield a straight line whose y-intercept is $1/V$. The two points given will determine a straight line of the form $y = ax + b$. We can either determine the y-intercept graphically by fitting a straight line through a plot of the two points, or determine it algebrically as follows:

For point 1 we have: $\dfrac{1}{2.72\times10^{-7}\text{ M/s}}=a\left(\dfrac{1}{1.4\times10^{-4}\text{ M}}\right)+b;$

and for point 2: $\dfrac{1}{4.03\times10^{-7}\text{ M/s}}=a\left(\dfrac{1}{2.2\times10^{-4}\text{ M}}\right)+b.$

Dividing the first equation by 2.2×10^{-4} M and the second equation by 1.4×10^{-4} M leads to

$$\frac{1}{(2.72\times10^{-7}\text{ M/s})(2.2\times10^{-4}\text{ M})}=\frac{a}{(1.4\times10^{-4}\text{ M})(2.2\times10^{-4}\text{ M})}+\frac{b}{2.2\times10^{-4}\text{ M}}\text{ and}$$

$$\frac{1}{(4.03\times10^{-7}\text{ M/s})(1.4\times10^{-4}\text{ M})}=\frac{a}{(2.2\times10^{-4}\text{ M})(1.4\times10^{-4}\text{ M})}+\frac{b}{1.4\times10^{-4}\text{ M}},$$

and subtracting the second equation from the first one allows us to cancel the a terms:

$$\frac{1}{(2.72\times10^{-7}\text{ M/s})(2.2\times10^{-4}\text{ M})}-\frac{1}{(4.03\times10^{-7}\text{ M/s})(1.4\times10^{-4}\text{ M})}=\frac{b}{2.2\times10^{-4}\text{ M}}-\frac{b}{1.4\times10^{-4}\text{ M}}.$$

Simplifying: $-1\times10^9\text{ M}^{-2}\cdot\text{s} = (-3\times10^3\text{ M}^{-1})\,b$, so $b = 4\times10^5\text{ M}^{-1}\cdot\text{s}$. This is the y–intercept, which corresponds to $1/V$, so $V = 3\times10^{-6}$ M/s..

20.83. The plot is left to the student. If done correctly, you should obtain a y-intercept of 20.0 $\text{mM}^{-1}\cdot\text{s}$, which corresponds to $1/V$. We therefore obtain $V = 0.0500$ mM/s. Because $V = k_2[E]_0$, we can determine k_2 from (0.0500 mM/s) / (0.010 mM) = 5.0 s^{-1}.

20.85. I_2 halogenation should proceed most easily because it has the weakest interhalogen bond, so it is most easily broken to start the chain reaction.

20.87. There are many possible reactions that can be devised, starting with initiation reactions of the formation of $H\cdot + \cdot C_2H_5$ radicals or 2 $\cdot CH_3$ radicals. There are no absolute, correct answers for a suggested mechanism.

20.89. In order of presentation, the reactions are initiation, propagation, branching, propagation, termination, and propagation.

20.91. If mechanism 1 of an oscillating reaction has the second step as the rate–determining step, then the rate law is rate = k [B] [C]. Invoking the steady–state approximation, the first step is at equilibrium: $K = \frac{[B]}{[A]}$. Therefore, $[B] = K[A]$, and rate = kK [A] [C] = k'[A] [C].

If mechanism 2 of an oscillating reaction has the second step as the RDS, then the rate law is

rate = k [B] [C]. Invoking the steady–state approximation, the first step is at equilibrium:

$K = \frac{[B]^2}{[A]}$. Therefore, $[B] = (K[A])^{1/2}$, and rate = k $(K\,[A])^{1/2}$ [C] = $k'[A]^{1/2}$ [C].

Note that the rate of mechanism 1 depends on the concentration of A, while the rate of mechanism 2 depends on the square root of the concentration of A.

20.93. Using c_0 = 1 M = 1 mol/L = 1000 mol/m^3, we rearrange equation 20.85:

$$e^{\Delta S^*/R} = \frac{Ac^\circ h}{ekT} = \frac{\left(3.77\times10^{6}\ \frac{\text{m}^3}{\text{mol}\cdot\text{s}}\right)\left(1000\ \text{mol/m}^3\right)\left(6.626\times10^{-34}\ \text{J}\cdot\text{s}\right)}{e\left(1.381\times10^{-23}\ \text{J/K}\right)\left(293.2\ \text{K}\right)} = 2.27\times10^{-4}.$$

Taking the logarithm of both sides, we can solve for ΔS^*:

$$\Delta S^* = \ln\left(2.27\times10^{-4}\right)R = -8.39\left(8.314\,\frac{\text{J}}{\text{mol}\cdot\text{K}}\right) = -69.8\,\frac{\text{J}}{\text{mol}\cdot\text{K}}.$$

20.95. As the temperature increases, A increases because higher and energy higher levels of the vibration that turns the activated complex into products are being populated. An increase in ΔS^* also increases A, because it favors the formation of the activated complex.

20.97. NaICl (probably structured as I–Na–Cl) would make a good transition state structure, linear or bent. To calculate k, we would need the partition functions of NaI, Cl, and NaICl, which in turn would require knowing their electronic, vibrational, rotational, and translational energies, along with the degeneracy of the ground state. We would also need to know which "vibration" ν* of NaICl represents the reaction pathway so that we can exclude it from the vibrational partition function for the transition state.

20.99. Compare $\exp\left(\frac{+1\,\frac{\text{J}}{\text{mol}\cdot\text{K}}}{8.314\,\frac{\text{J}}{\text{mol}\cdot\text{K}}}\right) = 1.128$ to $\exp\left(\frac{-1\,\frac{\text{J}}{\text{mol}\cdot\text{K}}}{8.314\,\frac{\text{J}}{\text{mol}\cdot\text{K}}}\right) = 0.8867$. This means that if ΔS^* changes from +1 to –1 J/(mol·K), the overall value of k (all else staying the same) will decrease by a little over 21%, slowing the reaction down by the same relative amount.

CHAPTER 21

The Solid State: Crystals

21.1. Ionic crystals are so brittle because at the atomic level, they are held together by strong interionic forces. However, if the planes of charged particles are moved by some force, they can now face like-charged particles that would repel, breaking the crystal apart.

21.3. Unit cells can be described for polycrystalline materials only if it is understood that each unit cell may apply only to a tiny portion of the overall solid, namely a single crystallite. Other than that, normal rules of unit cells apply.

21.5. Diamond has a face-centered cubic unit cell. With eight eighths atoms in the corners, six halves of atoms in the faces, and four complete atoms within the unit cell itself, there are a total of 8 atoms of carbon per unit cell.

21.7. **(a)** According to Table 21.1, a unit cell with only right angles is either cubic, tetragonal, or orthorhombic (3 possibilities).

(b) Only cubic and trigonal crystals have unit cell dimensions that are equal in all three dimensions (2 possibilities).

(c) All types of crystals except for triclinic crystals have at least one right angle (6 possibilities).

(d) Only triclinic crystals have no right angles; all of their unit cell dimensions are different in length, so there is only one possibility.

21.9. A proposed edge-centered cubic unit cell would actually be a simple cubic lattice, with some atoms located at tetrahedrally-oriented sites within the unit cell.

21.11. The largest atom that can fit in a body-centered cubic unit cell will have the corner atom touching the center atom, which will touch the opposite corner atom. Thus, going from one corner through the center of the unit cell to the opposite corner takes 4 radii of the atom. In terms of a, the unit cell parameter, that corner-to-corner distance is $\sqrt{3}\,a$.

Therefore, the maximum radius an atom in a bcc unit cell can have is $\frac{\sqrt{3}}{4}a$.

21.13. $1\text{ mL}\times\frac{1\text{ cm}^3}{1\text{ mL}}\times\frac{(1\text{ m})^3}{(100\text{ cm})^3}\times\frac{(10^{10}\text{ Å})^3}{(1\text{ m})^3}=1\times10^{24}\text{ Å}^3$.

21.15. The volume of this cubic unit cell is $V=\left(5.418\text{ Å}\right)^3=159.0\text{ Å}^3$. Therefore, the mass in each unit cell is

$\left(5.012\frac{\text{g}}{\text{cm}^3}\right)\times\left(159.0\text{ Å}^3\right)\times\frac{1\text{ u}}{1.6605\times10^{-27}\text{ kg}}\times\frac{1\text{ kg}}{1000\text{ g}}\times\frac{1\text{ cm}^3}{10^{24}\text{ Å}^3}=480.1\text{ u}$. We know that there are four formula units per unit cell, so each formula unit accounts for 120.0 u.

Iron has a mass of 55.845 u, and sulfur 32.066 u. The formula must contain at least one iron atom and one sulfur atom, which accounts for 87.911 u. The remaining mass is 120.0 u – 87.911 u = 32.1 u, which is one additional sulfur atom: the formula of pyrite is FeS_2.

21.17. The mass of one formula unit of SiO_2 is 28.0855 u + 2(15.9994 u) = 60.0843 u. Three of these units has a mass of 180.253 u. Now, we determine the volume of the unit cell, using equation 21.4:

$$V=(4.914\ \text{Å})(4.914\ \text{Å})(5.405\ \text{Å})\sqrt{1-\cos^2 90°-\cos^2 90°-\cos^2 120°+2\cos 90°\cos 90°\cos 120°}$$

$V = (130.5\ \text{Å}^3)(1\text{-}0\text{-}0\text{-}0.25\text{+}0)^{1/2} = 113.0\ \text{Å}^3$. Therefore, the density is:

$$d=\frac{180.253\ \text{u}}{113.0\ \text{Å}^3}\times\frac{10^{24}\ \text{Å}^3}{1\ \text{cm}^3}\times\frac{1.6605\times10^{-24}\ \text{g}}{1\ \text{u}}=2.648\ \text{g/cm}^3.$$

The experimental value is 2.648 g/cm^3, in perfect agreement with the calculated value.

21.19. Cubic crystal structures have $a = b = c$, so only one unit cell length needs to be specified. For the hexagonal crystal structure $a = b$, so only a needs to be specified. The missing angles for the hexagonal crystal structure are $\alpha = \beta = 90°$, $\gamma = 120°$ and for the monoclinic structure

$\alpha = \gamma = 90°$, as specified by Table 21.1.

21.21. Both hexagonal close-packed and face-centered cubic unit cells represent the most efficient use of space. One might speculate that in such crystals, interactions between individual atoms lead to an energy-minimum structure.

21.23. Using simple geometry, one can show that for an angle ϕ defined with respect to the line perpendicular to the crystal planes, the Bragg equation would be $n\lambda = 2d\cos\phi$. (This can also be seen by substituting $90° - \phi$ for θ in equation 21.5 and making use of the trigonometeric equality $\sin(90° - \phi) = \cos\phi$).

21.25. We use equation 21.5 in the form $\sin\theta=\frac{n\lambda}{2d}$ and solve for θ:

For first-order diffraction ($n = 1$): $\sin\theta=\frac{1(1.5511\ \text{Å})}{2(5.47\ \text{Å})}=0.142$, so $\theta = 8.15°$.

For second-order diffraction ($n = 2$): $\sin\theta=\frac{2(1.5511\ \text{Å})}{2(5.47\ \text{Å})}=0.284$, so $\theta = 16.5°$.

21.27. For a body-centered unit cell, there are 2 total atoms per cell (one in the center and eight eights in the corners). Assuming a cubic crystal structure, the volume of one unit cell is $(2.8664\ \text{Å})^3 = 23.551\ \text{Å}^3$. If the density is 7.8748 g/cm^3, then the number of grams per unit cell is

$\left(7.8748\ \frac{g}{cm^3}\right)\times\left(23.551\ Å^3\right)\times\frac{1\ cm^3}{10^{24}\ Å^3}=1.8546\times10^{-22}$ g. Because there are two atoms of Fe per unit cell, each atom has a mass of half of this, or 9.2730×10^{-23} g. Dividing the mass of 1 mol of ^{56}Fe by the mass of one iron atom gives us the number of atoms per mole (Avogadro's number): $\frac{55.93\ \frac{g}{mol}}{9.2730\times10^{-23}\ g}=6.031\times10^{23}\ mol^{-1}$.

21.29. Determining the ratio of the *d* spacings is aided with the appropriate diagrams. Figure 21.18 shows the (111) planes, while Figures 21.19a and 21.19b show (100) and (110) planes (in the latter case, we just swap the *b* and *c* labels).

For starters, we point out that the *d* spacing of the (100) planes is simply the lattice parameter *a*, whatever it is for a particular crystal.

For the (110) planes, we can see that the distance between planes is simply ½ of the length of the diagonal (which has a length of $\sqrt{2}\,a$), so $d=\frac{1}{2}\sqrt{2}\,a=\frac{a}{\sqrt{2}}$ or about $0.7071a$.

Figure 21.18 shows that the *d* spacing for the (111) planes is equal to the distance from one corner (take the origin, for example) to the center of the opposing triangle. Combine these four points, and you will find that they form a triangular pyramid (the corner being the apex, the triangle forming the base). The three edges starting at the apex have length *a*, the other three have the length of a diagonal, $s=\sqrt{2}\,a$. We want to calculate *d*, which equals the distance from the apex to the base of the triangular pyramid. To do this, look at the right triangle formed by the apex, the center of the triangle, and one of its corners. The hypotenuse has length *a*, and the distance from the center of the triangle to its corner is known to be $\frac{\sqrt{3}}{3}s=\frac{\sqrt{3}}{3}\sqrt{2}\,a$ (radius of the circumscribed circle to an equilateral triangle). Applying the Pythagorean theorem, we finally obtain $d^2=a^2-\frac{3}{9}\cdot2a^2=\frac{1}{3}a^2$, so $d=\frac{1}{\sqrt{3}}a$ or about $0.5774\,a$.

Therefore, the ratio of *d* spacings for (100), (110), and (111) planes would be $1:\frac{1}{\sqrt{2}}:\frac{1}{\sqrt{3}}$ (which is in agreement with equation 21.8) or 1 : 0.7071 : 0.5774.

21.31. The drawing is left to the student.

21.33. The given set of Miller indices is equivalent to the (111) plane.

21.35. Following Example 21 9 in the text we first determine the *d* spacings for each angle and look for the proper pattern. Using Bragg's law for first-order diffraction (equation 21.6):

$$d_1=\frac{\lambda}{2\sin\theta}=\frac{1.5418\ Å}{2\sin(15.7°)}=2.85\ Å;\ d_2=\frac{1.5418\ Å}{2\sin(18.2°)}=2.47\ Å;\ d_3=\frac{1.5418\ Å}{2\sin(26.1°)}=1.75\ Å;$$

$$d_4 = \frac{1.5418\ \text{Å}}{2\sin(31.1°)} = 1.49\ \text{Å};\ d_5 = \frac{1.5418\ \text{Å}}{2\sin(32.6°)} = 1.43\ \text{Å}.$$

Now we list the square of their reciprocals:

$1/(2.85\ \text{Å})^2 = 0.123\ \text{Å}^{-2}$; $1/(2.47\ \text{Å})^2 = 0.164\ \text{Å}^{-2}$; $1/(1.75\ \text{Å})^2 = 0.326\ \text{Å}^{-2}$;

$1/(1.49\ \text{Å})^2 = 0.449\ \text{Å}^{-2}$; $1/(1.43\ \text{Å})^2 = 0.488\ \text{Å}^{-2}$

The ratio of the two lowest reciprocals is 0.123 / 0.164 = 0.751. Thus, the crystal should be face-centered cubic. That means, by consulting Table 21.3, that the first diffraction is from the (111) plane. We can use equation 21.10 to determine the unit cell parameter:

$$a = \sqrt{h^2 + k^2 + \ell^2}\ d = \sqrt{1^2 + 1^2 + 1^2}\left(2.85\ \text{Å}\right) = 4.93\ \text{Å}.$$

21.37. While it is true that the first six diffraction patterns for a simple cubic unit cell occur for $(h^2 + k^2 + \ell^2)$ ratios of 1 : 2 : 3 : 4 : 5 : 6, those for a body-centered cubic unit cell follow the pattern 2 : 4 : 6 : 8 : 10 : 12, which is the exact same ratio! In order to decide which of these two cases is correct, the crystallographer has several possibilities. He could extend the measurement to the next diffraction pattern; for the simple cubic unit cell, the value 7 is missing, while it is present (14) for the body-centered case. Another possibility is to take the ratio of the seventh and eight $1/d^2$ values and check whether it is 0.875 (body-centered cubic) or 0.889 (simple cubic). He could also calculate the density from the derived unit cell parameters and compare it to the experimental value.

21.39. The drawing is left to the student. It should show that successive planes diffract light of the right wavelength in a destructive-interference fashion, effectively eliminating the diffraction.

21.41. The sample could be MgO, as we would expect these isoelectronic ions to have similar scattering abilities. The other ionic compounds are composed of ions of very different size, which should be reflected in a pattern of varying intensities.

21.43. Use Table 21.4 to obtain the ionic radii of the ions involved, and take the ratio of the smaller to larger ions.

(a) For titanium sulfide: $\frac{r_{\text{smaller}}}{r_{\text{larger}}} = \frac{0.68\ \text{Å}}{1.84\ \text{Å}} = 0.37$. Therefore, TiS_2 should have the rutile structure (tetragonal unit cell).

(b) For barium fluoride: $\frac{r_{\text{smaller}}}{r_{\text{larger}}} = \frac{1.33\ \text{Å}}{1.34\ \text{Å}} = 0.993$. Therefore, BaF_2 should have fluorite structure (face-cetered-cubic unit cell).

(c) For potassium sulfate: $\frac{r_{\text{smaller}}}{r_{\text{larger}}} = \frac{1.33\ \text{Å}}{2.30\ \text{Å}} = 0.578$. Therefore, K_2SO_4 should have the rutile structure (tetragonal unit cell).

21.45. Carbon typically exists as two types of solids: a covalent-network solid (as diamond), and a planar hexagonal lattice (as graphite). In graphite, the planar sheets are far enough apart that carbon isn't close-packed.

21.47. Referring to Figure 21.29, we see that for fluorite, the coordination number for the Ca^{2+} ions is 8, while the coordination number for the F^- ions is 4. For rutile, the coordination number for the Ti^{4+} ions is 6 while the coordination for the O^{2-} ions is 3. The coordination numbers are different because these aren't 1:1 ionic compounds; they are 1:2 ionic compounds.

21.49. The reactions whose energy change represent the lattice energy are:

(a) $K^+ (g) + F^- (g) \rightarrow KF (s)$

(b) $Mg^{2+} (g) + Se^{2-} (g) \rightarrow MgSe (s)$

(c) $2\ Na^+ (g) + O^{2-} (g) \rightarrow Na_2O (s)$

(d) $2\ Na^+ (g) + O_2^{2-} (g) \rightarrow Na_2O_2 (s)$

21.51. (a)

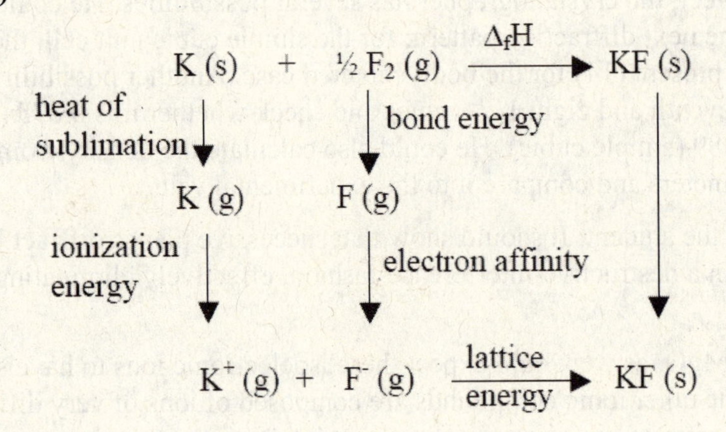

(b)

$\Delta_f H$

Mg (s) + Se (s) → MgSe (s)

heat of sublimation ↓ | heat of sublimation ↓

Mg (g)   Se (g)

1st and 2nd ionization energy ↓ | 1st and 2nd electron affinity ↓

Mg^{2+} (g) + Se^{2-} (s) —lattice energy→ MgSe (s)

(c)

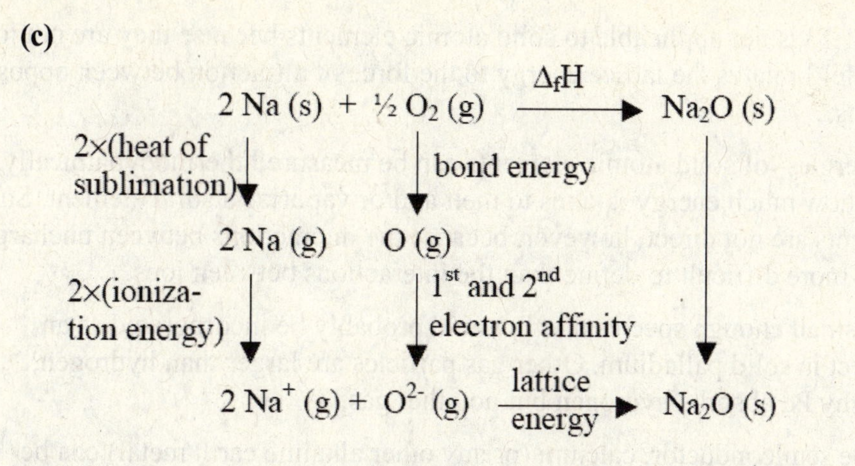

(d)

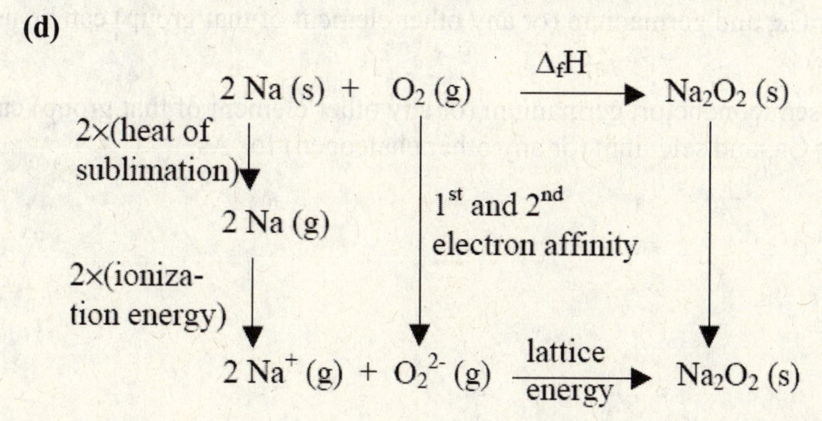

21.53. We will assume that KCl has sodium chloride structure (it is actually borderline because $\frac{r_{\text{smaller}}}{r_{\text{larger}}} = \frac{1.33\ \text{Å}}{1.81\ \text{Å}} = 0.735$). Using Table 21.6 and equation 21.13:

$$\text{lattice energy} = \frac{\left(6.022\times10^{23}\ \text{mol}^{-1}\right)\left(1.74756\right)\cdot1^2\cdot\left(1.602\times10^{-19}\ \text{C}\right)^2}{4\pi\left(8.854\times10^{-12}\ \frac{\text{C}^2}{\text{J}\cdot\text{m}}\right)\left(0.319\times10^{-9}\ \text{m}\right)}\left(1-\frac{0.326\ \text{Å}}{3.14\ \text{Å}}\right).$$

This evaluates to 682×10^3 J/mol or 682 kJ/mol, which is close to the experimental value of 701 kJ/mol (from Table 21.5).

21.55. KI has the same crystal structure as NaCl, a face-centered cubic (as expected from $\frac{r_{\text{smaller}}}{r_{\text{larger}}} = \frac{1.33\ \text{Å}}{2.20\ \text{Å}} = 0.605$). Therefore, we use 1.74756 as the Madelung constant. We rearrange equation 21.13 to solve for $(1 - \rho / r)$:

$$\left(1-\frac{\rho}{r}\right) = \frac{\left(627.2\times10^3\ \text{J/mol}\right)\cdot4\pi\left(8.854\times10^{-12}\ \frac{\text{C}^2}{\text{J}\cdot\text{m}}\right)\left(3.533\times10^{-10}\ \text{m}\right)}{\left(6.022\times10^{23}\ \text{mol}^{-1}\right)\left(1.74756\right)\cdot1^2\cdot\left(1.602\times10^{-19}\ \text{C}\right)^2} = 0.9129.$$

Therefore, the ratio ρ / r is 0.0871, which, again with r = 3.533 Å, gives ρ = 0.308 Å.

21.57. **(a)** Equation 21.13 is not applicable to solid atomic elements because they are not ions. Equation 21.13 relates the lattice energy to the force of attraction between oppositely-charged ions.

(b) 'Lattice energies' of solid atomic elements can be measured thermodynamically, by measuring how much energy it takes to melt and/or vaporize a solid element. Such measurements are not direct, however, because the interactions between uncharged species are more difficult to define than the interactions between ions.

21.59. Hydrogen is a small enough species that it would probably be incorporated as an interstitial defect in solid palladium. Other gas particles are larger than hydrogen, rationalizing why Pd absorbs hydrogen but no other gas.

21.61. **(a)** For a p-type semiconductor, calcium (or any other alkaline earth metal) can be substituted for Ga, and germanium (or any other element of that group) can be used to substitute for As.

(b) For an n-type semiconductor, germanium (or any other element of that group) can be substituted for Ga, and selenium (or any other chalcogen) for As.

CHAPTER 22

Surfaces

22.1. Because there are unbalanced forces at a liquid interface, work is always performed (against a force pointing towards the bulk) when increasing the surface area of that liquid. Recall that work is a form of energy, implying that energy is always required to increase the surface area of a liquid.

22.3. It is evident from Table 22.1 that the surface tension decreases with increasing temperature. This is not surprising, given that a higher temperature favors greater translational motion, which weakens the forces that hold the molecules of the liquid together.

22.5. With 1 dyn/cm = 1 erg/cm^2 and 1 erg equaling 1×10^{-7} joules, a surface tension of 27.1 dyn/cm corresponds to 27.1×10^{-7} J/cm^2.

(a) Therefore increasing the surface area by 50.0 cm^2 takes an energy of $(27.1\times10^{-7}\ \text{J/cm}^2)\ (50.0\ \text{cm}^2) = 1.36\times10^{-4}$ J.

(b) A *film* of chloroform with an area of 0.010 m^2 has a surface area of twice that amount, so $(27.1\times10^{-7}\ \text{J/cm}^2)\ (0.020\ \text{m}^2)\ (100\ \text{cm/m})^2 = 5.4\times10^{-4}$ J.

22.7. The force acting on a 2.00 cm bar drawn across acetone at 20 °C (Table 22.1) is $F = \gamma \cdot \ell = (23.7\ \text{dyn/cm})\ (2.00\ \text{cm}) = 47.4$ dyn or 47.4×10^{-5} N if the bar is increasing the surface area (as shown in Fig. 22.2).

22.9. The left-hand side of equation 22.6 is the surface tension, which has units of N/m. The right-hand side is energy per surface area, with units of J / m^2. Recall that 1 J = 1 N · m, so

$(\text{N} \cdot \text{m}) / \text{m}^2 = \text{N} / \text{m}$.

22.11. The volume per droplet is $V = \frac{4}{3}\pi r^3 = \frac{4}{3}\pi\left(1.00\times10^{-7}\ \text{cm}\right)^3 = 4.19\times10^{-21}\ \text{cm}^3$, meaning that the total volume of both droplets is 8.38×10^{-21} cm^3. Their total surface area is

$A = 2\cdot 4\pi r^2 = 8\pi\left(1.00\times10^{-7}\ \text{cm}\right)^2 = 2.51\times10^{-13}\ \text{cm}^2$.

Now the radius of the new droplet needs to be calculated so we can determine its surface area. We know that its volume equals 8.38×10^{-21} cm^3, so the radius can be calculated from

$8.38\times10^{-21}\ \text{cm}^3 = \frac{4}{3}\pi r^3$, which results in $r = 1.26\times10^{-7}$ cm. Therefore, the new surface area is $A = 4\pi r^2 = 4\pi\left(1.26\times10^{-7}\ \text{cm}\right)^2 = 1.99\times10^{-13}\ \text{cm}^2$.

The amount of surface area lost is therefore 2.51×10^{-13} cm^2 – 1.99×10^{-13} cm^2 = 5.2×10^{-14} cm^2.

Because surface area was *lost*, we can calculate that amount of energy released using the surface tension of water: (72.75 erg/cm^2) (5.2×10^{-14} cm^2) = 3.8×10^{-12} erg. Because 1 erg equals

1×10^{-7} joules, this is an energy of 3.8×10^{-19} J – a very small amount of energy, but remember that the amount of water is also very tiny.

We now need to know the heat capacity of liquid water in order to calculate by how much exactly the temperature of the droplet will rise. The specific heat capacity of liquid water at

20.0 °C can be approximated by its value at 25 °C, which is c = 4.184 J per g and per K (see Table 2.1). Using a density of 1.00 g/cm^3, 8.38×10^{-21} cm^3 of water correspond to 8.38×10^{-21} g. If we assume that all of the released energy will be put into increasing the droplet's temperature, the temperature change can be calculated from $q = m\cdot c\cdot\Delta T$:

$\Delta T = \dfrac{q}{m\cdot c} = \dfrac{3.8\times10^{-19}\ \text{J}}{\left(8.38\times10^{-21}\ \text{g}\right)\left(4.184\ \frac{\text{J}}{\text{g}\cdot\text{K}}\right)} = 11\ \text{K}$. So, despite the tiny amount of energy involved, it's still enough to increase the temperature of the water from 20.0 °C to 31 °C.

22.13. Work is being done by the bubble. As it decreases its surface area, it is releasing surface energy to the surroundings. Therefore, work is being done by the bubble on the surroundings.

22.15. Equation 22.9 is simply following the possible energy changes around an interface. It is dividing the interface into three regions and – showing that the first law of thermodynamics holds in this circumstance – requires that the overall infinitesimal energy change be zero in accordance with the first law.

22.17. If surface tension has units of energy per area, say J/m^2, we have

$\dfrac{\text{J}}{\text{m}^2}\left(\dfrac{\text{m}^2}{\text{m}^3}\right) = \dfrac{\text{J}}{\text{m}^3} = \dfrac{\text{N}\cdot\text{m}}{\text{m}^3} = \dfrac{\text{N}}{\text{m}^2} = \text{Pa}$ (which is a unit of pressure, as required).

22.19. Because of the inverse relationship between Δp and r, the smaller the radius, the greater the pressure difference across the interface. Thus, small droplets of deuterium-tritium mixes may be preferable for fusion research because of higher pressures inside the droplets.

22.21. Ethanol has a surface tension of 22.8 dyn/cm at 20 °C. Thus, the pressure difference is

$\Delta p = \dfrac{2(22.8\ \text{dyn/cm})}{1.00\ \text{cm}} = 45.6\ \text{dyn/cm}^2$. Recall that 1 dyn = 1×10^{-5} N and 1 cm^2 = 1×10^{-4} m^2, so this is 45.6×10^{-1} N/m^2 or 4.56 Pa (or, using yet another pressure unit: 4.56×10^{-5} bar).

Similarly, for 1.00 mm = 0.100 cm: $\Delta p = \dfrac{2(22.8\ \text{dyn/cm})}{0.100\ \text{cm}} = 456\ \text{dyn/cm}^2$ or 45.6 Pa.

22.23. Here, we must use equation 22.15. Again, with 1 erg/cm^2 = 1 dyn/cm:

$$\Delta p = \frac{4(24.3 \text{ dyn/cm})}{1.75 \text{ cm}} = 55.5 \text{ dyn/cm}^2$$. Recall that 1 dyn = 1×10^{-5} N and 1 cm^2 = 1×10^{-4} m^2, so this is 55.5×10^{-1} N/m^2 or 5.55 Pa (which is 5.55×10^{-5} bar or 0.0417 torr).

22.25. As temperature increases, the vapor pressure of the liquid phase increases, which must stem from an increase of the pressure inside the liquid drop. However, the pressure of the surrounding gas increases even more rapidly, so that Δp becomes smaller. This is consistent with a lowering of surface tension as temperature increases.

22.27. **(a)** Because r is in the denominator of the Kelvin equation, the value of the fraction increases as r gets smaller. Thus, the equilibrium vapor pressure of the droplet increases as the droplet gets smaller. This suggests that small droplets won't form spontaneously from the gas or vapor phase; rather, that some other influence is necessary for droplets to form from the gas phase. (Cloud seeding is one application of this concept.)

(b) According to Table 22.1, the surface tension of water at 20 °C is 72.75 erg/cm^2. Using this value for a temperature of 298 K and a density of 1.00 g/cm^3 for water (which gives a molar volume of 18.0 cm^3 per 1 mol, or 18.0152 g, of water):

$$\ln\left(\frac{p_{\text{vapor}}}{p^{\text{o}}_{\text{vapor}}}\right) = \frac{2(72.75 \text{ erg/cm}^2)(18.0 \text{ cm}^3\text{/mol})}{(20.0\times10^{-7} \text{ cm})\left(8.314 \frac{\text{J}}{\text{mol}\cdot\text{K}}\right)(298 \text{ K})} = 5.29\times10^{5} \text{ erg/J}$$. Recall that one erg equals 1×10^{-7} joules, so the units cancel and we get simply 5.29×10^{-2} = 0.0529. Then: $\frac{p_{\text{vapor}}}{p^{\text{o}}_{\text{vapor}}} = e^{0.0529} = 1.05$, so p_{vapor} = 1.05 (23.77 mmHg) = 25.1 mmHg.

22.29. Capillary action is not seen in cylinders with large radii because capillary action is inversely proportional to r. The greater the radius R of the cylinder, the greater $r = R / \cos\theta$, and the smaller h in equation 22.18.

22.31. We use equation 22.19: $h = \frac{2(485.5 \text{ erg/cm}^2)\cos(115.5°)}{(13.6 \text{ g/cm}^3)(9.81 \text{ m/s}^2)(0.200 \text{ cm})} = -15.7 \frac{\text{erg}}{\text{g}\cdot\text{m}\cdot\text{s}^{-2}}$.

The units in the numerator are erg, which is 1×10^{-7} J, or 1×10^{-7} kg · m^2 / s^2. This leads to

$$h = -15.7 \frac{\text{erg}}{\text{g}\cdot\text{m}\cdot\text{s}^{-2}} \times \frac{1\times10^{-7} \text{ kg}\cdot\text{m}^2/\text{s}^2}{1 \text{ erg}} \times \frac{1000 \text{ g}}{1 \text{ kg}} = -0.00157 \text{ m}.$$

The capillary depression is 1.57 mm.

22.33. Many surfaces may be covered with silicones adsorbed from the vapor phase because the silicones have very low surface energies.

22.35. **(a)** Because the (100) plane contains one of the faces of the unit cell, the Na^+-Na^+ distance in this plane equals the distance between a corner Na^+ and a facial Na^+, which is $\frac{1}{2}\sqrt{2}(5.640\ \text{Å}) = 3.988\ \text{Å}$. (See the fcc unit cell in Figure 21.9 to better visualize this.) Closer inspection shows that all adjacent Na^+ have this distance.

(b) The (110) plane goes through opposite edges of the unit cell, cutting the unit cell into two 45º-45º-90º prisms. Along the diagonal direction, the (110) plane contains corner and facial Na^+ ions, so the closest Na^+-Na^+ distance in this direction is again 3.988 Å. In the c direction, however, the spacing of adjacent Na^+ ions is 5.640 Å.

(c) The (111) plane contains the Na^+ ions at the corners and face diagonals. This means that the closest Na^+–Na^+ distance is again 3.988 Å. Closer inspection shows that all Na^+ are 3.988 Å apart, similar to the (100) plane. The difference is that in the (100) plane, each ion has four closest neighbors, while it has six in the (111) plane.

22.37. A clean surface would be a surface that terminates in atoms or molecules of the bulk material. However, most surfaces are far from clean at the atomic or molecular level, due to adsorbed species that interact with the dangling bonds of the surface atom.

22.39. Almost immediately upon breakage, whatever dangling bonds that are formed interact with each other (slightly rearranging the atomic layer at the surface) or interact with gas-phase species in the air. Either way, those bonds are no longer available to interact with the corresponding atoms on the other piece, and additional material (i.e. glue) is needed to connect the two pieces back together.

22.41. We will use the ideal gas law. 1 torr is exactly 1/760 atm and 1 cm^3 = 1 mL = 10^{-3} L, so

$$n = \frac{pV}{RT} = \frac{(1.00\times10^{-8}/760\ \text{atm})(10^{-3}\ \text{L})}{\left(0.08205\frac{\text{L}\cdot\text{atm}}{\text{mol}\cdot\text{K}}\right)(273\ \text{K})} = 5.87\times10^{-16}\ \text{mol}.$$ 1 mol contains 6.022×10^{23} particles, so there are $(5.87\times10^{-16}\ \text{mol})(6.022\times10^{23}\ \text{mol}^{-1}) = 3.54\times10^{8}$ gas particles per cubic centimeter at 273 K in this ultra high vacuum.

22.43. $1.00\ \text{bar}\cdot\text{s}\times\frac{1\ \text{atm}}{1.01325\ \text{bar}}\times\frac{760\ \text{torr}}{1\ \text{atm}}\times\frac{1\ \text{L}}{1\times10^{-6}\ \text{torr}\cdot\text{s}} = 7.50\times10^{8}\ \text{L}$ (remember that L in this context means langmuir). This is such a high exposure that for all practical purposes any surface is completely contaminated with adsorbed species, and "exposure" in terms of monolayers of adsorbed species loses its meaning.

22.45. Starting with equation 22.26, $\theta = \frac{k_{ads}\cdot[\text{gas}]}{k_{ads}\cdot[\text{gas}]+k_{des}}$, we first multiply both the numerator and denominator by $1/k_{des}$:

$$\theta = \frac{k_{ads}\cdot[\text{gas}]}{k_{ads}\cdot[\text{gas}]+k_{des}}\times\frac{1/k_{des}}{1/k_{des}} = \frac{k_{ads}\cdot[\text{gas}]/k_{des}}{k_{ads}\cdot[\text{gas}]/k_{des}+k_{des}/k_{des}} = \frac{(k_{ads}/k_{des})\cdot[\text{gas}]}{(k_{ads}/k_{des})\cdot[\text{gas}]+1}$$

Now, we substitute the definition of the equilibrium constant $K = k_{ads} / k_{des}$:

$$\theta = \frac{K \cdot [\text{gas}]}{K \cdot [\text{gas}] + 1},$$ which is equation 22.27.

22.47. The plot of 1/coverage versus 1/pressure is left to the student. If done properly, you should get a straight-line fit with a slope of 4.04×10^{-8} torr, which corresponds to a value of K of approximately 2.5×10^{7} torr^{-1}.

22.49. If this step requires energy, it would simply add to the requirement that the exothermic steps in the mechanism must supply the energy for any endothermic processes. It is difficult to conceive that the desorption of species from a surface be exothermic (after all, energy is always required to break a chemical bond, so all bond-breaking processes are endothermic). Thus, the desorption processes should be as minimally endothermic as possible to support the overall catalytic process.